T0139853

Lecture Notes in Networks and Systems

Volume 256

Series Editor

Janusz Kacprzyk, Systems Research Institute, Polish Academy of Sciences, Warsaw, Poland

Advisory Editors

Fernando Gomide, Department of Computer Engineering and Automation—DCA, School of Electrical and Computer Engineering—FEEC, University of Campinas—UNICAMP, São Paulo, Brazil

Okyay Kaynak, Department of Electrical and Electronic Engineering, Bogazici University, Istanbul, Turkey

Derong Liu, Department of Electrical and Computer Engineering, University of Illinois at Chicago, Chicago, USA; Institute of Automation, Chinese Academy of Sciences, Beijing, China

Witold Pedrycz, Department of Electrical and Computer Engineering, University of Alberta, Alberta, Canada; Systems Research Institute, Polish Academy of Sciences, Warsaw, Poland

Marios M. Polycarpou, Department of Electrical and Computer Engineering, KIOS Research Center for Intelligent Systems and Networks, University of Cyprus, Nicosia, Cyprus

Imre J. Rudas, Óbuda University, Budapest, Hungary

Jun Wang, Department of Computer Science, City University of Hong Kong, Kowloon, Hong Kong

The series "Lecture Notes in Networks and Systems" publishes the latest developments in Networks and Systems—quickly, informally and with high quality. Original research reported in proceedings and post-proceedings represents the core of LNNS.

Volumes published in LNNS embrace all aspects and subfields of, as well as new challenges in, Networks and Systems.

The series contains proceedings and edited volumes in systems and networks, spanning the areas of Cyber-Physical Systems, Autonomous Systems, Sensor Networks, Control Systems, Energy Systems, Automotive Systems, Biological Systems, Vehicular Networking and Connected Vehicles, Aerospace Systems, Automation, Manufacturing, Smart Grids, Nonlinear Systems, Power Systems, Robotics, Social Systems, Economic Systems and other. Of particular value to both the contributors and the readership are the short publication timeframe and the world-wide distribution and exposure which enable both a wide and rapid dissemination of research output.

The series covers the theory, applications, and perspectives on the state of the art and future developments relevant to systems and networks, decision making, control, complex processes and related areas, as embedded in the fields of interdisciplinary and applied sciences, engineering, computer science, physics, economics, social, and life sciences, as well as the paradigms and methodologies behind them.

Indexed by SCOPUS, INSPEC, WTI Frankfurt eG, zbMATH, SCImago.

All books published in the series are submitted for consideration in Web of Science.

More information about this series at http://www.springer.com/series/15179

Rajiv Misra · Rudrapatna K. Shyamasundar ·
Amrita Chaturvedi · Rana Omer
Editors

Machine Learning and Big Data Analytics (Proceedings of International Conference on Machine Learning and Big Data Analytics (ICMLBDA) 2021)

 Springer

Editors
Rajiv Misra
Indian Institute of Technology Patna
Patna, India

Rudrapatna K. Shyamasundar🆔
Indian Institute of Technology Bombay
Mumbai, India

Amrita Chaturvedi
Indian Institute of Technology (BHU)
Varanasi, India

Rana Omer
Cardiff University
Cardiff, UK

ISSN 2367-3370   ISSN 2367-3389   (electronic)
Lecture Notes in Networks and Systems
ISBN 978-3-030-82468-6   ISBN 978-3-030-82469-3   (eBook)
https://doi.org/10.1007/978-3-030-82469-3

© The Editor(s) (if applicable) and The Author(s), under exclusive license
to Springer Nature Switzerland AG 2022
This work is subject to copyright. All rights are solely and exclusively licensed by the Publisher, whether
the whole or part of the material is concerned, specifically the rights of translation, reprinting, reuse of
illustrations, recitation, broadcasting, reproduction on microfilms or in any other physical way, and
transmission or information storage and retrieval, electronic adaptation, computer software, or by similar
or dissimilar methodology now known or hereafter developed.
The use of general descriptive names, registered names, trademarks, service marks, etc. in this
publication does not imply, even in the absence of a specific statement, that such names are exempt from
the relevant protective laws and regulations and therefore free for general use.
The publisher, the authors and the editors are safe to assume that the advice and information in this
book are believed to be true and accurate at the date of publication. Neither the publisher nor the
authors or the editors give a warranty, expressed or implied, with respect to the material contained
herein or for any errors or omissions that may have been made. The publisher remains neutral with regard
to jurisdictional claims in published maps and institutional affiliations.

This Springer imprint is published by the registered company Springer Nature Switzerland AG
The registered company address is: Gewerbestrasse 11, 6330 Cham, Switzerland

Preface

This book Machine Learning and Big Data Analytics presents the proceedings of the International Conference on Machine Learning and Big Data Analytics (ICMLBDA) 2021 that was held in collaboration with Indian Institute of Technology (BHU), Varanasi, India; California State University, San Bernardino, USA; Emlyon Business School, France; and International Association of Academicians (IAASSE), USA.

The ICMLBDA 2021 provided a platform for researchers and professionals to share their research and reports of new technologies and applications in Machine Learning and Big Data Analytics like biometric recognition systems, medical diagnosis, industries, telecommunications, AI Petri Nets Model-Based Diagnosis, gaming, stock trading, intelligent aerospace systems, robot control, law, remote sensing and scientific discovery agents and multiagent systems, and natural language and Web intelligence.

The ICMLBDA 2021 tried to bridge the gap between these non-coherent disciplines of knowledge and fosters unified development in next generation computational models for machine intelligence. This conference basically focused on advanced automation, computational optimization of machine learning in all engineering-based applications as well as includes specific plenary sessions, invited talks and paper presentations focusing on the applications of ML and BDA in the fields of computer/electronics/electrical/mechanical/chemical/textile engineering, healthcare and agriculture, business and social media and other relevant domains.

Due to the outbreak of COVID-19, this year's conference was organized as fully virtual conference. This was an incredible opportunity to experiment with a conference format that we plan to continue in the future. In order to prepare ICMLBDA 2021, the organizing committees, reviewers, session chairs as well as all the authors and presenters have made a lot of efforts and contributions. Thank you very much for always being supportive to the conference. We could not have pulled off this convention without all of your hard work and dedication.

Organization

Organizing Committee

Chairman of Organizing Committee

Rajiv Misra   IIT Patna, India

Members of Organizing Committee

Omer Rana	Cardiff University, UK
Muttukrishnan Rajarajan	City University, London, UK
Bharadwaj Veeravalli	National University of Singapore, Singapore
R. K. Shyamsunder	IIT Bombay, India

Scientific Committee

Chairperson of Scientific Committee

Amrita Chaturvedi   IIT (BHU), Varanasi, India

Members of Scientific Committee

Imene Brigui	Emlyon Business School, France
Nishtha Kesswani	CURAJ, India
Sriparna Saha	IIT Patna, India
Ahmed Zobaa	Brunel University London, UK
Pramod Mishra	Banaras Hindu University, India
Doaa Wafik	BUC University, Egypt
Alboaie Lenuta	Alexandru Ioan Cuza University of Iasi, Romania
Sugata Sanyal	IIT Guwahati, India
Sami Bedra	Université Abbes Laghrour-Khenchela, Algeria
Khashayar Yazdani	University of Science and Technology, Malaysia
Fareeda Khodabocus	University of Mauritius, Mauritius

Shailen Gungah	University of Mauritius, Mauritius
Lo Man-fung	Hong Kong Polytechnic University, Hong Kong
Joseph Vella	University of Malta, Malta
Rajib Ghosh	NIT Patna, India
Yasha Hasija	DTU Delhi, India
Abu Kaisar Mohammad	DIU, Bangladesh
Nada Matta	Université de Technologie de Troyes, France
Sarachandran	Muscat College, Oman
Nader Nashat Nashed	Université Française d'Egypte, Egypt
Şaban Öztürk	Amasya Üniversitesi, Turkey

Contents

Analysis on Applying the Capabilities of Deep Learning Based Method for Underwater Fish Species Classification 1
Suja Cherukullapurath Mana and T. Sasipraba

Bug Assignment Through Advanced Linguistic Operations 12
Shubham Kumar, Saurabh Kumar, Rahul Agrawal, Nitesh Goyal,
Shubham Kumar, and Vivek Kumar

**An Empirical Framework for Bangla Word Sense Disambiguation
Using Statistical Approach** 22
Monisha Biswas, Omar Sharif, and Mohammed Moshiul Hoque

Engagement Analysis of Students in Online Learning Environments . . . 34
Sharanya Kamath, Palak Singhal, Govind Jeevan, and B. Annappa

**Static and Dynamic Hand Gesture Recognition for Indian
Sign Language** 48
A. Susitha, N. Geetha, R. Suhirtha, and A. Swetha

**An Application of Transfer Learning: Fine-Tuning BERT
for Spam Email Classification** 67
Amol P. Bhopale and Ashish Tiwari

Concurrent Vowel Identification Using the Deep Neural Network 78
Vandana Prasad and Anantha Krishna Chintanpalli

**Application of Artificial Intelligence to Predict the Degradation
of Potential mRNA Vaccines Developed to Treat SARS-CoV-2** 85
Ankitha Giridhar and Niranjana Sampathila

**Explainable AI for Healthcare: A Study for Interpreting
Diabetes Prediction** 95
Neel Gandhi and Shakti Mishra

**QR Based Paperless Out-Patient Health and Consultation Records
Sharing System** 106
S. Thiruchadai Pandeeswari, S. Padmavathi, and S. S. Srilakshmi

**Searching Pattern in DNA Sequence Using ECC-Diffie-Hellman
Exchange Based Hash Function: An Efficient Approach** 117
M. Ravikumar, M. C. Prashanth, and B. J. Shivaprasad

**Integrated Micro-Video Recommender Based on Hadoop
and Web-Scrapper** 128
Jyoti Raj, Amirul Hoque, and Ashim Saha

**Comparison of Machine Learning Techniques to Predict Academic
Performance of Students** 141
Bhavesh Patel

**Misinformation–A Challenge to Medical Sciences:
A Systematic Review** 150
Arpita Sharma and Yasha Hasija

**Comparative Analysis Grey Wolf Optimization Technique & Its
Diverse Applications in E-Commerce Market Prediction** 160
Shital S. Borse and Vijayalaxmi Kadroli

**Applying Extreme Gradient Boosting for Surface EMG Based Sign
Language Recognition** 175
Shashank Kumar Singh, Amrita Chaturvedi, and Alok Prakash

**Automated Sleep Staging System Based on Ensemble Learning
Model Using Single-Channel EEG Signal** 186
Santosh Kumar Satapathy, Hari Kishan Kondaveeti,
and Ravisankar Malladi

**Histopathological Image Classification Using Ensemble
Transfer Learning** 203
Binet Rose Devassy and Jobin K. Antony

**A Deep Feature Concatenation Approach for Lung
Nodule Classification** 213
Amrita Naik, Damodar Reddy Edla, and Ramesh Dharavath

**A Deep Learning Approach for Anomaly-Based Network Intrusion
Detection Systems: A Survey and an Objective Comparison** 227
Shailender Kumar, Namrata Jha, and Nikhil Sachdeva

Review of Security Aspects of 51 Percent Attack on Blockchain 236
Vishali Aggarwal and Gagandeep

Transfer Learning Based Convolutional Neural Network (CNN)
for Early Diagnosis of Covid19 Disease Using Chest Radiographs 244
Siddharth Gupta, Avnish Panwar, Sonali Gupta, Manika Manwal,
and Manisha Aeri

Review of Advanced Driver Assistance Systems and Their
Applications for Collision Avoidance in Urban Driving Scenario 253
Manish M. Narkhede and Nilkanth B. Chopade

Segregation and User Interactive Visualization of Covid-19 Tweets
Using Text Mining Techniques 268
Gauri Chaudhary and Manali Kshirsagar

Sparse Representation Based Face Recognition Using VGGFace 280
Jitendra Madarkar and Poonam Sharma

Detection and Classification of Brain Tumor Using Convolutional
Neural Network (CNN) 289
Smita Deshmukh and Divya Tiwari

Software Fault Prediction Using Data Mining Techniques
on Software Metrics 304
Rakesh Kumar and Amrita Chaturvedi

MMAP: A Multi-Modal Automated Online Proctor 314
Aumkar Gadekar, Shreya Oak, Abhishek Revadekar, and Anant V. Nimkar

A Survey on Representation Learning in Visual
Question Answering 326
Manish Sahani, Priyadarshan Singh, Sachin Jangpangi,
and Shailender Kumar

Evidence Management System Using Blockchain and Distributed
File System (IPFS) 337
Shritesh Jamulkar, Preeti Chandrakar, Rifaqat Ali, Aman Agrawal,
and Kartik Tiwari

Author Index 361

Analysis on Applying the Capabilities of Deep Learning Based Method for Underwater Fish Species Classification

Suja Cherukullapurath Mana[1]($\boxtimes$) and T. Sasipraba[2]

[1] Department of CSE, Sathyabama Institute of Science and Technology, Chennai 603103, India
suja.cse@sathyabama.ac.in
[2] Sathyabama Institute of Science and Technology, Chennai 603103, India

Abstract. Fish species classification is an important part of fish conservation studies. Identifying the fish species from underwater images is a difficult task because of the facts like complexities of environment, high noise in the image and low illumination. Similarly the number images produced by unmanned underwater equipments are huge so that manual processing will not be suitable. Advances in machine learning field leads to deep learning based networks which can effectively extract features without much preprocessing of the image. Deep learning networks can be effectively utilized in marine species classification studies. This study analyses some of the deep learning and computer vision based methods to classify marine species.

Keywords: Deep learning · Convolution neural network · Fish species classification

1 Introduction

Species classification is an important part of any conservation studies. If we can clearly classify various species then studying about it will be easier. Manual classification of species is a tedious task and possibilities of errors are more in such process. With advancement in areas of machine learning, classification of species can be done effectively with less amount of time and more accuracy using the machine learning algorithms. Deep learning is a branch of machine learning in which artificial neural networks are being used to construct the model. In deep learning models a network similar to the human neuron connection is being designed. This model consist of many hidden layers connected together similar to how the neural cells of our body is being connected. These deep networks can effectively process information and extract features. These features can be utilized to classify various fish species.

This paper progresses in a way that in the upcoming sections authors gives an overview of deep learning and computer vision techniques. Further section describes some of the deep learning based models which can be effectively used for fish species classification, followed by conclusion and future work sections.

© The Author(s), under exclusive license to Springer Nature Switzerland AG 2022
R. Misra et al. (Eds.): ICMLBDA 2021, LNNS 256, pp. 1–11, 2022.
https://doi.org/10.1007/978-3-030-82469-3_1

2   Overview of Deep Learning

Deep learning networks are designed similar to the network of human neurons. Due to the increase in the amount of data needs to be processed every day, importance of models like deep learning is increasing in all scientific fields. The diagram (Fig. 1) displays deep learning which as a subfield of machine learning

Fig. 1. Deep learning as a subfield of machine learning

Deep learning is used in a wide range of areas like medical diagnosis, species classification in ecology, gaming etc. (Amiri et al. 2019). Because of the feature learning capability of deep learning networks the amount of preprocessing required is very less. This property of deep learning networks makes it beneficial for many applications. Deep learning models have the capability to learn complex features from large datasets (Amiri et al. 2019). The deep learning network is a deep artificial neural network consists of one or more hidden layers as shown in Fig. 2. Here the first hidden layer receive data from input layer process it and pass it to the second hidden layer. That hidden layer processes the received input and finally output will be given to the output layer.

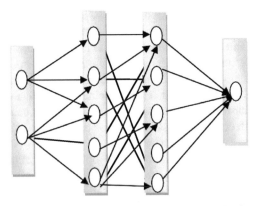

Fig. 2. An artificial neural network with two hidden layers

There are different types of deep learning models. Following sections briefly describe about some of the deep learning algorithms like convolution neural networks, recurrent neural networks, auto encoders and generative adversial networks.

Convolution Neural Networks (CNN)

Convolution neural network is an artificial neural network and the connectivity pattern in a convolution neural network is designed from the organization of visual cortex. Convolution neural network is highly effective in image processing applications (Hui and Yu-jie 2018; Zhang and Jing 2018; Yang and Li 2017). The convolution operation is defined between two matrices. There is a kernel defined which will slide over the matrix representation of input image. Specially designed kernels can be utilized for special purpose applications. Convolution neural network are very effective in the marine species classification because it can effectively process the images from under water. The amount of preprocessing required will be very less as the convolution neural network can effectively extract features and these features can be utilized to classify the animal species. Architecture of convolution neural network is given below.

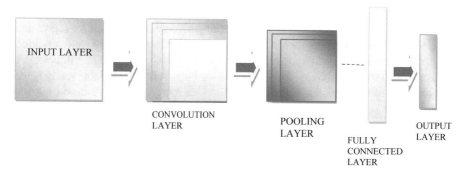

Fig. 3. Convolution neural network architecture

Figure 3 shows the architecture of convolution neural networks. The convolution layer consists of different number of hidden layers. There are pooling layer in between the convolution layer and these pooling layer will help to extract prominent features at each layers. These extracted features will be passed to next layer and will continue the same process in further hidden layers also. The input data will be classified based on these extracted features.

Recurrent Neural Networks (RNN)

Recurrent neural networks are another class of artificial neural networks. The prominent feature of recurrent neural network is the memory. Because of these memory RNN is suitable for applications like speech recognition, handwriting identification etc. (Yang et al. 2016). This model architecture has a recurrent loop which allows it to remember previous information. Long short term memory (LSTM) is a type of recurrent neural network which is highly successful in speech recognition type of application.

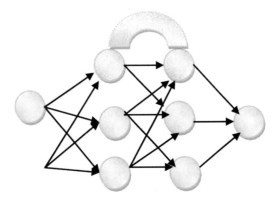

Fig. 4. Recurrent neural network architecture

Figure 4 shows recurrent neural network architecture. It has a loop between the hidden layers which is responsible for the memory elements of the network.

Auto Encoders
Auto encoder is an unsupervised machine learning algorithm. It is used for representational learning. Auto encoders' uses approximation functions to process and reproduce the input. Auto encoders will be sensitive to those data which is required for the model to precisely build reconstructions. Auto encoders are successfully used in anomaly detection, information retrieval, and data de noising applications. Structure od an auto encoder is given below.

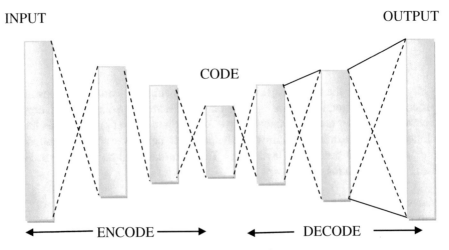

Fig. 5. Auto encoder architecture

Figure 5 shows architecture of an auto encoder.

3   Review of Deep Learning Based Fish Classification Models

This section analyses various deep learning based methods for fish species classification. The analysis of fish abundance, distribution and types of fish species is important to the fish conservation studies. Manual method to classify fish species are highly difficult, time consuming and expensive. With the development of machine learning and computer vision technologies latest algorithms can be effectively utilized for fish species classification. This session describes some of the deep learning based methods to classify fish species.

Convolution Neural Network Based Model
In the paper named 'Underwater Fish Species Classification using Convolution Neural Network and Deep Learning '(Dhruv Rathi et al. 2017) authors describe about a convolution neural network based model for fish species detection. This model tries to reduce the problems like presence of other water bodies, distortion of image and high noise level etc. in underwater image processing. The proposed model in the paper has an accuracy of 96.2%. The model uses convolution neural network based application to identify and classify the fish species in underwater images. The main advantages of using convolution neural network are that the amount of preprocessing needed is very less. The layers in the convolution neural network are capable to extracting features efficiently. Each layer will extract features and these features will be passed to the next hidden layers. Finally based on these features the model can effectively classify the fish species (Dhruv Rathi et al. 2017). Fish4knowlege dataset is used for this study.

The architecture of a fish classification model is given below (Fig. 6).

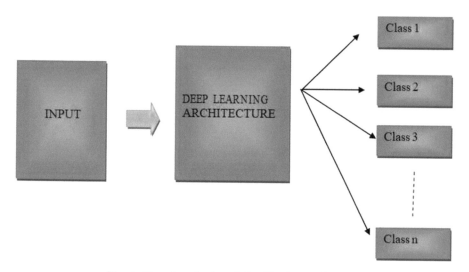

Fig. 6. Deep learning based classification architecture

Initial step is the preprocessing. In preprocessing step the noise will be removed and this preprocessed image will be given as an input to the model. Otsu's method is a preprocessing method used by this model. In Otsu's thresholding a threshold value will be set and it will compare the pixel value with the threshold value. If the pixel value is greater than the threshold value then it will be taken as white. Otherwise the pixel will be taken as black. As a result of this preprocessing method a grey level histogram will be created (Dhruv Rathi et al. 2017). The binarized image will then undergo erosion and dilation operations. After the preprocessing, the input will be given to convolution network. The convolution neural network consists of number of hidden layer with pooling layers in between. The convolution layers extract the features and only the important features will be passed to the next hidden layer. The same process will be repeated in the second layer also. Finally we have the fully connected layer where the trained output will be received. Pooling layers are placed in between the convolution layers and these pooling layers will extract the prominent features by max pooling, min pooling or average pooling. In the proposed model (Dhruv Rathi et al. 2017) three activation functions namely ReLU, tanh and softmax are being used. Relu is giving an overall accuracy of 96.29%, tanh is giving an accuracy of 77.26% and softmax is giving an accuracy of 61.91%.

In another study by Muhammad Ather Iqbal et al. (2019) suggest a modified AlexNet model for fish species classification. This model shows an accuracy of 90.48%. The performance is compared in terms of accuracy, number of hidden layers, number of fully connected layers etc. A dropout layer also introduced in this model which results in better accuracy. Fish species under different environments are taken for experimentation. The inclusion of dropout layer followed by the softmax layer resulted in better accuracy.

Hybrid Model Using Convolution Neural Network and Support Vector Machine
Support vector machine algorithm can be utilized along with the convolution neural network to improve the output. The output of the convolution neural network is given to the support vector machine algorithm (SVM) and this SVM is utilized to classify the data into various classes (Vikram Deep and Dash 2019). This is a hybrid approach in which the convolution neural networks will be utilized for feature extraction and support vector machine algorithm is utilized for classification. The architecture diagram of this hybrid model is given below.

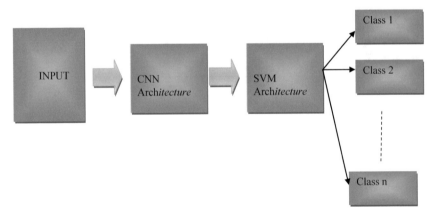

Fig. 7. Hybrid model using CNN and SVM

Figure 7 shows the architecture of the hybrid model. Here the input data is given to the CNN architecture. CNN architecture will do the preprocessing and feature extraction. The extracted features will be given to the SVM based model. The SVM model will perform the classification. The experiment is carried out with the help of fish4knowledge dataset and the model provides an accuracy of 98.32% (Vikram Deep and Dash 2019).

Hybrid Model Using Convolution Neural Network and KNN
Another hybrid model is by combining convolution neural network and K nearest neighbor (KNN) algorithm (Vikram Deep and Dash 2019). In this implementation the output of the convolution neural network model is given to the KNN model. Here the convolution neural network based model is used for feature extraction and classification is done by the KNN algorithm. KNN is a supervised machine learning algorithm. It will identify K nearest neighbor classes and will classify the input data into any one of these nearest neighboring classes. Here the input is given to the convolution neural network based model. This model will do the feature extraction and then that extracted features will be given to the KNN model. The KNN based model will do the classification. Figure 7 shows the general architecture of this hybrid model.

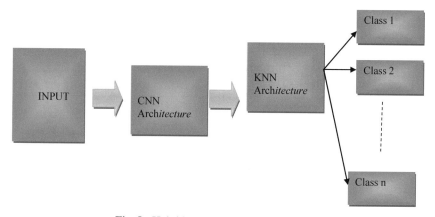

Fig. 8. Hybrid model using CNN and SVM

As shown in the diagram (Fig. 8) the input is given to the CNN model and CNN model will do the feature extraction. The pooling layer in between the hidden layer will apply the filter over the data. It will extract features and pass to the next hidden layer. Finally the output of the CNN model is given to the KNN model and it will do the classification. The final output is the different species of fishes (Vikram Deep and Dash 2019). In the study by Vikram Deep and Ratnakar Dash, this model is used for a dataset from the fish4knowledge and it is giving an accuracy of 98.79%.

4 Performance Evaluation

In order to increase the trustworthiness of the model performance evaluation is necessary. Some of the criteria for performance evaluation are precision, accuracy, recall, fscore and root mean square error.

Accuracy measures how many predictions are correct. Suppose AP is the correct prediction and TP is the total predictions then the formula for accuracy is given below.

$$\text{Accuracy} = (AP)/(TP)$$

Precision is the ratio of true positive (tp) to the total positive prediction which is the sum of true positive and false positive (fp). The formula for precision is given below.

$$\text{Precision} = (tp)/(tp + fp)$$

Recall is the ratio of true positive value to the sum of true positive and false negative (fn) values.

$$\text{Recall} = (tp)/(tp + fn)$$

The formula for fscore is given below.

$$\text{Fscore} = 2 . (\text{precision} * \text{recall})$$
$$(\text{precision} + \text{recall})$$

The CNN based model and hybrid models are tested with the fish4knowledge data set and their performance measures are discussed here (Vikram Deep and Dash 2019). The deepCNN model is giving an accuracy of 98.65%, the hybrid model with CNN and SVM is giving an accuracy of 98.32% and hybrid model with CNN and KNN is giving an accuracy 98.79% (Vikram Deep and Dash 2019). The graph showing the accuracy level is given below (Fig. 9).

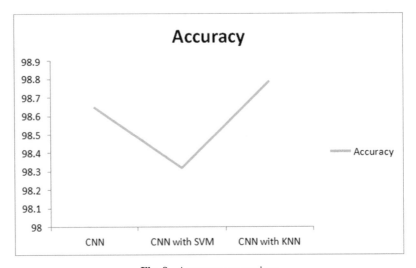

Fig. 9. Accuracy comparison

The precision of the CNN based model is 92.11%, hybrid CNN-SVM based model is 96.27 and hybrid CNN-KNN based model is 98.74. Similarly average recall for CNN

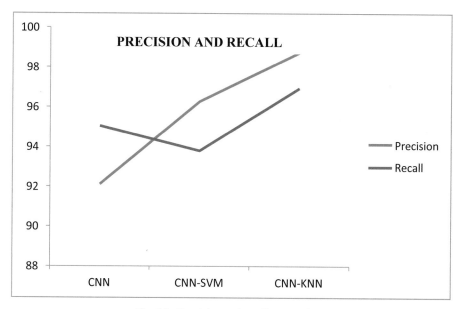

Fig. 10. Precision and recall comparison

based model is 95.04 %, hybrid CNN- SVM based model is 93.79% and CNN-KNN based model is 96.94%. The above measure is displayed in graphical format (Fig. 10)

The fscore of the model for the three architecture is shown in the graph below (Fig. 11).

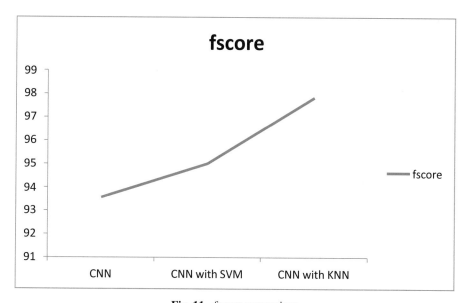

Fig. 11. fscore comparison

As shown in the graph (Fig. 11) the fscore of the CNN based model is 93.56%. The fscore of the hybrid model with CNN-SVM is 95.01% and 97.83% respectively (Vikram Deep and Dash 2019).

These performance evaluations clearly shows that the hybrid model with CNN-KNN outperform other models in accuracy, precision, recall and fscore values.

5   Conclusion and Future Work

This paper studies about various deep learning models suitable for fish species classification. Convolution neural network based models are highly successful in feature extraction and classification of fish species. In one of the studies (Vikram Deep and Dash 2019) this model when tested on the fish4knowledge dataset gives an accuracy of 98.65%. Hybrid models are formed by combining the CNN model with support vector machine algorithm and KNN algorithm. These hybrid models also highly effective in identifying and classifying the fish species. This study reveals the efficiency of deep learning based methods in fish classification and conservation. Manual process of fish classification is very much inefficient and time consuming. Scientists have to spend a lot of time to manually classify the fish species. Deep learning technology can be effectively utilized by the marine scientist to improve the efficiency of their research. Deep learning technologies can contribute a lot in the conservation studies also. This study is an initial step towards the implementation of an efficient deep learning based system to classify marine species. Authors are also working on the implementation of a mobile based application which can be conveniently used by marine scientist for species identification.

References

Amiri, S., Salimzadeh, S., Belloum, A.S.Z.: A Survey of scalable deep learning frameworks. In: 2019 15th International Conference on eScience (eScience), San Diego, CA, USA, pp. 650–651 (2019). https://doi.org/10.1109/eScience.2019.00102

Hui, L., Yu-jie, S.: Research on face recognition algorithm based on improved convolution neural network. In: 2018 13th IEEE Conference on Industrial Electronics and Applications (ICIEA), Wuhan, pp. 2802–2805 (2018). https://doi.org/10.1109/ICIEA.2018.8398186

Zhang, Y., Jing, L.: Convolution neural network application for glioma image processing. In: 2018 2nd IEEE Advanced Information Management, Communicates, Electronic and Automation Control Conference (IMCEC), Xi'an, pp. 2629–2632 (2018). https://doi.org/10.1109/IMCEC.2018.8469348

Yang, J., Li, J.: Application of deep convolution neural network. In: 2017 14th International Computer Conference on Wavelet Active Media Technology and Information Processing (ICCWAMTIP), Chengdu, pp. 229–232 (2017). https://doi.org/10.1109/ICCWAMTIP.2017.8301485

Yang, T., Tseng, T. Chen, C.: Recurrent neural network-based language models with variation in net topology, language, and granularity. In: 2016 International Conference on Asian Language Processing (IALP), Tainan, pp. 71–74 (2016). https://doi.org/10.1109/IALP.2016.7875937

Billa, J.: Dropout approaches for LSTM based speech recognition systems. In: 2018 IEEE International Conference on Acoustics, Speech and Signal Processing (ICASSP), Calgary, AB, pp. 5879–5883 (2018). https://doi.org/10.1109/ICASSP.2018.8462544

Atmaja, B.T., Akagi, M.: Speech emotion recognition based on speech segment using LSTM with attention model. In: 2019 IEEE International Conference on Signals and Systems (ICSigSys), Bandung, Indonesia, pp. 40–44 (2019). https://doi.org/10.1109/ICSIGSYS.2019.881108

Rathi, D., Jain, S., Indu, S.: Underwater fish species classification using convolutional neural network and deep learning. In: 2017 Ninth International Conference on Advances in Pattern Recognition (ICAPR), Bangalore, pp. 1–6 (2017)

Vikram Deep, B., Dash, R.: Underwater fish species recognition using deep learning techniques. In: Proceedings of 6th International Conference on Signal Processing and Integrated Networks (SPIN) (2019)

Iqbal, M.A., Wang, Z., Ali, Z.A., Riaz, S.: Automatic fish species classification using deep convolutional neural networks. Wireless Pers. Commun. **116**(2), 1043–1053 (2019). https://doi.org/10.1007/s11277-019-06634-1

Bug Assignment Through Advanced Linguistic Operations

Shubham Kumar$^{(\boxtimes)}$ ⓘ, Saurabh Kumar, Rahul Agrawal, Nitesh Goyal,
Shubham Kumar, and Vivek Kumar

Samsung Research Institute, Noida, India
sh3.kumar@samsung.com
http://samsung.com

Abstract. In this era we are blessed with huge amount of data, we come across a lot of ways by which we can reduce human efforts in common day to day work. With the availability of new and updated algorithms we can improve accuracy in our tasks. Bug assignment has been a huge pain for companies since years. It takes a lot of human effort and expertise to assign a reported bug to its correct owner. Resolution time for issues depends upon how quickly the reported bug was assigned to the correct owner. Moreover, developers waste a lot of time in bouncing the reported bugs to one another after analysing huge log files. Lots of time is wasted in analysing the issues which have already been fixed by some other developer. In this paper we will focus on approaches we can take to create some intelligent machine learning models for automatic bug assignment with high accuracy. We are trying to make an automated system which assign the issues directly to the developer. We have used two approaches to achieve this. First we have made an Artificial Neural Network (ANN) and Recurrent Neural Network (RNN) model on 2M bugs and by ensemble of these two Neural networks we have framed a final model with the help of optimum α. By applying this approach we have achieved accuracy of 83% @top-1 and 89% @top-2. To achieve more accuracy we have used transfer learning as we didn't have a huge dataset to make a large neural network so we have use state-of-art language model ULMFiT which is trained on 103 million words corpus. We fine-tuned that model with our dataset and used it for prediction .With this approach we achieved an accuracy of 89% @top-1 and 95% @top-2. So as soon as a new bug reported the models will predict the suitable developer to resolve that bug.

Keywords: Bug · Automated · Language model · Accuracy

1 Introduction

Big companies receive millions and millions of bugs and the worst part is to distribute bugs to the developer. Bug assignment activity wastes significant man-hour per day and leads to non-scalable process [1].

S. Kumar—ICMLBDA.

© The Author(s), under exclusive license to Springer Nature Switzerland AG 2022
R. Misra et al. (Eds.): ICMLBDA 2021, LNNS 256, pp. 12–21, 2022.
https://doi.org/10.1007/978-3-030-82469-3_2

There are few literature which address similar problem statement and provide different approaches. For a text document, IR algorithms like topic modeling [2] helps in inferring the inherent latent topics. A topic model converts terms in a document to topics which help in dealing with synonyms and polysemy problems. T-REC [3] provides a simple solution by comparing the text-similarity score and predicting top-5 suitable technical group for the bug. Similarily, there are few approaches [4–6] which takes care of the specialized technical groups. But for platforms like GSD [7] where number of developers is quite large, redundancy of technical expertise is quite common so it becomes difficult for the system to choose a particular developer for the issue.

We propose a robust solution which assigns the reported bug to the correct developer by using state-of-the-art deep learning models, taking care of technical group architecture and similarity of text as well. Figure 1 shows general overview of the proposed solution. We also consider expertise redundancy by the help of load balancing in various technical groups.

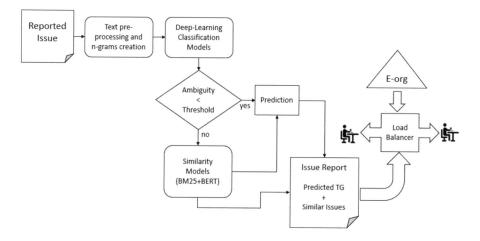

Fig. 1. Overview of solution architecture

Deep learning algorithms [8] have shown a lot of breakthroughs in Image Processing as well as Natural Language Processing tasks. In this paper we will see how we can use this intelligence for our bug assignment task. For this purpose we will use a robust architecture which will provide correct assignment as well as analysis of log files (future task) and description of similar issues for the reported bug. The assignment process takes the output generated by all these models into its consideration to minimize the scope of error and then we take care of the load balancing aspect also. However, the main core model is for categorization of reported issues into the correct group of developers in the organization and we use the rest of the models as plugins.

2   Our Work

In the upcoming sub-sections our approach will be discussed. Our work starts with data management part and then we create various models on the basis of the cleaned data.

2.1   Data

We collected 2 M rows of labelled data of resolved issues containing textual information of the issues as well as some categorical fields. Each row is labelled with its "Category", "Resolver Name" and "Resolver Department". For each Resolver Name, we got his working department details and with time duration for which he has been working in the department.

2.2   Data Pre-processing

To classify of the issues into its categories, we apply ML algorithms. For NLP tasks pre-processing of data is a crucial task. Pre-processing of the data can be described in two steps:

Basic Sanitation of the Text Data and Removal of Undesired Tokens. In this step, we apply basic pre-processing of text like lowering the case of the characters, stop words removal, expansion of abbreviations, removal of numeric digits, special characters removal, spelling correction etc.

In using advance state-of-the-art ML models, the performance of the models also depends upon the number of unique tokens we are using in the text data. So removal of tokens which do not have any semantic relation with the issue description becomes important. To achieve this efficiency, we created an Information Retrieval (IR) based algorithm which calculates term-frequency and inverted document-frequency for each token present in the corpus. Then these values are used to calculate a score for each token which describes the irrelevancy for that token with the corpus.

$$Score = (\alpha * tf + \beta * df)/(1 + tf * df) \tag{1}$$

Adding 1 in the denominator eliminate the chance of having unbounded score. α and β are two hyper-parameter fractions which respectively emphasise on the weightage we want to provide to tf and idf for finding the irrelevancy score of the tokens. α and β are model parameters not token parameters, it means both must be tuned for the whole corpus not for single tokens.

We picked up 1000 tokens with highest irrelevant scores and removed them from them form the corpus. Selection of the number of tokens to be removed should be chosen precisely as this create a speed-accuracy trade-off.

Creating Useful N-Gram Tokens. N-grams is defined as the collection of N number of tokens in a sequence. The idea behind creating n-gram tokens is that the language used in the issue description in IT companies is little different from natural language we use in our day to day tasks. So to use transfer learning on this data, creating new tokens provides learning the embedding (vector representation of a token in multi-dimensional plane) from scratch for the new tokens rather than using the embedding which were trained on natural data like Wikipedia. To understand this better, consider these two tokens "parental" & "control". If a model is trained on general language then its understanding will be different with respect to the developers working on "Parental Control" module in IT companies. So to provide better understanding of the tokens to the model we create a new token "parental_control". Using this idea yields quite a better accuracy (+2%) with same ML model.For creating new tokens we generated Part of Speech (POS) tagging for all the tokens in the corpus. POS tags of the tokens become important because the frequency of prepositions, pronouns, verbs are higher than nouns and adjectives. Then we count the occurrence of trigrams and bigrams which follow noun_adjective_noun, noun_noun_noun, adjective_noun and noun_noun pattern. If the frequency of these bigrams and trigrams are above the threshold then these tokens get replaced with a single token having '_' between them. As bigrams are more common than trigrams, so the threshold for the bigrams should be set higher than that of trigrams.

2.3   Classification Models

For classification of reported issues into its categories, we created a few machine learning models. We used models like Naive Bayes (NB), Support Vector Machine (SVM), Word embedding & ANN [9], Word embedding & RNN [10], Model Ensembling [11] and state-of-the-art models like Universal Language Model Fine-tuning (ULMFiT) and Bidirectional Encoder Representations from Transformers. In this paper we will discuss two models which provided better accuracies over other models.

Proposed Ensemble Model of ANN & RNN with Word Embedding. Word Embedding [12] is the vector representation of the word in multi-dimensional space. It provides a shallow relationship between words which are semantically similar. We created two separate deep learning models, one with three layered ANN architecture and other with two layered RNN architecture with LSTM flavour. Both the models output a 300-D vector, which is concatenated and followed by two relu layers and a softmax layer at the end. The softmax layer provides n number of probabilities corresponding to n number of categories of the issue. As shown in Fig. 2, on the top of the classification models we put the embedding layer. For the input to the model, a vocabulary of the tokens present in the corpus is created and the model is fed with the index of the tokens in the input sentences. The embedding of the words are also trainable parameters for the model. During the training of the complete model, sparse categorical cross entropy loss function was used to calculate the loss. The ensemble of the ANN & RNN models provided and accuracy of 83% @top-1 and 89% @top-2.

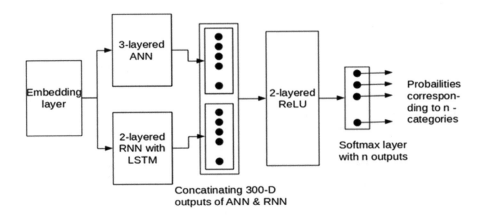

Fig. 2. Proposed ensemble model of ANN & RNN with word embedding

Language Modeling. Main task of a Language model is to predict the next word in a sequence. However we used inductive transfer learning approach in which source task is "next word prediction in a sequence" while target task is "classification of text in different categories". UMLFiT [13] is a state-of-the-art LM, which is trained on WikiText-103 dataset. We can fine tune this model with our dataset.

ULMFiT contains a three step process, i) General purpose LM pre-training, ii) LM fine-tuning and iii) Target task. The general purpose pre-trained models are available online which is trained on WikiText-103 dataset, a huge text corpus of 103 million words. However, it is important to fine tune the model on our dataset, as our data can contain tokens which are previously unseen to the model. ULMFiT models have an embedding layer on the top of three RNN layers of LSTM flavour. For next word prediction, the output of the LSTM layers is passed through a softmax layers which contains output units equal to the number of the words present in the vocabulary. But to use this model for our target task (i.e., classification of the issues) the last layer (softmax layer) of the language model is removed from the model and this is replaced by a relu layer followed by a softmax layer which output the probabilities of the issue lying in different categories.

ULMFiT model uses a three-layered RNN architecture, as shown in Fig. 3. On testing the model we found, there was a remarkable improvement and we took a jump of 6% in the accuracy to 89% @top-1 and 95% @top-2.

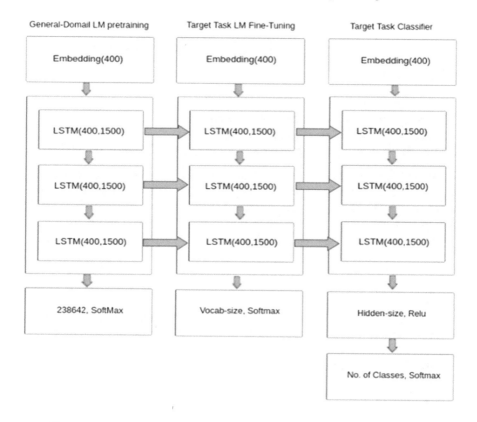

General-Domain LM pretraining   Target Task LM Fine-Tuning   Target Task Classifier

Fig. 3. Model architecture of ULMFiT along with shape of weight matrix

2.4   Similar Issues

For developers, the information of similar issues can be very helpful for resolving the reported issues and auto-resolving the issues with similarity score higher than threshold. For this task we considered textual features of the issue like Problem details, Reproduction root etc.

To generate the similarity score, we averaged the output of two algorithms: IR based BM25 [14] with some modification and ii) Cosine distance between BERT embedding matrix.

BM25. BM25 is a ranking algorithm used by various search engines in order to generate relevance score of documents to a given query. It uses bag-of-words approach based on the query terms and the terms appearing in each document. Equation 2 calculates the BM25 score corresponding to previously resolved issues (D, considered as documents) with respect to reported issue (Q, considered as query).

$$score(D,Q) = \sum_{i=1}^{n} IDF(q_i) * \frac{f(q_i, D) * (k_1 + 1)}{f(q_i, D) + k_1 * (1 - b + b * \frac{|D|}{avgdl}))} \qquad (2)$$

Where,

q_i are the query terms.

$|D|$ is the count of the words in the document D.

$avgdl$ is the average length of the documents.

b and k_1 are the hyper-parameters to amplify the effect of $avgdl$ and q_i

However for any query Q, its relevance or similarity score with document D has unbounded upper limit. But, to use this algorithm along with our second algorithm this score need to be bounded between 0 to 1. So to cast the BM25 score in this range we took best five results and normalized their score.

Final normalized score for j_{th} ranked document, D_j is given as,

$$Score_{norm}(D_j, Q) = Score_{BM25}(D_j, Q) / \sum_{i=1}^{5} Score_{BM25}(D_i, Q) \qquad (3)$$

Thus we get the similarity score of top five similar issues on the basis of bag-of-words method. But it is important to notice that bag-of-words approach lacks the semantic information of the text. So along with BM25, we also calculate the similarity score with the help of BERT [15] model.

BERT. BERT is the state-of-the-art model for many NLP tasks. It is a huge model basically based on transformer architecture. Explanation of BERT will be out of the scope for this paper. However, the idea is to fine tune this huge model on our dataset and then extract the Embedding of the words. BERT provides a 700 dimensional word-embedding vector for each word in the sentence. These word-embedding are normalized to get a document-embedding of same shape. Now, cosine distance between query sentence (newly reported issue, Q) and each document sentence (previously resolved issues) is calculated. Inverse of the cosine distance is the cosine similarity score between the query and the document sentence.

$$Score_{BERT}(D,Q) = 1/cos\_dist(Emb_{BERT}(Q), Emb_{BERT}(D)) \qquad (4)$$

Issues with top 5 scores are selected and then these scores are added to those of the top 5 similar issues selected by BM25 algorithm. If any issue is present in only one list then we add 0 to them and then all these values are divided by 2 to be normalized. At the end, final top 5 similar issues get selected on the basis of their normalized similarity score.

2.5   Load Balancing

Classification of issues provides the group to which the issue belongs. However, for assigning the issue we need to select a developer within this group. It is important to insure the involvement of every developer in some task as well as we also need to tackle the severity of the issue.

As discussed in data section, we are provided data for every employee like his date of joining, his man-hours in the modules he has worked for, total number of resolved issues, total number of open issues etc. We used some of the information to get a better and effective way for issue distribution among the members of the module. Severity (1 - Normal, 2 - High, 3 - Critical) of the newly reported issue plays an important role in finding the most suitable developer for it. For every new reported issue, we calculate a score corresponding to every developer in the module. To calculate this score we used the expertise (man-hours of the employee in that module) and total count of open issues of the employees. Before jumping to the formula, this is important to understand that the values of man-hours and the open issues count must be normalized to get a bounded result and the hyper-parameter can be adjustable.

$$X_{norm} = (X_i - X_{min})/(X_{max} - X_{min}) \qquad (5)$$

As negative value of man-hours and negative count of open issues makes no sense, so in the numerator we subtracted Xi with Xmin rather than Xavg, the normalized score will never be negative. Now, the score for each employee with respect to newly reported issue with severity S is calculated as:

$$Score = Severity * MH_{norm} - \beta * OPEN\_CT_{norm} \qquad (6)$$

Here, β is adjustable. Suppose in a module, new developers are recruited. As the count of developers increased, more number of issues should be assigned to developers which have low count of open issues. In this case β should be decreased to increase the chance (score) of assigning the issues to the newly recruited developers. The reported issue gets assigned to the developer with highest score.

3 Accuracy

We tried various models like Naive Bayes, SVM, TF-IDF vectorizer models in the start and then we chose our two best performing models and fine-tuned the models for better accuracy. ANN and RNN based Ensembling model yie 3lded an accuracy of 83% @top-1 and 89% @top-2. However the model which was picked for the production was the fine-tuned ULMFiT model with an accuracy of 89% @top-1 and 95% @top-2. Accuracies with all the models tried and tested has been shown in the Table 1.

Accuracies of the selected models was tracked on the production before selecting the final model. Ensemble model and ULMFiT model performed as shown in Fig. 4 on the testing data.

Table 1. Accuracies of different models

Models	% Accuracy @top-1	% Accuracy @top-2
Naive bayes	66	70
SVM	72	76
Ensemble	83	89
ULMFiT	89	95

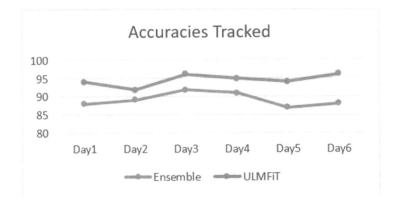

Fig. 4. Comparison of tracked accuracies of two best models.

4   Conclusion

This paper is intended to contribute in automation of bug assignment process. We are using Machine Learning algorithms to predict the suitable developer so that bug can be resolved in less time and reducing human effort. We were able to get quite a remarkable accuracy with the ULMFiT model.

However we are still trying to get higher accuarcy and tuning the model. We are also trying to develop a model for log analysis & prediction and plugging it to the current architecture.

Acknowledgment. We would like to thank Samsung's PLM team for providing great support to us from data collection to providing APIs for logs downloading. We sincerely thank our management team for providing great infrastructure support and valuable assets whenever required without which this project was far from achieving.

References

1. Zhang, T., Jiang, H., Luo, X., Chan, A.T.S.: A literature review of research in bug resolution: tasks, challenges and future directions. Comput. J. **59**(5), 741–773 (2016). https://doi.org/10.1093/comjnl/bxv114

2. Steyvers, M., Griffiths, T.: Probabilistic topic models. In: Landauer, C.T., Mcnamara, D., Dennis, S., Kintsch, W. (eds.) Latent Semantic Analysis: A Road to Meaning. Laurence Erlbaum, Hillsdale (2007)

3. Pahins, C.A.D.L., D'Morison, F., Rocha, T.M., Almeida, L.M., Batista, A.F., Souza, D.F.: T-REC: towards accurate bug triage for technical groups. In: 2019 18th IEEE International Conference On Machine Learning and Applications (ICMLA), Boca Raton, FL, USA, pp. 889–895 (2019). https://doi.org/10.1109/ICMLA.2019.00154

4. Xia, X., Lo, D., Ding, Y., Al-Kofahi, J.M., Nguyen, T.N., Wang, X.: Improving automated bug triaging with specialized topic model. IEEE Trans. Softw. Eng. **43**(3), 272–297 (2017). https://doi.org/10.1109/TSE.2016.2576454

5. Mani, S., Sankaran, A., Aralikatte, R.: Deeptriage: exploring the effectiveness of deep learning for bug triaging. In: Proceedings of the ACM India Joint International Conference on Data Science and Management of Data, ACM (2019). https://doi.org/10.1145/3297001.3297023

6. Yujian, L., Bo, L.: A normalized levenshtein distance metric. IEEE Trans. Pattern Anal. Mach. Intell. **29**(6), 1091–1095 (2007). https://doi.org/10.1109/TPAMI.2007.1078

7. Conchúir, E.O., Ågerfalk, P., Olsson, H., Fitzgerald, B.: Global software development: where are the benefits? Commun. ACM **52**(8), 127–131 (2009). https://doi.org/10.1145/1536616.1536648

8. Shrestha, A., Mahmood, A.: Review of deep learning algorithms and architectures. IEEE Access **7**, 53040–53065 (2019)

9. Mishra, M., Srivastava, M.: A view of artificial neural network. In: 2014 International Conference on Advances in Engineering and Technology Research (ICAETR - 2014), Unnao, India, pp. 1-3 (2014). https://doi.org/10.1109/ICAETR.2014.7012785

10. Sherstinsky, A.: Fundamentals of recurrent neural network (RNN) and long short-term memory (LSTM) network. Physica D: Nonlinear Phenom. **404**,(2020). https://doi.org/10.1016/j.physd.2019.132306

11. Opitz, D., Maclin, R.: Popular ensemble methods: an empirical study. J. Artif. Intell. Res. **11**, 169–198 (1999). https://doi.org/10.1613/jair.614

12. Tomas, M.: Efficient estimation of word representations in vector space. arXiv:1301.3781 (2013)

13. Faltl, S., Schimpke, M., Hackober, C.: ULMFiT: state-of-the-art in text analysis. https://humboldt-wi.github.io/blog/research/information_systems_1819/group4_ulmfit/. Accessed 10 Apr 2021

14. Jones, S.: Okapi BM25: A Non-Binary Model, Online, pp. 219–235. Cambridge University, London (2009)

15. Devlin, J., Chang, M.W., Lee, K., Toutanova, K.: BERT: pre-training of deep bidirectional transformers for language understanding. arXiv:1810.04805 (2019)

An Empirical Framework for Bangla Word Sense Disambiguation Using Statistical Approach

Monisha Biswas, Omar Sharif⬥, and Mohammed Moshiul Hoque$^{(\boxtimes)}$⬥

Department of Computer Science and Engineering, Chittagong University
of Engineering and Technology, Chittagong, Bangladesh
{omar.sharif,moshiul_240}@cuet.ac.bd

Abstract. Recently, word sense disambiguation has gained increased attention by NLP practitioner due to its various potential applications in language technology. This paper proposes a Naïve Bayes classifier for resolving lexical ambiguities of Bangla words with the help of a Bangla sense annotated corpus. At the initial stage, a Bangla sense annotated corpus is generated from a raw text corpus for serving as a training dataset. For a given input Bangla sentence, ambiguous words detection is done first and then Bayes probability theorem is applied to calculate the posterior probability that an ambiguous word belongs to a particular sense class. The values of posterior probability of several senses of the detected ambiguous word finally train the Naïve Bayes classifier to classify a closest sense of the ambiguous word. Experimental outcome reveals that the proposed method outdoes existing techniques by achieving the highest F1-score of 90% on the test data.

Keywords: Natural language processing · Word sense disambiguation · Sense annotated corpus · Ambiguous word · Bangla language

1 Introduction

Word Sense Disambiguation (WSD) is one of the significant research issues in NLP [1]. It is concerned the task of selecting the real sense of a word in context if it carried numerous interpretations. Words that have different meanings in different contexts are called ambiguous words. A human being can quickly identify these ambiguities, but for machines, it is difficult to detect the real sense of an ambiguous term in a context. Due to huge morphological complexities, many Bangla words have different meanings in different contexts, and these are called Bangla ambiguous words. For example, the Bangla word "হাত [Hand]" has different meanings in different contexts based on different feature words such as (i) তার হাত টানের অভ্যাস আছে [He has a habit of pulling hands] (ii) শিশুটির আজ হাতে খড়ি [Chalk in the child's hand today]. In (i) the word hand means চুরির অভ্যাস [The habit of stealing] and in (ii), it possesses the meaning

© The Author(s), under exclusive license to Springer Nature Switzerland AG 2022
R. Misra et al. (Eds.): ICMLBDA 2021, LNNS 256, pp. 22–33, 2022.
https://doi.org/10.1007/978-3-030-82469-3_3

প্রাথমিক শিক্ষা [Primary education] because of the feature word খড়ি [Chalk]. The determination of the real sense of a word is beneficial in various applications including machine translation, information retrieval, lexicography, knowledge mining, and semantic web. Developing an automated tool for WSD is a very complicated task for highly inflected languages like Bangla due to the substantial morphological variations of root words. Thus, selecting the real senses of the words in a given context is the critical challenge in WSD. There are about 70% of words are inflected in Bangla and due to this high level of inflections result in difficulty to develop Bangla language processing tools such as machine translation and WSD. Traditional dictionary-based techniques are not useful for any inflected language due to the existence of word-level inflections. To address the problem of Bangla WSD, this paper proposes a statistical based technique to identify the ambiguous Bangla words and find their real senses. The key contributions are:

- Develop a Bangla sense annotated corpus for WSD.
- Present a statistical based framework for disambiguation of Bangla ambiguous words by proposing a sense selection and ambiguity checking techniques.
- Investigate the performance of the proposed model by employing the comparative analysis of existing techniques with error analysis.

2   Related Work

There are three kinds of strategies have been explored to deals with problem of word sense disambiguation such as knowledge based, unsupervised and supervised [1]. The knowledge-based methodology [2] concerns on exterior resources such as semantic lexicon, thesauri, and wordnet to extract sense definitions of the lexical constituents. Samhith et al. [3] proposed a knowledge based WSD system, where lexical category of the given word is determined first, and then with the help of English WordNet, the correct sense of ambiguous word is found in a given context. A recent work on WSD used Bangla WordNet, where a knowledge-based disambiguation technique is used to extract sense of Bangla ambiguous word [4]. The system was tested on 9 common Bengali ambiguous words and achieved an accuracy rate of 75%. Haque et al. [5] proposed a dictionary based approach, where a machine-readable dictionary is used to detect ambiguous word. The system tested on limited size of dictionary and obtained 82.40% accuracy. In unsupervised method, first the sentences are clustered and these are tagged with appropriate senses [6]. Then, a distance-based similarity measure approach is utilized to determine the minimum distance of a test data with the sense-tagged clusters. A most recent work on unsupervised BWSD is proposed in [7] which achieved an accuracy of 61%.

Supervised WSD approaches require annotated training dataset in order to train the classification algorithms [8,9]. Several supervised statistical WSD methods have been developed for English and other languages [10,11]. A most recent work on supervised WSD system for Bangla language is proposed in [12], where four supervised methods such as decision tree, support vector machine, artificial neural network and Naïve Bayes are used as baseline methods. Among all techniques, Naive Bayes achieved the highest accuracy of 80.23%. Nazah et al. [13] proposed a statistical Bangla word sense disambiguation system which achieved of 82% accuracy.

3   Dataset Preparation

Bengali raw texts accumulated from several sources such as Bengali articles, short stories, news portals and blogs. Initially search operations were performed on the web to collect top links of sources based on a few popular Bangla ambiguous words. To extract raw texts from HTML contents; "JavaScript DOM API" is used and basic sanitization was done. About 90% of data of raw corpus is used to develop sense annotated training corpus and remaining 10% is preserved for testing purpose. Several activities are performed to prepare the sense annotated corpus, such as preprocessing, tokenization, stemming, POS tagging, a lexicon of ambiguous words creation, ambiguous word selection, sense selection.

Pre-processing: The collected raw text is normalized in python platform by (i) remove all numbers using str.translate() function, (ii) remove punctuation (!" %'()*+,-/:;<=>?@[]) using translate() method with string.punctuation, (iii) remove leading and ending white space by using strip() function, and (iv) convert the whole texts into single Unicode compatible Bangla font (i.e., Vrinda). The raw corpus contains 13241 Bangla sentences with 358850 words from which a sense annotated corpus is to be built. Figure 1 shows a sample of the raw corpus.

ছেলে পরীক্ষায় পাশ করে পিতার মুখ রাখল, আজ সে তার পরিবারের মান রাখল। কথা বলে খুব ভাল লাগল ছেলেটিকে।তাই যাওয়ার বেলায় তাকে জিজ্ঞেস করেছিলাম, তোমার শখ কী? উত্তরে সে বলেছিল, 'আংকেল মাঝেমধ্যে ছবি আঁকি, আর অবসরে মাঝে মাঝে বই পড়তে পছন্দ করি।

Fig. 1. A sample of the raw text

Tokenization: All sentences of the raw corpus divided into individual words or tokens by importing the built-in word tokenizing library using NLTK module. For example, the tokenizer generates output for the sentence, "তিনি গ্রামের মাথা ছিলেন [He was the head of the villge]" ; as "তিনি[He]" , "গ্রামের [Of Village]" , "মাথা [Head]" , "ছিলেন [Was]". Tokens are stored in an iterative list to convert them into their constituent root parts by stemming operation.

Stemming and POS Tagging: Stemming operation is done to convert inflected Bengali words into their constituent roots. The suffix part is stripped off from the inflected word and then validity of the remaining part is checked using a machine-readable dictionary. If the valid root word is not found in the dictionary, then the affix part should be stripped off again. After stemming, corresponding POS category for each root word is assigned with the help of an available machine-readable online Bangla dictionary. After assigning appropriate POS category to each word stem, the sample output of stemming and POS tagging process is shown in the following:

তিনি[He]/Pro/ তিনি [He], গ্রামের [of Village]/ Noun /গ্রাম [Village], মাথা [Head]/ Noun/ মাথা[Head]ছিলেন [Was]/ Aux /ছিলেন [Was].

Lexicon of Ambiguous Words Creation: A database of 48 popular Bangla ambiguous words with several senses are stored in a lexicon. Some sample ambiguous words with their different meanings or senses is shown in Table 1.

Table 1. A fragment of a lexicon of Bangla ambiguous words

Ambiguous word	Feature word	Sense word
হাত	টান	চুরির অভ্যাস
হাত	নেই	প্রভাব
হাত	পাতা	সাহায্য চাওয়া
হাত	দেয়া	কাজ শুরু করা
হাত	করা	বশীভূত করা
হাত	তুলা	প্রহার কর
হাত	Null	শরীরের অঙ্গ বিশেষ
মাথা	আছে	বুদ্ধি
মাথা	ঠেকান	প্রনাম করা
মাথা	কাটা	লজ্জা পাওয়া
মাথা	গরম	চটে যাওয়া
মাথা	গ্রাম	প্রধান ব্যক্তি
মাথা	ব্যথা	শারীরিক অসুস্থতা
মাথা	Null	শরীরের অঙ্গ বিশেষ

Ambiguous Word Detection and Sense Selection: With the help of lexicon, individual root word of each sentence (S_i) is checked one by one in order to find whether it is ambiguous or not. If an ambiguous word is found, its collocating left and right feature words in the sentence are compared with the corresponding feature word list to find the actual sense of the ambiguous word. For example, consider the sentence, "তিনি[He]/Pro/তিনি[He] গ্রামের [of Village]/Noun/গ্রাম [Village] মাথা [Head] "/ Noun/ মাথা [Head] ছিলেন[Was]/ Aux /ছিলেন [Was]";

From Table 1, it is observed that মাথা [Head] is stored as an ambiguous word in the lexicon. Thus, the system will detect "মাথা[Head]" as an ambiguous word. After ambiguous word detection, it will compare its left and right collocating root words "গ্রাম" and "ছিলেন" with the corresponding feature word list of ambiguous word "মাথা". There are five feature words ((আছে ,ঠেকান,কাটা, গরম, গ্রাম)) in the lexicon for the ambiguous word "মাথা" (as shown in Table 1). As the feature word "গ্রাম" is matched; the ambiguity detector will provide "প্রধান ব্যক্তি [The main person]" as the actual sense of the ambiguous word "মাথা" for the above sentence. If no ambiguous word is found from a sentence, that sentence will be discarded and not considered to store into the sense annotated corpus.

3.1 Bangla Sense Annotated Corpus

Bangla sense annotated corpus is a collection of validated words tagged with features such as POS types, root word forms and detection of Bangla ambiguous words tagged with their actual senses in different sentences. After ambiguous word detection and sense selection in the above step, each sentence is stored in the resultant sense annotated corpus and the ambiguous words of each sentence are tagged with their corresponding senses. A sample snapshot of our developed Bangla sense annotated corpus is illustrated in Fig. 2. The details procedure of creating sense annotated corpus described in [14]. The developed corpus contains 5028 Bangla sentences with sense annotation of 15 Bangla ambiguous words.

[1] করিম/Noun/করিম সাহেব/Noun/সাহেব গাঁয়ের/Noun/গাঁয়ের [aw_id=1, sense2=প্রধান] মাথা/Noun/মাথা

[2] পরিবারের/Noun/পরিবার [aw_id=2, sense13=সম্মান রাখা] মুখ/Noun/মুখ রাখা/verb/রাখা আমাদের/Noun/আমাদের সকলের/adv/সকল দায়িত্ব/Noun/দায়িত্ব

[3] তোমার/pron/তোমার [aw_id=1, sense1=অঙ্গ বিশেষ] মাথা/Noun/মাথা

[4] [aw_id=2, sense17=রাগ করা] মুখ/Noun/মুখ ভার/adj/ভার করে/verb/করা আর/adv/আর কতদিন/adv/কতদিন থাকবে/verb/থাকা

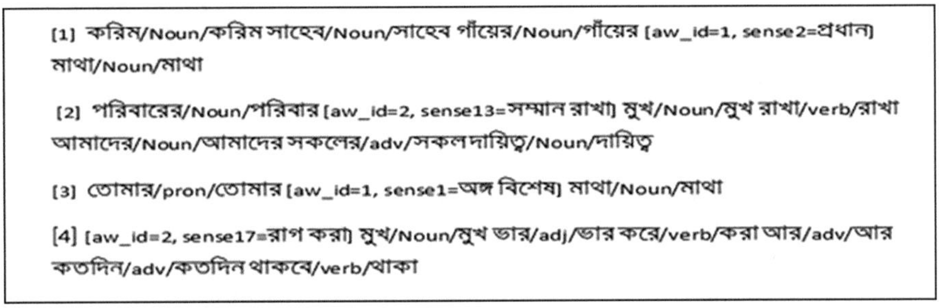

Fig. 2. Sample structure of the Bangla sense annotated corpus

4 Proposed BWSD Framework

The Bangla word sense disambiguation (BWSD) system consists of four major phases: corpus creation, ambiguity checker, training classifier and sense prediction. The corpus generation procedure is already stated above which is prerequisite to perform disambiguation process. The overall procedure of the proposed statistical BWSD system is illustrated in Fig. 3. After developing the sense annotated corpus, the procedure of our word sense disambiguation process can be described by the Algorithm 1. Each sentence s_i with ambiguous words is tokenized to create individual token list $W[w_1, w_2, ..., w_n]$. After that, stemming operation is performed to convert each word w_j into its corresponding root word r_k. Now, we check whether r_k exists in the predefined ambiguous word list CR. If r_k is in CR classifier model is trained with Bayesian probability estimation and r_k is disambiguated with closest sense S'. For a input sentence to the system, tokenization, POS tagging and stemming operations will be performed in similar manner as described in corpus generation section. After that, all root word list of a given test sentence are taken collectively to treat as input for the ambiguity checker program.

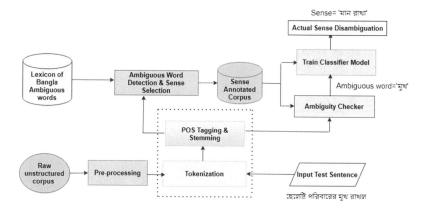

Fig. 3. Proposed framework of statistical based BWSD system

Algorithm 1: Sense Selection

---

1 $CR \leftarrow$ predefined ambiguous words list;
2 $S \leftarrow$ set of sentences with ambiguous words;
3 **for** $s_i \in S$ **do**
4 $\quad$ $W \leftarrow$ tokenized word list;
5 $\quad$ $W[w_1, w_2, ..., w_n] = tokenize(s_i)$;
6 $\quad$ **for** $w_j \in W$ **do**
7 $\quad\quad$ $r_k = stemming(w_j)$;
8 $\quad\quad$ **if** $(r_k \ in \ CR)$ **then**
9 $\quad\quad\quad$ train classifier using Bayesian probability estimation;
10 $\quad\quad\quad$ Disambiguate r_k with closest sense S';
11 $\quad\quad$ **else**
12 $\quad\quad\quad$ End disambiguation process;
13 $\quad\quad$ **end**
14 $\quad$ **end**
15 **end**

4.1 Ambiguity Checker

In Bengali language, a root word may has 20 or more types of inflections. Thus, most of the Bengali words may carry certain level of ambiguity. However, to simplify the disambiguation task, the most ambiguous Bangla words from a test instance should be identified by the system rather than considering minimal ambiguity of all words. An ambiguity checker function is developed to check whether any word of input sentence is found as ambiguous or not. Algorithm 2 illustrates the process of checking ambiguity of words. For each word r_i in root word list R check whether r_i present in the existing sense annotated corpus SAC. If word r_k found in SAC it is marked as ambiguous. Variable $flag$ used to determine whether ambiguous words present in R. If $flag$ is set to 1 then ambiguous words exists otherwise not. For example, if we consider one test sentence "ছেলেটির হাত টানের অভ্যাস আছে (The boy has a habit of pulling hands)"; then based on the information of the sense annotated corpus, the ambiguity detector will select "হাত [Hand]" as an ambiguous word.

Algorithm 2: Ambiguity checking

1 $SAC \leftarrow$ sense annotated corpus;
2 $R \leftarrow$ set of root words;
3 $flag = 0$;
4 **for** $r_i \in R$ **do**
5   **if** $(r_i\ in\ SAC)$ **then**
6   r_i is ambiguous;
7   flag=1;
8   **end**
9 **end**
10 **if** $(flag==0)$ **then**
11   no ambiguous words found;
12 **end**

4.2   Train Classifier

The words surrounding the ambiguous words in the sense annotated corpus are considered as a feature word set $F = (f_1, f_2, ..., f_n)$. For all feature words, ambiguous words and senses of ambiguous words of the sense annotated corpus, several parameters are considered to train disambiguation classifier (Table 2). A sample of the training corpus developed based on these parameters is shown in Table 3. From the table, it is shown that the ambiguous word "হাত [Hand]" is found in total 622 times in the whole corpus where it appears as sense, $S_i=$ "প্রহার করা [To beat]" for 36 times in presence of feature word "তোলা [lift]" and for 60 times in presence of feature word "দেয়া [Give]". So the total occurrence of the sense $S_i=$ "প্রহার করা [To beat]" is 96.

Table 2. Parameters considered for training classifier

Parameters	Description
Count(S_i)	Total occurrence of sense S_i for an ambiguous word w
Count(w)	Total occurrence of ambiguous word w
Count($F_j|S_i$)	No. of appearances of feature word F_j in a context of sense S_i

Table 3. Sample of statistical training model

Ambigu-ous word	Count(w)	Feature word	Corresponding sense	Count$(F_j|S_i)$	Count(S_i)
মাথা	615	গ্রাম	প্রধান ব্যক্তি	39	141
		গাঁয়ের	প্রধান ব্যক্তি	41	141
		এলাকা	প্রধান ব্যক্তি	38	141
		শহর	প্রধান ব্যক্তি	23	141
		পাতা	সম্মত হওয়া	38	40
		আছে	বুদ্ধি	35	37
		ঠেকান	প্রণাম করা	32	38
		আসা	বুঝতে পারা	33	33
		ব্যথা	শারীরিক অসুস্থতা	32	32
		আমার	শরীরের অঙ্গ বিশেষ	21	63
হাত	622	টান	চুরির অভ্যাস	28	28
		নেই	প্রভাব	38	45
		পাতা	সাহায্য চাওয়া	48	48
		আমার	শরীরের অঙ্গ বিশেষ	21	39
		খড়ি	প্রাথমিক শিক্ষা	38	38
		তোলা	প্রহার করা	36	96
		দেয়া	প্রহার করা	60	96

4.3 Actual Sense Prediction for Disambiguation

Bayes theorem is the most suitable statistical learning technique to acquire the probability of a target word concerning surrounding words [15]. As the multiple occurrences of words in the sense annotated corpus significantly impact the training model preparation, Multinomial Naive Bayes (MNB) is selected as a classifier which gives better classification performance [12]. MNB follows multinomial distribution and uses Bayes theorem to classify discrete features. For text disambiguation using Bayes rule, the conditional probability $P(S_i|F_j)$ that an ambiguous word (w) belongs to a particular sense class (S_i) in presence of feature word (F_j) can be estimated by using the Eq. 1.

$$P(S_i|F_j) = \frac{P(S_i) * P(F_j|S_i)}{P(F_j)} \qquad (1)$$

where, $P(S_i)$ denotes prior probability of sense S_i of an ambiguous word w, $P(F_j|S_i)$ means conditional probability of a feature word F_j in a given sense class S_i and $P(F_j)$ is the occurrence probability of feature word F_j which is constant. It is assumed that all feature words around an ambiguous word are conditionally independent. Thus, $P(F_j)$ is constant in the corpus and it does not influence the value of $P(S_i|F_j)$. Thus, Eq. 1 can be rewritten as Eq. 2.

$$P(S_i|F_j) = P(S_i) * P(F_j|S_i) \qquad (2)$$

here, $P(S_i)$ and $P(F_j|S_i)$ can be computed using Eqs. 3–4.

$$P(S_i) = \frac{Count(S_i)}{Count(w)} \tag{3}$$

$$P(F_j|S_i) = \frac{Count(F_j|S_i)}{Count(S_i)} \tag{4}$$

The classifier returns the class c' out of all classes $c \in C$ which has the maximum posterior probability in the given context. For a detected ambiguous word (w) in the given input sentence, the candidate classes $(S=S_1, S_2, ..., S_n)$ that represent the possible senses of the ambiguous word and the feature word set $(F = F_1, F_2, ..., F_n)$ in which the ambiguous word occurs in the given context, our WSD classifier will find out the proper sense class (S') that maximizes the posterior probability $P(S_i|F_j)$ by the Eq. 5.

$$S' = \arg\max_S P(S_i|F_j = F_1, F_2, ...F_n) \tag{5}$$

5   Experiments

Due to the smaller number of instances, we used 90% data for training and 10% data for testing purpose. To evaluate the proposed system, precision (P), recall (R) and F1-score are used as the performance measures.

5.1   Results

The different values of precision, recall and F1-score for few ambiguous words (concerning the test dataset) are illustrated in Table 4. Out of 210 test sentences, the proposed system can correctly predict the senses of ambiguous words of 187 sentences, which leads to an average accuracy of 90%. Figure 4a illustrates a sample snapshot of actual sense prediction for the ambiguous word: হাত (Hand).

Table 4. Performance of the proposed BWSD system

Ambiguous word	No. of inputs	No. of disam-biguation	No. of correct disam-biguation	P (%)	R (%)	F1-score
মাথা[Head]	55	52	48	92.30	87.27	0.89
বুক[Chest]	11	10	9	90.00	81.82	0.85
হাত[Hand]	51	49	47	96.00	92.16	0.94
মন[Mind]	31	29	28	96.55	90.32	0.93
মুখ[Face]	29	29	26	89.66	89.66	0.89
কড়া[Strict]	11	10	9	90.00	81.82	0.85
উঠা[Get Up]	9	9	8	88.89	88.89	0.89
পা[Leg]	13	13	12	92.30	92.30	0.92
Total	210	201	187	93.03	89.05	0.90

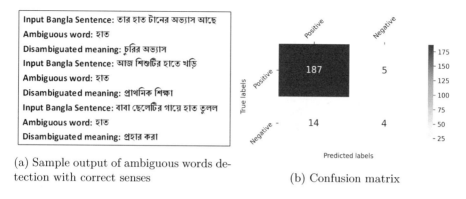

(a) Sample output of ambiguous words detection with correct senses

(b) Confusion matrix

Fig. 4. Sample output and confusion matrix of the proposed BWSD

A detail error analysis is carried out using the confusion matrix to investigate more insights concerning the model's performance. In Fig. 4b, the class *positive* denotes to the event when an ambiguous word is detected as well as sense disambiguation is performed and the class *negative* denotes the case when no ambiguity is found or no sense disambiguation is performed. The system gives false positive and false negative outputs for 14 and 5 input sentences respectively out of 210 sentences. Several reasons behind the misclassification by the proposed system are stated in Table 5 with some misclassified examples. The overall reason of these inappropriate classifications is mainly the scarcity of vast semantic information of Bengali ambiguous words in the sense annotated corpus as it is in developing phase and not a complete reference for disambiguating all types of Bengali ambiguous words.

5.2 Comparison with Previous Techniques

The proposed BWSD system is also compared with the previous techniques [9,13]. Due to the unavailability of standard dataset, comparison performed based on our developed dataset. Table 6 shows the detail results of the comparison based on precision, recall and F1-scores. Results reveal that the proposed system outperforms the previous techniques by achieving highest precision (93%), recall (89%) and F1-score (0.90). The use of well-developed sense annotated corpus in the training phase has increased performance of the proposed BWSD system than previous techniques. Figure 5 shows a comparative analysis between the proposed and existing techniques. The proposed method achieved the highest average accuracy of 89.03% compared to GRNN [13] (83.3%) and k-NN [9] (81.9%).

Table 5. Few examples of inappropriate classification by the proposed system

Input Sentence	Ambiguous word	Actual class	Predicted class	Reason of misclassification
তাতে তোমার এত মাথা ব্যথা কিসের? [Why are you so anxious for that?]	মাথা [Head]	দুশ্চিন্তা [Anxiety]	শারীরিক অসুস্থতা [Physical illness]	Surrounding same feature word "ব্যথা" in semantically dissimilar sentences
ঘটনার বিস্তারিত শুনে চোখের পাতা ভারী হয়ে উঠল [Hearing the details of the incident, the eyelids became heavy]	পাতা [Leaf]	কান্না আসা [To cry]	No ambiguity detected	Missing semantic ambiguity information of lexical word "পাতা" in sense annotated training corpus
ভীষণ কড়া! [Very strict]	কড়া [Strict]	কঠিন [Tough]	Null	Containing insufficient information for sense prediction as no contexual similarity is found for feature word "ভীষণ" with ambiguous word "কড়া"

Table 6. Comparison with existing BWSD systems

Methods	Input sentences	Disambiguated sentences	Correctly predicted sentences	P (%)	R (%)	F1-score
Proposed system	210	201	187	93	89	0.90
GRNN [13]	210	192	175	91	83	0.86
k-NN [9]	210	190	172	90	81	0.85

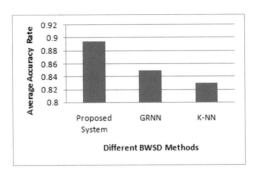

Fig. 5. Average accuracy of various WSD methods in Bengali

6   Conclusion

In this article, a framework of word sense disambiguation in Bengali language has been suggested based on the Naive Bayes technique. A sense selection process is used to detect the ambiguous words and to find out its appropriate sense in the sentence. A sense annotated corpus is also introduced to train the classifier model. The results revealed that the proposed statistical based WSD system outperforms previous systems. The size of the corpus can be extended to handle various context. Consideration of the function and content words detection, singular and conjugate forms detection and sense divergence may improve the performance of the Bengali word sense disambiguation system.

References

1. Pal, A.R., Saha, D., Das, N.S., Pal, A.: Word sense disambiguation in Bangla language using supervised methodology with necessary modifications. J. Inst. Eng. India Ser. B **99**(5), 519–526 (2018)
2. Parameswarappa, S., Narayana, V.N.: Kannada word sense disambiguation using decision list. Int. J. Emerg. T. Tec. Comput. Sci. **2**(3), 272–278 (2013)
3. Samhith, K., Tilak, A.S., Panda, G.: Word sense disambiguation using WordNet lexical categories. In: International Conference on SCOPES, pp. 1664–1666. India (2016)
4. Pal, R.A., Saha, D., Naskar, K.S.: Word sense disambiguation in Bengali: a knowledge based approach using Bengali WordNet. In: International Conference on ICECCT, pp. 1–5. India (2017)
5. Haque, A., Hoque, M.M.: Bangla word sense disambiguation system using dictionary based approach. In: Proceedings of the ICAICT (2016)
6. Pedersen, T.: In: Agirre, E., Edmonds, P. (Eds.): Word Sense Disambiguation: Algorithms and Applications. Springer (2007)
7. Pal, R.A., Saha, D.: Word sense disambiguation in Bengali language using unsupervised methodology with modifications. Sadhana **44**(168), 1–13 (2019)
8. Màrquez, L., Escudero, G., Martínez, D., Rigau, G: Supervised corpus-based methods for WSD. In: TLTB, vol. 33. Springer (2007)
9. Pandit, R., Naskar, K.S.: A memory based approach to word sense disambiguation in Bengali using k-NN method. In: International Conferences on Recent Trends in Information Systems, pp. 383–386. India (2015)
10. Brown, F.P., Pietra, D.J.V., Mercer, L.R.: Word sense disambiguation using statistical methods. In: Annual Meeting of the ACL, pp. 264–270. USA (1991)
11. Soltani, M., Faili, H.: A statistical approach on Persian word sense disambiguation. In: International Conference on Informatics and Systems, pp. 1–6. Cairo, Egypt (2010)
12. Pal, A.R., Saha, D., Dash, N.S., Naskar, S.K., Pal, A.: A novel approach to word sense disambiguation in Bengali language using supervised methodology. Sādhanā **44**(8), 1–12 (2019). https://doi.org/10.1007/s12046-019-1165-2
13. Nazah, S., Hoque, M.M., Hossain, R.: Word sense disambiguation of Bangla sentences using statistical approach. In: Proceedings of the ECCE, pp. 1–6. Bangladesh (2017)
14. Biswas, M., Hoque, M.M.: Development of a Bangla sense annotated corpus for word sense disambiguation. In: Proceedings of the ICBSLP, pp. 1–6. Sylhet, Bangladesh (2019)
15. Dawn, D.D., Shaikh, H.S., Pal, K.R.: A comprehensive review of Bengali word sense disambiguation. In: Artificial Intelligence Review, pp. 4183–4213. Springer (2019)

Engagement Analysis of Students in Online Learning Environments

Sharanya Kamath$^{(\boxtimes)}$, Palak Singhal, Govind Jeevan, and B. Annappa

Department of Computer Science, National Institute of Technology Karnataka, Mangalore, India

Abstract. Engagement rate is considered a metric that measures the extent of engagement a particular content is receiving from the audience. In e-learning settings, educators want to observe the level of interest of learners to appropriately modify their courses and make the educational process more effective. In this paper, an ensemble approach is proposed to detect student engagement levels while watching an e-learning video. The ensemble model consists of a deep convolutional neural network (DCNN) for facial expression recognition and a deep recurrent neural network (DRNN) for establishing a relationship between eye-gaze and engagement intensity. OpenFace 2.0 toolbox abilities are leveraged for feature extraction. Experimental results on the test datasets give an accuracy of 55.64% on DAiSEE and an MSE of 0.0598 on Engagement in the Wild Dataset.

1 Introduction

As the world moves rapidly towards digitization, swift advancements in technologies in the educational sector are observed, as online learning and Massive Open Online Courses (MOOCs) start becoming common across the world (Chatterjee and Nath 2014). This paper focuses on a comprehensive multi-model approach that aims at judging the level of attentiveness of students while attending an online lecture. The results provided would help the lecturers to appropriately study their audience, and modify the content according to the level of interest.

There have been studies that delve into the calculation of student engagement based on venues given to the audience to communicate with the speaker. For example, in the case of Wu He (2013), student online interaction is examined based on their online chat messages, and interaction with the teacher. In the case of Burch *et al.* (2015), surveys were developed for students to fill. The attentiveness data gained through student reviews is not comparable to the information gained by a quick glance over the classroom in an offline environment. Thus, this paper focuses on calculating engagement intensity on the basis of facial expressions and eye-gaze movements of the students while they watch an online lecture, in the hope to provide a similar kind of knowledge about the students' interests as in an offline classroom.

Equal contribution–All authors

Proceedings of the 38th International Conference on Machine Learning, PMLR 139, 2021. Copyright 2021 by the author(s).

© The Author(s), under exclusive license to Springer Nature Switzerland AG 2022
R. Misra et al. (Eds.): ICMLBDA 2021, LNNS 256, pp. 34–47, 2022.
https://doi.org/10.1007/978-3-030-82469-3_4

People usually rely on verbal and non-verbal cues to detect engagement. In some cases this is particularly easy, such as during driver attention detection (Nguyen et al. 2015). But monitoring an audience in a classroom environment is difficult. Student reactions may be subtle and quick. Manually identifying these can be tedious and prohibitive. Notable progress in computer vision has enabled the development of recent tools that can make the task of identifying facial landmarks easier. One such tool is OpenFace 2.0 (Baltrusaitis et al. 2018), which this paper uses to analyze input videos and obtain eye-gaze landmarks, which is further used as input to the model.

By leveraging the universality of webcam technology, this paper presents a solution to calculate student engagement by utilizing computer vision techniques. For this an ensemble approach is presented, comprising of two deep convolutional neural networks - one for analyzing the subject's facial expressions, and the other for tracking the subject's eye gaze movements.

An analysis of engagement analysis methods using computer vision techniques is covered in Sect. 2. In Sect. 3, 3.2 and 3.3 of the paper the proposed solution is explained in detail. Section 4 presents the performance comparison of this approach with existing approaches, followed by conclusions in Sect. 5.

2 Related Work

The computer vision based methods for learners' engagement detection can be divided into 3 categories- facial expressions, postures and gestures, and gaze movement, based on the modalities used for detection. As shown in Table 1, some studies use these modalities separately, while others combine two or more to achieve better accuracy.

Xuesong et al. (2018) used GRU networks over a feature set of facial action units, gaze, and head pose. Chang et al. (2018) proposed an ensemble model using classic machine learning techniques such as AdaBoost and K-Means, and deep learning techniques such as Bi-LSTM, while also tracking features such as hand and body fidget movements. Whitehill et al. (2014) studied the participants of a Cognitive Skills Training experiment. Their approach included facial analysis using Boost filter, SVM (Gabor) techniques, and CERT toolbox to determine the engagement of the participants.

Mustafa et al. (2018) suggested a multi-instance learning method utilizing Local Binary Patterns from Three Orthogonal Planes (LBP-TOP) while analyzing videos of students viewing MOOCs. Thomas et al. (2017) calculated engagement intensity of students based on their facial behavior indicators such as pose, gaze, and occurrences of facial AUs. Bidwell et al. (2011) came up with a framework to analyze individual behavior by using gaze targets as feature points.

This paper explains an approach that utilizes MTCNN for face detection, and a deep convolutional neural network (DCNN) along with the Haar Cascade module of OpenCV to perform facial expression recognition. It further utilizes OpenFace 2.0 to extract features corresponding to eye gaze, and a deep recurrent neural network (DRNN) with LSTM layers for establishing a relationship between gaze and engagement intensity.

Table 1. A systematic review of engagement analysis methods using computer vision techniques

Paper	Dataset used	Modalities considered	Methodology	Advantages	Disadvantages
(Thomas and Jayagopi 2017)	IIITB Facial Behaviour dataset	Eye gaze, head movement and facial AU	OpenFace to extract features. ENN to remove noise. Under-sampling of dataset is done to tackle imbalanced data	Correlation analysis effectively removes features which are negatively correlated with engagement	Occlusions over the face result in face detection failure. AUs are not robust
(Dewan et al. 2019)	SDMATH, DAiSEE, HBCU, In the Wild	Facial AUs, gestures, postures, tracking eye movement	A review of pre-existing methods used for engagement detection. Datasets and performance evaluation metrics are benchmarked	Removing images having large label disagreement helps in improving engagement detection results	Incomparable results due to lack of training data, large subject differences, differing environments
(Nezami et al. 2018)	FER-2013 dataset	Gaze behaviour and facial emotions	VGGNet model layers are optimized using a trained CNN	2 dimensions considered - behavior and emotion	Occluded faces have to be manually removed. Annotation isn't robust
(Hernandez et al. 2013)	MIT Media Lab TV audience behaviour dataset	Face angles/distances, head position, head size, head roll	Viola and Jones face detector. LIBSVM for classification	Forward Feature Selection (FFS) effectively finds the best subset of features	Expensive computation algorithms and invasive sensors are required
(Whitehill et al. 2014)	HBCU dataset	Eyes, nose, and mouth positions using CERT	CERT toolbox used for capturing features. Binary classification using Boost (Box Filter), SVM (Gabor) and MLR (CERT)	SVM is robust in generalisation in cases varied ethnicity subjects	The Box Filter overfits and CERT doesn't perform well
(Bidwell and Fuchs 2011)	FPG child engagement dataset	Intersection of gaze direction with manually defined objects in classroom	PittPatt SDK to extract individual gaze. HMM and Viterbi algorithm to classify sequences	The first to analyze automated gaze tracking as an engagement indicator	Low frequency behaviors of well-behaved students are difficult to track
(Chang et al. 2018)	EmotiW 2018	Eye gaze, head pose, facial AUs, hand and body fidget movements	An ensemble of K-Means clustering method and RNN model using BLSTM-Attention	Postural behaviour dynamics are analysed in a comprehensive manner	Model overfits due to lack of large dataset
(Niu et al. 2018)	EmotiW 2018	Gaze, head pose, facial AUs, Gaze-AU-Pose (GAP) feature	OpenFace 2.0 is used for feature extraction. Prediction model uses Gated Recurrent Unit (GRU) layers	The novel GAP feature is robust and captures engagement cues effectively	GAP + LBP-TOP performs worse due to subject-dependent data distribution

3　Proposed Work

3.1　Facial Expression Recognition

3.1.1　Dataset

The model for facial expression which classifies the users into 4 engagement levels based on expressions is trained using the "DAiSEE" dataset (Gupta et al. 2016). Figure 2 shows a few samples of frames from the DAiSEE dataset. The dataset consists of 9068 videos involving 112 users out of which 80 are male and 32 are female. The videos are taken in an unconstrained setting such as a library, lab, hostel room, etc. where the illumination level varies from light to dark. They are captured with a video camera mounted on a computer while the users' watch some educational videos. All the subjects in the dataset are Asians. The annotation for videos is done for the four labels namely: engagement, boredom, confusion, and frustration. Each of these labels has four intensities ranging from 0–3, signifying levels low to high.

However, in this paper, only labels corresponding to the users' engagement are required. The engagement label intensities range from 0–3, where 0 signifies disengaged and 3 denotes highly engaged. The best frame for each video from all the videos above are identified based on the keyframe extraction approach explained in (Huang et al. 2009). The resulting frames are then passed to an MTCNN (Zhang et al. 2016) model to detect and crop the faces. The detected faces are resized to 48 × 48 size. The number of images that could be used from the DAiSEE dataset after removing all the images containing more than one face (noisy images) is 6988 images for training and 1302 for testing purposes. However, all the images are further pre-processed before passing them to the model by applying Haar Cascade (Wilson 2006) framework of OpenCV which is used for object detection. It can accurately classify images with and without a face. Since a lot of images obtained from DAiSEE are occluded and the face isn't clearly visible, these samples are removed to make the dataset noise-free, and for better classification. However, this results in fewer images than the actual number of images present in DAiSEE. The final distribution of the dataset used is 5381 images for training, 1607 images for validation, and 853 images for testing. The training images are then passed to the proposed Facial Expression Recognition model for training.

3.1.2　Proposed Model for Facial Expression Recognition

The model is a DCNN based on the work of Gudi (2015), which mainly focuses on identifying the basic 7 emotions namely happy, sad, anger, fear, disgust, surprise, and neutral, trained using the FERC-2013 dataset (Goodfellow et al. 2013). Thus, this model is used as a baseline for the proposed work. The proposed model adopts the architecture presented by (Krizhevsky et al. 2012). The proposed network as shown in Fig. 1, begins with an input layer that takes images of the size 48 × 48. Following this layer is a convolution layer, a local contrast normalization layer, and finally a max-pooling layer. Two more convolution layers are added to detect fine-grained features. A fully connected layer is added,

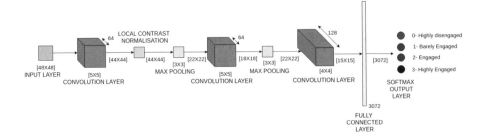

Fig. 1. Engagement intensity through facial expression model architecture

Fig. 2. Examples of frames from the DAiSEE dataset (Gupta et al. 2016)

followed by a final softmax output layer which predicts labels for four different classes namely disengaged, barely engaged, engaged, and highly engaged. Dropout is applied to all layers to prevent overfitting and the ReLu activation function is used. The predicted labels which range from 0–4 are then normalized to 0–1. This float value is thresholded to give an engagement level prediction, as explained in Sect. 3.3.

The model is tinkered more based on (Correa et al. 2016) and another max-pooling layer is added to reduce the parameters. It greatly lowers the computational intensity with very little loss in accuracy (around 1 to 2%). Furthermore, momentum is used to monitor the learning rate as the change in gradient slope is almost negligible.

3.2 Gaze Detection

Survey of previous work (Chang et al. 2018; Kaur et al. 2018; Niu et al. 2018) had shown that level of engagement had a high correlation with eye-gaze movement. Thus, OpenFace is used to track eye-gaze landmarks and use a deep recurrent neural network (DRNN) to predict engagement intensity based on it.

3.2.1 Dataset

The model to detect engagement level via gaze is trained on the "Engagement in the wild dataset" (Kaur et al. 2018), which is used in the Student Engagement prediction task, one of the challenges in EmotiW 2019 (Dhall 2019). Figure 3 shows a few samples of frames from the Engagement in the wild dataset. The dataset consists of 75 subjects - 25 females and 50 males, reacting to educational

Fig. 3. Examples of frames from Engagement in the Wild dataset (Kaur et al. 2018)

videos that were 5 min in duration. Their reactions are captured using a webcam, with the environments varying between classrooms, homes, canteens, and open grounds. In total, 102 videos are compiled, and manual annotation is done for four engagement levels. A group of five annotators rated the videos based on the subjects' eye gazes and classified them into engagement levels 0 to 3. The engagement level mapping with the corresponding gaze attributes is as follows:

- Level 0: completely disengaged. The viewer's gaze is always away from the screen.
- Level 1: barely engaged. The viewer views the screen intermittently with partially closed eyes.
- Level 2: engaged. The viewer looks at the screen and seems to be involved in the content.
- Level 3: highly engaged. The viewer doesn't take their eyes off the screen.

3.2.2 Proposed Model for Eye Gaze Detection

The architecture of this model is based on the approach suggested in (Thong Huynh et al. 2019). The input video is first processed by OpenFace(Baltrusaitis et al. 2018), an open-source toolkit for the analysis of facial components. This toolkit employs MTCNN (Zhang et al. 2016) for detection of faces from the video frames and makes use of techniques mentioned by (Baltrusaitis et al. 2013; Zadeh et al. 2017) for detection, extraction, and tracking of facial landmarks. It adopts the methodology by (Wood et al. 2015) for eye gaze estimation and tracking, and (Baltrušaitis et al. 2015) for facial action unit detection.

For data preprocessing, the input video (10 s duration) is divided into 15 video segments and fed into OpenFace. The output consists of 300 feature points,

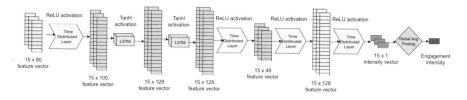

Fig. 4. Engagement intensity through gaze detection model architecture

out of which 60 feature points shown in Table 2 are utilized. The change in gaze direction of the subject overall the frames are captured by calculating the standard deviation and mean of head pose, 2D and 3D gaze landmarks, as well as gaze direction and angle. The eye location in the frames is one of the factors responsible for the change in the gaze direction. Thus, the eye-to-camera distance feature is also taken into consideration. This results in 60 feature points for each video segment, which results in a 15 × 60 feature vector for the input video.

Table 2. OpenFace 2.0 feature set summary

Feature type	Feature information	No. of feature points
Gaze direction	standard deviation, mean	16
2D, 3D eye landmarks, eye to camera distance	mean, standard deviation, coefficient of variation, min-max ratio	32
Head pose	standard deviation, mean	12
Total no. of feature points		60

The 15 × 60 input feature vector is passed through a DRNN as shown in Fig. 4 which consists of 2 LSTM layers and 3 fully connected layers, and a global average pooling layer. A dense layer is placed right before the LSTM layers to be able to learn any unseen representation of data. It returns a 15 × 100 feature vector, which is passed forward. The engagement intensity varies over time, and it depends on the current content being viewed along with the previous. These variations are analyzed by the LSTM layers. A 15 × 128 feature vector is obtained, which is fed through 3 time distributed fully connected layers to get the output separately in timesteps. Each of the LSTM layers applies the TanH activation function, while the fully connected layers apply ReLU activation.

These layers learn and capture the interrelation among adjacent input segments, and help in producing an engagement intensity prediction for each of the 15 input video segments as the output of the final fully connected layer. This output is a 15 × 1 vector in which each value is a float between 0 and 1. The final pooling layer takes an aggregate of these values and outputs the engagement intensity for the entire 10-s input video.

3.3   Ensemble Model

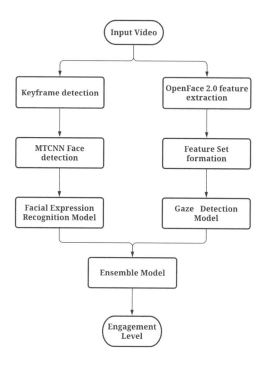

Fig. 5. Engagement level prediction system pipeline

The proposed approach also uses ensemble modeling, as shown in Fig. 5, in which multiple diverse models - Facial Expression Recognition model and Gaze Detection model are created to obtain better predictive performance than any single constituent model. As explained in the previous sections, this paper uses two different modeling algorithms, as well as different training datasets. The predictions P_1 and P_2 mentioned in Eq. 1 are aggregated, and the output is a float value in the range [0, 1].

$$engagement\ intensity = 0.5 \times (P_1 + P_2)$$
$$; P_1, P_2 \subseteq [0, 1] \qquad (1)$$

In the proposed approach, engagement intensity threshold is defined as:

- Level 0: 0.0–0.4
- Level 1: 0.4–0.6
- Level 2: 0.6–0.8
- Level 3: 0.8–1.0

Through the process of experimental trial and error, it is concluded that these threshold values give the best results of engagement analysis for the proposed facial expression and gaze detection ensemble approach.

4 Results

The results of the individual models are shown in Table 3. The detailed performance report of the ensemble model with percentage accuracy and mean squared error metrics corresponding to each engagement level is shown in Table 4. To see how well the ensemble model performed compared to existing models, a comparison analysis is done and the results are shown in Table 5 and 6. Since two datasets are used for training the models, a performance comparison is done using both datasets - DAiSEE and Engagement in the Wild. Performance on the DAiSEE dataset is evaluated by using Eq. 2 accuracy as a metric.:

$$accuracy = \frac{correctly\ predicted\ class}{total\ testing\ class} \times 100\% \tag{2}$$

Performance on the Engagement in the Wild dataset is evaluated using Mean Square Error (MSE) among the values predicted from the test set, and the ground truths. The MSE over all four levels is obtained by using the Eq. 3, where Y denotes predicted engagement intensity, $\bar{Y}$ denotes ground truth intensity, and n is the total number of samples.

$$OverallMSE = \frac{\sum_{i=0}^{n}(Y_i^2 - \bar{Y}_i^2)}{n} \tag{3}$$

Table 3. Results of DCNN and DRNN models

Number of participants: 1		
Model	Accuracy	MSE
DCNN (Facial Expression)	45.87%	0.105
DRNN (Gaze Detection)	41.30%	0.112
Ensemble (DCNN + DRNN)	46.32%	0.102
Number of Participants: 10		
DCNN (Facial Expression)	55.71%	0.051
DRNN (Gaze Detection)	50.35%	0.044
Ensemble (DCNN + DRNN)	57.23%	0.047
Number of Participants: 20		
DCNN (Facial Expression)	53.22%	0.083
DRNN (Gaze Detection)	48.50%	0.060
Ensemble (DCNN + DRNN)	55.64%	0.059

Table 4. Detailed Performance Report of Ensemble Model

DAiSEE dataset (Number of Participants: 20)					
Accuracy	Level 0	Level 1	Level 2	Level 3	Overall
Validation %	51.8%	65.64%	55.45%	60.65%	60.16%
Testing %	47.43%	60.81%	53.67%	57.81%	**55.64%**
Validation MSE	0.0508	0.0411	0.0831	0.0465	0.0501
Testing MSE	0.0486	0.0515	0.0798	0.0545	**0.0622**
In The Wild dataset (Number of Participants: 20)					
MSE	Level 0	Level 1	Level 2	Level 3	Overall
Validation %	46.23%	50.35%	49.68%	51.92%	51.56%
Testing %	46.41%	53.03%	50.98%	48.66%	**50.48%**
Validation MSE	0.1848	0.0201	0.1564	0.0482	0.045
Testing MSE	0.2204	0.0297	0.1033	0.0353	**0.0598**

Table 5. Performance comparison on DAiSEE dataset

Method	Accuracy
InceptionNet Video Level (Gupta et al. 2016)	46.4%
InceptionNet Frame Level (Gupta et al. 2016)	47.1%
C3D (Gupta et al. 2016)	48.6%
EmotioNet (Gupta et al. 2016)	51.07%
Proposed ensemble model	55.64%

Table 6. Performance comparison on in the wild dataset

Method	MSE
SVR + mean (Dhall et al. 2018)	0.15
DNN (Kaur 2018)	0.1416
GRU + GAP feature (Niu et al. 2018)	0.0724
Deep MIL (Chang et al. 2018)	0.0593
Proposed ensemble model	0.0598

The results of Table 5 and 6 indicate that the ensemble model outperforms various other methods that have been used for predicting student engagement - InceptionNet, EmotioNet, SVR, GRU, etc. The success of this proposed framework is the result of a combination of gaze action units and facial expression landmarks, along with a deep multi-model architecture. The performance improvements are also due to the diversity in the feature type as well as feature representation.

The proposed approach predicts the engagement intensity with a low error even in varying conditions of illuminations and environment settings because of the robustness of the OpenFace toolbox to capture features.

4.1   Real-Time Engagement Analysis Capability

The computation load on testing is minimal because the models are pre-trained and the weights are stored. Since OpenFace2.0 feature retrieval is also efficient, the proposed approach of engagement analysis can operate on webcam input feed in real-time. The ensemble model makes an engagement prediction at every 10-second interval of the video received.

Table 7. Real-time performance evaluation

Number of participants: 10					
Size	Level 0	Level 1	Level 2	Level 3	Overall
10 s	44.2%	46.51%	42.66%	42.9%	43.51%
30 s	48.91%	53.27%	50.01%	53.65%	50.95%
1 min	55.45%	55.1%	45.6%	56.33%	55.12%
2 min	52.62%	61.64%	58.77%	60.5%	58.78%

Table 7 shows the performance evaluation results of the real-time testing. For this testing, 10 participants were presented with e-learning videos clips of sizes 10 s to 2 min. At the end of each clip, a survey was filled by the participants to rate their level of engagement. The overall accuracy is calculated by taking the weighted average of the accuracies at each level. The optimal number of participants the model can tolerate in real-time is 10. A significant delay is observed for more than 10 participants.

5   Conclusion

This paper proposed an ensemble approach to detect student engagement levels on the basis of their facial expressions and gaze movements while watching an e-learning video. This approach presented the following novel improvements: (a) the use of combined datasets for training- DAiSEE and In the Wild datasets, leading to a better generic model which works in the real world, (b) fast engagement level predictions with temporal dynamics which enables real-time engagement analysis, (c) a model ensemble for robust and accurate predictions, and (d) a notable improvement against the baseline models using accuracy and MSE as evaluation metrics.

The proposed approach of analyzing the interest levels of learners has great potential for educational data mining. Future studies should work upon making the model robust to facial occlusions, as well as including the context of the tutoring session while making an engagement analysis. Additional features such as audio can also be examined. Other traits of facial expressions, such as how quickly an expression appears and how rapidly it disappears can also be analyzed.

References

Baltrusaitis, T., Robinson, P., Morency, L.: Constrained local neural fields for robust facial landmark detection in the wild. In: 2013 IEEE International Conference on Computer Vision Workshops, pp. 354–361 (2013)

Baltrusaitis, T., Zadeh, A., Lim, Y.C., Morency, L.: Openface 2.0: facial behavior analysis toolkit. In: 2018 13th IEEE International Conference on Automatic Face Gesture Recognition (FG 2018), pp. 59–66 (2018)

Baltrušaitis, T., Mahmoud, M., Robinson, P.: Cross-dataset learning and person-specific normalisation for automatic action unit detection. In: 2015 11th IEEE International Conference and Workshops on Automatic Face and Gesture Recognition (FG), vol. 06, pp. 1–6 (2015)

Bidwell, J., Fuchs, H.: Classroom analytics: measuring student engagement with automated gaze tracking, November 2011

Burch, G.F., Heller, N.A., Burch, J.J., Freed, R., Steed, S.A.: Student engagement: developing a conceptual framework and survey instrument. J. Educ. Bus. **90**(4), 224–229 (2015). https://doi.org/10.1080/08832323.2015.1019821

Chang, C., Zhang, C., Chen, L., Liu, Y.: An ensemble model using face and body tracking for engagement detection. In: Proceedings of the 20th ACM International Conference on Multimodal Interaction, ICMI 2018, pp. 616–622. Association for Computing Machinery, New York (2018). ISBN 9781450356923. https://doi.org/10.1145/3242969.3264986

Chatterjee, P., Nath, A.: Massive open online courses (MOOCs) in higher education – unleashing the potential in India. In: 2014 IEEE International Conference on MOOC, Innovation and Technology in Education (MITE), pp. 256–260 (2014)

Correa, E., Jonker, A., Ozo, M., Stolk, R.: Emotion recognition using deep convolutional neural networks. Tech. Rep. IN4015 (2016)

Dewan, M.A.A., Murshed, M., Lin, F.: Engagement detection in online learning: a review. Smart Learn. Environ. **6**(1), 1–20 (2019). https://doi.org/10.1186/s40561-018-0080-z

Dhall, A.: Emotiw 2019: automatic emotion, engagement and cohesion prediction tasks. In: 2019 International Conference on Multimodal Interaction, ICMI 2019, pp. 546–550. Association for Computing Machinery, New York (2019). ISBN 9781450368605. https://doi.org/10.1145/3340555.3355710

Dhall, A., Kaur, A., Goecke, R., Gedeon, T.: Emotiw 2018: audio-video, student engagement and group-level affect prediction. In: Proceedings of the 20th ACM International Conference on Multimodal Interaction, ICMI 2018, pp. 653–656. Association for Computing Machinery, New York (2018). ISBN 9781450356923. https://doi.org/10.1145/3242969.3264993

Goodfellow, I.J.: Challenges in representation learning: a report on three machine learning contests. In: Lee, M., Hirose, A., Hou, Z.-G., Kil, R.M. (eds.) Neural Information Processing, pp. 117–124. Springer, Heidelberg (2013). ISBN 978-3-642-42051-1

Gudi, A.: Recognizing semantic features in faces using deep learning. *ArXiv*. arXiv:abs/1512.00743 (2015)

Gupta, A., Jaiswal, R., Adhikari, S., Balasubramanian, V.: DAISEE: dataset for affective states in e-learning environments. CoRR. arXiv:abs/1609.01885 (2016)

He, W.: Examining students' online interaction in a live video streaming environment using data mining and text mining. Comput. Hum. Behav. **29**(1), 90–102 (2013). ISSN 0747-5632. https://doi.org/10.1016/j.chb.2012.07.020. Including Special Section Youth, Internet, and Wellbeing

Hernandez, J., Zicheng Liu, Hulten, G., DeBarr, D., Krum, K., Zhang, Z.: Measuring the engagement level of tv viewers. In: 2013 10th IEEE International Conference and Workshops on Automatic Face and Gesture Recognition (FG), pp. 1–7 (2013)

Huang, M., Shu, H., Jiang, J.: An algorithm of key-frame extraction based on adaptive threshold detection of multi-features. In: 2009 International Conference on Test and Measurement, vol. 1, pp. 149–152 (2009)

Kamath, A., Biswas, A., Balasubramanian, V.: A crowdsourced approach to student engagement recognition in e-learning environments. In: 2016 IEEE Winter Conference on Applications of Computer Vision (WACV), pp. 1–9 (2016)

Kaur, A.: Attention network for engagement prediction in the wild. In: Proceedings of the 20th ACM International Conference on Multimodal Interaction, ICMI 2018, pp. 516–519. Association for Computing Machinery, New York (2018). ISBN 9781450356923. https://doi.org/10.1145/3242969.3264972

Kaur, A., Mustafa, A., Mehta, L., Dhall, A.: Prediction and localization of student engagement in the wild. In: 2018 Digital Image Computing: Techniques and Applications (DICTA), pp. 1–8 (2018)

Krizhevsky, A., Sutskever, I., Hinton, G.E.: ImageNet classification with deep convolutional neural networks. In: Pereira, F., Burges, C.J.C., Bottou, L., Weinberger, K.Q. (eds.) Advances in Neural Information Processing Systems, vol. 25, pp. 1097–1105. Curran Associates, Inc. (2012)

Nezami, O.M., Dras, M., Hamey, L., Richards, D., Wan, S., Paris, C.: Automatic recognition of student engagement using deep learning and facial expression (2018)

Nguyen, T.P., Chew, M.T., Demidenko, S.: Eye tracking system to detect driver drowsiness. In: 2015 6th International Conference on Automation, Robotics and Applications (ICARA), pp. 472–477 (2015)

Niu, X., Han, H., Zeng, J., Sun, X., Shan, S., Huang, Y., Yang, S., Chen, X.: Automatic engagement prediction with gap feature. In: Proceedings of the 20th ACM International Conference on Multimodal Interaction, ICMI 2018, pp. 599–603. Association for Computing Machinery, New York (2018). ISBN 9781450356923. https://doi.org/10.1145/3242969.3264982

Thomas, C., Jayagopi, D.B.: Predicting student engagement in classrooms using facial behavioral cues. In: Proceedings of the 1st ACM SIGCHI International Workshop on Multimodal Interaction for Education, MIE 2017, pp. 33–40. Association for Computing Machinery, New York (2017). ISBN 9781450355575. https://doi.org/10.1145/3139513.3139514

Thong Huynh, V., Kim, S.-H., Lee, G.-S., Yang, H.-J.: Engagement intensity prediction with facial behavior features. In: 2019 International Conference on Multimodal Interaction, ICMI 2019, pp. 567–571. Association for Computing Machinery, New York (2019). ISBN 9781450368605. https://doi.org/10.1145/3340555.3355714

Whitehill, J., Serpell, Z., Lin, Y., Foster, A., Movellan, J.R.: The faces of engagement: automatic recognition of student engagement from facial expressions. IEEE Trans. Affect. Comput. **5**(1), 86–98 (2014)

Wilson, P.I., Fernandez, J.: Facial feature detection using HAAR classifiers. J. Comput. Sci. Coll. **21**(4), 127–133 (2006). ISSN 1937-4771

Wood, E., Baltrusaitis, T., Zhang, X., Sugano, Y., Robinson, P., Bulling, A.: Rendering of eyes for eye-shape registration and gaze estimation (2015)

Zadeh, A., Lim, Y., Baltrusaitis, T., Morency, L.: Convolutional experts constrained local model for 3D facial landmark detection. In: 2017 IEEE International Conference on Computer Vision Workshop (ICCVW), Los Alamitos, pp. 2519–2528. IEEE Computer Society, October 2017. https://doi.org/10.1109/ICCVW.2017.296

Zhang, K., Zhang, Z., Li, Z., Qiao, Y.: Joint face detection and alignment using multitask cascaded convolutional networks. IEEE Signal Process. Lett. **23**(10), 1499–1503 (2016). ISSN 1558-2361. https://doi.org/10.1109/lsp.2016.2603342

Static and Dynamic Hand Gesture Recognition for Indian Sign Language

A. Susitha, N. Geetha$^{(\boxtimes)}$, R. Suhirtha, and A. Swetha

Coimbatore Institute of Technology, Tamilnadu, Coimbatore, India
geetha.n@cit.edu.in

Abstract. Sign language recognition offers better types of assistance to the hard of hearing as it avoids the gap of communication between the deaf and mutes and the remaining people in the society. Hand signals, the essential mode of communication via gestures correspondence, plays a critical part in improving communication through gestures. Approaches for image detection, analysis and classification are available in glut, but the distinction between such approaches continues to be esoteric. It is essential that proper distinctions between such techniques should be interpreted and they should be analyzed. Standard Indian Sign Language (ISL) images of a person's hand photographed under several different environmental conditions are taken as the dataset. In this work, the system has been designed and developed which can recognize gestures in front of a web camera. The main aim is to acknowledge and classify hand gestures to their correct which means with the most accuracy doable. A unique approach for same has been planned and a few different wide standard models have compared with it. The novel model is made using canny edge detection, dilation, threshold and ORB. The preprocessed information is passed through many classifiers to draw effective results. The accuracy of the new models has been found considerably higher than the prevailing model.

Keywords: Computer vision · Gesture recognition · Feature extraction · Image processing · Indian sign language

1 Introduction

In recent years, the research in the development of natural, intuitive user interfaces have been developed. Such interfaces should be invisible to the users and allow them to interact with the application without using any of the specialized equipment. There should be a support for natural interaction and it has to be easily adaptable to the user. The application need to satisfy the role in real time with greater accuracy and supply robust against background clutter. These requirements and complexity makes it more challenging for the researchers.

© The Author(s), under exclusive license to Springer Nature Switzerland AG 2022
R. Misra et al. (Eds.): ICMLBDA 2021, LNNS 256, pp. 48–66, 2022.
https://doi.org/10.1007/978-3-030-82469-3_5

Hand gestures are considered as an intuitive and convenient means of communication between the human and the machines. This has led the research community to make more development and advancement in the hand gesture technologies. The significant ability of the natural user interface is the capability to identify the real time hand gestures.

As the technology is developing day by day, various methods of interaction with the computers have been developed. Traditionally started with a keyboard mouse and then joysticks, trackpads and electromagnetic gloves etc., have been used. Apart from these methods gesture recognition has been considered as a natural mode of interaction as it portrays the normal communication with human being. So it provides a better interface for r the Human Computer Interaction (HCI). Gesture recognition has many applications such as computer interaction, sign language recognition, robotic device manipulation etc., Gestures are identified as meaningful way to interact with the environment. These gestures can be either static or dynamic. In dynamic recognition it is necessary to identify both the spatial and temporal movements of hand and this paper proposes a method for dynamic gesture recognition. For static recognition, features extracted from the hand pose were considered for recognition whereas for dynamic gestures features extracted from the gesture trajectory were used. Feature selection has been considered as an important method for recognizing the gestures correctly.

Understanding sign language is associate degree arduous task and it's a talent that needs to be learned with observe. However, with this paper, we have a tendency to aim to supply many schemes for understanding such letters while not learning the language. We focus totally on the event of recent procedures to know language, and to seek out variations between the approaches and best technique of recognition of the language. There are many difficulties in developing a higher methodology for sign recognition like, in world the pictures captured are therefore to a fault screaky that top level of preprocessing is needed, the datasets offered on-line are usually noiseless, that working on them leads to the development of models trained only to handle images with less or nearly no noise, hence being impractical for real-life application. Thus, it's imperative to form a model which will handle noisy pictures and even be able to turn out positive results.

Indian sign language is found to be the widely used language for physically impaired (as shown in Fig. 1).

The model is built to acknowledge sign gesture pictures of the hand, that utilizes Oriented FAST and Rotated BRIEF (ORB) as a feature detector having effectiveness and performance higher than wide used feature detectors like SIFT and SURF, etc. The model utilizes a mixture of many alternative techniques and classifiers.

2   Related Work

Automatic sign language recognition gives a better service to the deaf as it bridges the existing communication between the rest of the society. Hand gesture is the primary mode of communication for sign language which plays a vital role in enhancing the sign language recognition [7]. There are various techniques which can be used to implemented for classification and recognition of images using machine learning. Work has also been done in the field of depth camera sensing and processing of videos apart from recognizing static images. Various processes embedded in the system was developed using different programming languages to implement the procedural techniques for the final system's maximum efficacy [5]. The Hand gesture recognition system focus on thresholding approach and skin color model along with an effective template matching using principal component analysis. The set of input frames are preprocessed for correct hand region extraction mistreatment numerous image process techniques. The preprocessing techniques such as face detection and elimination, hand segmentation, connected component extraction, and noise removal are considered for better hand segmentation [10].

The hand region of the image is segmented from the whole image using mask images available in open access dataset. The mask image is inverted and it is convolved with the hand image which results in the convolved image. The gray-level threshold is applied on the convolved image which produced hand region segmented image [3]. It is necessary to develop feature selection technique to obtain the optimal features. Feature Detection and Extraction is done through Oriented Fast and Rotated Brief (ORB). ORB is made up of two well-known descriptors namely FAST (Features from Accelerated and Segments Test) and BRIEF (Binary Robust Independent Elementary Features) with various modifications to improve the overall performance [5]. ORB feature detector is used to detect patches from the image and a 32-dimensional vector for each of the patches generated [5]. Canny edge detection is a good and widely used technique for the detection of the edges of a picture. It uses a multi-stage algorithm to differentiate sharp discontinuities or edges. This helps in reducing the ground noise in order that further techniques are often effectively applied [5]. Subsequent feature extraction can only be performed when the gesture is stable [1].

For dynamic gestures, hand gestures recognition depends most significantly on motion features, instead of skin color features, based on silhouette or shapes or edges and their variations over time because of its movements. Even though variations of skin color play little role, the information collection has been done by taking extreme care to incorporate participants with maximum variations to review the dependency of gesture recognition performance on human skin color [7].

3   Dataset and Methods

3.1   Description of Data Set

The dataset is taken with an aim to develop a benchmark for static and dynamic hand gesture recognition and the corresponding results of the classification are used as a reference for the future enhancements in the Indian Sign Language (ISL) recognition.

Fig. 1. Gestures for numbers (0–9) and alphabets

The dataset we are using for the model contains 1199 images for each of the alphabet and numbers. The total number of gestures were 35 which includes Indian sign language for numbers 1 to 9 and alphabets A to Z. In Fig. 1, it has been shown.

3.2 Methods

There are various models involved in implementation of each module starting from image pre-processing to classification of gesture. The system flow for classification is shown in the Fig. 2. Initially the hand movement is captured through web camera. Then the image is preprocessed in various methods and further feature extraction and feature selection is made using the suitable algorithm. Then the suitable testing and training of the images is done and the classification is made to recognize the given hand gesture input. The model is trained in such a way that maximum accuracy is obtained.

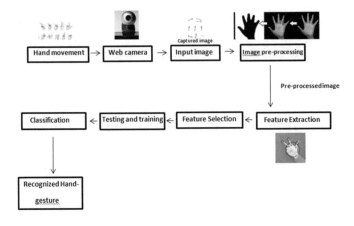

Fig. 2. Proposed architecture

4 Experimental Model

4.1 Image Preprocessing

The aim of pre-processing is to supply improvement of the image information that suppresses unsought distortions or enhances some image options relevant for more processing and analysis task. Image preprocessing use the redundancy in images. The images taken from the video will be preprocessed. A standard dataset has been used. The dataset contains 0 to 9 numbers and A to Z alphabets 1199 images each.

Fig. 3. Capturing of image using web camera

Capturing of Image

Initially the image is captured using the web camera and the captured image is saved in the local disk and further preprocessed as shown in Fig. 3.

RGB to Gray Conversion

Gray scaling is the process of converting an image from other color to gray shades. It varies between complete black and complete white. Since we want to convert our original image from the BGR color space to gray, we use the code COLOR_BGR2GRAY. This gray scale image is also saved. This has been shown in Fig. 4. The use of this conversion is to reduce the dimension and complexity reduction.

Image Thresholding

Thresholding is a method in OpenCV, which will be assigning pixel values according to the threshold value provided. In thresholding, each pixel value is compared with the value of threshold. The function used is cv2.threshold. OpenCV gives various thresholding methods and it is based on the fourth parameter of the function. Here the thresholding style used is THRESH_BINARY. Figure 5 shows the image thresholding.

Fig. 4. Conversion of RGB image to grayscale

Fig. 5. Image thresholding

Image Dilation

Then the image is further dilated by the process of dilation. This is done to increase the object area and to accentuate features. It increases the white region in the image or results in increase in size of foreground image. This is performed using the CV2.DIALATE function. Figure 6 shows the dilated image after image dilation.

Fig. 6. Dilated image after image dilation

4.2 Image Segmentation

Canny Edge Detection

On the converted gray scale image, canny edge detection is used here to generate only strong edges present in the image. Canny edge detection is an efficient and widely used technique for the detection of the sides in a picture. It uses a multi-stage algorithm to differentiate sharp discontinuities or edges. This helps in reducing the background noise so that further techniques can be effectively applied. Figure 7 shows the image segmentation. The function used is cv2.Canny.

4.3 Feature Extraction

Feature Detection and Extraction is done through Oriented Fast and Rotated Brief (ORB). ORB is a better feature detection and matching algorithm to SIFT or SURF. ORB is combination of two well-known descriptors FAST (Features from Accelerated and Segments Test) and BRIEF (Binary Robust Independent Elementary Features) with several modifications to spice up the performance. It firstly uses the FAST key point detector technique to compute key points along with the orientation which is calculated by the direction of the vector from the located corner point to the centroid of the patch. The orientation is not a part of FAST features so that ORB uses a multi-scale image pyramid. The key

Fig. 7. Image Segmentation using canny edge detection

points are effectively located at each pyramid level which contains the down sampled version of the image. ORB feature detector is used to detect patches (as shown in Fig. 8) from the image and a 32-dimensional vector for each of the patches generated. Thus, for every image belonging to a set of a single class of sign images, features are produced.

Algorithm-Oriented Fast and Rotated Brief
In this paper, Oriented Fast and Rotated Brief algorithm is used for feature selection and extraction. ORB is a combination of FAST key point detector and BRIEF descriptor with many modifications to improve the performance. First it uses FAST to seek out key points, then apply Harris corner measure to seek out top N points among them. It also uses pyramid to supply multiscale-features. It calculates the intensity weighted centroid of the patch located at center. The direction of the vector from this corner point to centroid gives the orientation. Thus, the features of the hand are detected using Oriented Fast and Rotated Brief algorithm. ORB is much faster compared to SURF and SIFT and ORB descriptor is found to be working better than SURF.

Algorithm 1 Oriented Fast and Rotated Brief

Begin
Step S1: Input the detected figure and carries out the SURF feature point detection;
Step S2: Determines unique point coordinate;
Step S3: Set up figure picture pyramid;
Step S4: Remove the unique point at picture edge;
Step S5: Calculate ORB unique point descriptor;
Step S6: Adopt K nearest neighbor algorithm to carry out feature points matching;
Step S7: Screening feature points matching to and output detections.
End

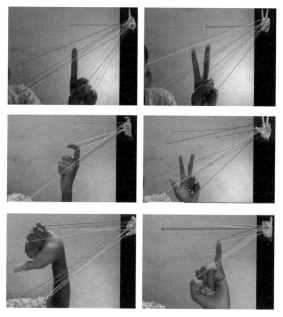

Fig. 8. Feature extraction

4.4 Classification

Before classification the dataset images is classified a 80% for training data and 20% for validation data. Then the labels are converted from categories to vectors. Label binarizer is used for binary classification. Here is type of model built is Sequential. Then two hidden layers are built using the sigmoid function. Then finally the output layer is built using the softmax function. Then the epochs are run to train the model. Here the optimizer used is SGD (Stochastic Gradient Descent Optimizer). In this project, binary cross entropy is used for the classification. Figure 12 shows the classification result.

Label Binarizer
Label Binarizer is from the SciKit Learn class that which will be accepting categorical data as input and returns an Numpy array as an output. Label Encoder, encodes the information into dummy variables for showing the presence of a specific label or not.

Stochastic Gradient Descent Optimizer
The word 'stochastic' means a system or a process that's linked with a random probability. In Stochastic Gradient Descent, a couple of samples are choosen randomly rather than choosing the entire data set for every iteration. In Gradient Descent, there's a term called "batch" which denotes the entire number of samples from a dataset that's used for calculating the gradient for every iteration. SGD has been employed to largescale problems often encountered in text classification and natural language processing. The advantage of using this algorithm is that it is more efficient and it is easy for implementation.

Binary Cross Entropy Classification
Binary cross entropy may be a loss function that's utilized in binary classification tasks. Sigmoid is that the only activation function compatible with the binary cross entropy loss function. Figure 9 shows the flowchart for classification.

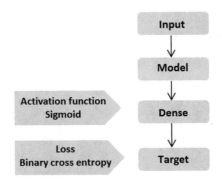

Fig. 9. Flowchart for classification

Training Loss and Validation Loss
Training loss is the error on the training set. Validation loss is the error after running the validation set of data through the trained network. Figure 10 shows this.

1) Underfitting.
    This is the case where the validation loss is less than the training loss, then the model is said to be underfit.

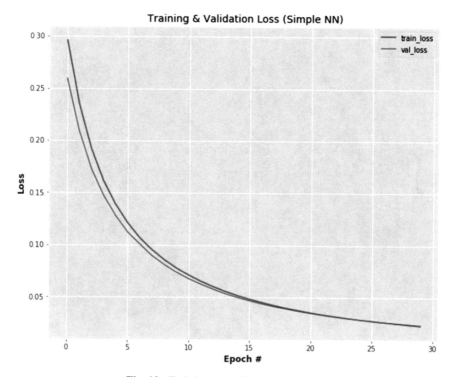

Fig. 10. Training and validation loss graph

2) Overfitting

This is the case where the validation loss is greater than the training loss, then the model is said to be underfit. This means that model is fitting very nicely for the training data but not for the validation data.

3) Perfect fitting

This is the case where the validation loss is equal to the training loss, then the model is said to be perfectly fit. If both values are found to be roughly the same and also if the values are converging then chances are very high

Training Accuracy and Validation Accuracy

Training Accuracy

The accuracy of a model on examples it was constructed on.

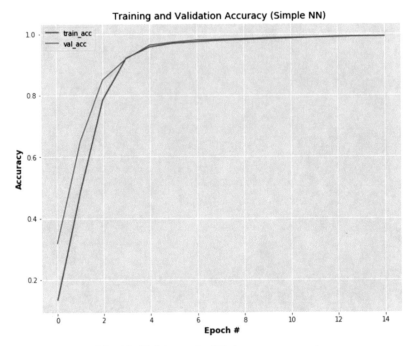

Fig. 11. Training and validation accuracy graph

Validation Accuracy

The test accuracy often refers to the validation accuracy, the accuracy calculated on the data set which is not used for training, but used for training for validating (or "testing") the generalization ability of your model or for "early stopping". Figure11 shows the training and validation accuracy graph.

Training of Samples

The overall classification performance of the network is high. In this paper, 80% of data is used for training and 20% of the data is used for validation. Feature extraction creates new transformed data from the base dataset, and the new images are added into the dataset to improve the model's gesture recognition accuracy. Figure 12 and Figure 13 shows the accuracy after running in number of epochs. Figure 14 shows the classification accuracy.

4.5 Prediction and Output Generation

Predictions are made using the batch size of 32 and the classification report is generated. Training and validation loss graph is generated. In addition to that training and validation accuracy graph is also plotted. Finally, the summary of the model is obtained and the model is saved to the disk. After saving the model the prediction is made and the output is obtained with the probability on the output image (shown in Table 1). The final output is achieved by the processing of above-mentioned model in an efficient way and the maximum accuracy is met with.

```
Train on 24960 samples, validate on 6240 samples
Epoch 1/30
24960/24960 [==============================] - 38s 2ms/step - loss: 0.2961 - ac
curacy: 0.9808 - val_loss: 0.2592 - val_accuracy: 0.9827
Epoch 2/30
24960/24960 [==============================] - 39s 2ms/step - loss: 0.2349 - ac
curacy: 0.9824 - val_loss: 0.2090 - val_accuracy: 0.9843
Epoch 3/30
24960/24960 [==============================] - 40s 2ms/step - loss: 0.1926 - ac
curacy: 0.9843 - val_loss: 0.1732 - val_accuracy: 0.9857
Epoch 4/30
24960/24960 [==============================] - 40s 2ms/step - loss: 0.1620 - ac
curacy: 0.9859 - val_loss: 0.1479 - val_accuracy: 0.9877
Epoch 5/30
24960/24960 [==============================] - 39s 2ms/step - loss: 0.1389 - ac
curacy: 0.9875 - val_loss: 0.1280 - val_accuracy: 0.9886
Epoch 6/30
24960/24960 [==============================] - 39s 2ms/step - loss: 0.1212 - ac
curacy: 0.9889 - val_loss: 0.1123 - val_accuracy: 0.9899
Epoch 7/30
24960/24960 [==============================] - 41s 2ms/step - loss: 0.1071 - ac
curacy: 0.9901 - val_loss: 0.1011 - val_accuracy: 0.9902
Epoch 8/30
24960/24960 [==============================] - 45s 2ms/step - loss: 0.0957 - ac
curacy: 0.9914 - val_loss: 0.0901 - val_accuracy: 0.9912
Epoch 9/30
24960/24960 [==============================] - 49s 2ms/step - loss: 0.0861 - ac
curacy: 0.9921 - val_loss: 0.0814 - val_accuracy: 0.9920
Epoch 10/30
24960/24960 [==============================] - 47s 2ms/step - loss: 0.0782 - ac
curacy: 0.9929 - val_loss: 0.0741 - val_accuracy: 0.9936
Epoch 11/30
24960/24960 [==============================] - 47s 2ms/step - loss: 0.0713 - ac
curacy: 0.9934 - val_loss: 0.0678 - val_accuracy: 0.9941
Epoch 12/30
24960/24960 [==============================] - 46s 2ms/step - loss: 0.0655 - ac
curacy: 0.9940 - val_loss: 0.0627 - val_accuracy: 0.9939
Epoch 13/30
24960/24960 [==============================] - 46s 2ms/step - loss: 0.0604 - ac
curacy: 0.9942 - val_loss: 0.0580 - val_accuracy: 0.9944
Epoch 14/30
24960/24960 [==============================] - 44s 2ms/step - loss: 0.0559 - ac
curacy: 0.9952 - val_loss: 0.0538 - val_accuracy: 0.9949
Epoch 15/30
24960/24960 [==============================] - 50s 2ms/step - loss: 0.0520 - ac
curacy: 0.9955 - val_loss: 0.0502 - val_accuracy: 0.9955
Epoch 16/30
24960/24960 [==============================] - 47s 2ms/step - loss: 0.0485 - ac
curacy: 0.9960 - val_loss: 0.0471 - val_accuracy: 0.9957
Epoch 17/30
24960/24960 [==============================] - 44s 2ms/step - loss: 0.0455 - ac
curacy: 0.9960 - val_loss: 0.0442 - val_accuracy: 0.9960
Epoch 18/30
24960/24960 [==============================] - 43s 2ms/step - loss: 0.0426 - ac
curacy: 0.9962 - val_loss: 0.0417 - val_accuracy: 0.9965
Epoch 19/30
24960/24960 [==============================] - 46s 2ms/step - loss: 0.0401 - ac
curacy: 0.9967 - val_loss: 0.0395 - val_accuracy: 0.9965
Epoch 20/30
24960/24960 [==============================] - 46s 2ms/step - loss: 0.0378 - ac
curacy: 0.9969 - val_loss: 0.0372 - val_accuracy: 0.9965
```

Fig. 12. Accuracy after running epochs

```
Epoch 21/30
24960/24960 [==============================] - 44s 2ms/step - loss: 0.0358 - ac
curacy: 0.9970 - val_loss: 0.0353 - val_accuracy: 0.9968
Epoch 22/30
24960/24960 [==============================] - 43s 2ms/step - loss: 0.0339 - ac
curacy: 0.9976 - val_loss: 0.0336 - val_accuracy: 0.9970
Epoch 23/30
24960/24960 [==============================] - 37s 1ms/step - loss: 0.0322 - ac
curacy: 0.9979 - val_loss: 0.0319 - val_accuracy: 0.9973
Epoch 24/30
24960/24960 [==============================] - 39s 2ms/step - loss: 0.0306 - ac
curacy: 0.9982 - val_loss: 0.0304 - val_accuracy: 0.9974
Epoch 25/30
24960/24960 [==============================] - 40s 2ms/step - loss: 0.0291 - ac
curacy: 0.9983 - val_loss: 0.0292 - val_accuracy: 0.9974
Epoch 26/30
24960/24960 [==============================] - 42s 2ms/step - loss: 0.0278 - ac
curacy: 0.9984 - val_loss: 0.0281 - val_accuracy: 0.9981
Epoch 27/30
24960/24960 [==============================] - 45s 2ms/step - loss: 0.0266 - ac
curacy: 0.9986 - val_loss: 0.0267 - val_accuracy: 0.9986
Epoch 28/30
24960/24960 [==============================] - 46s 2ms/step - loss: 0.0254 - ac
curacy: 0.9989 - val_loss: 0.0257 - val_accuracy: 0.9986
Epoch 29/30
24960/24960 [==============================] - 45s 2ms/step - loss: 0.0244 - ac
curacy: 0.9990 - val_loss: 0.0248 - val_accuracy: 0.9989
Epoch 30/30
24960/24960 [==============================] - 42s 2ms/step - loss: 0.0234 - ac
curacy: 0.9993 - val_loss: 0.0238 - val_accuracy: 0.9989
Time taken: 1300.0 seconds
```

Fig. 13. Accuracy after running epochs

```
[INFO] evaluating network...
              precision    recall  f1-score   support

           A       1.00      1.00      1.00       226
           B       1.00      1.00      1.00       226
           C       0.98      1.00      0.99       244
           D       1.00      1.00      1.00       258
           E       1.00      1.00      1.00       235
           F       1.00      1.00      1.00       231
           G       1.00      1.00      1.00       218
           H       1.00      1.00      1.00       234
           I       1.00      1.00      1.00       227
           J       1.00      1.00      1.00       242
           K       1.00      1.00      1.00       235
           L       1.00      1.00      1.00       235
           M       1.00      1.00      1.00       263
           N       1.00      1.00      1.00       225
           O       1.00      1.00      1.00       247
           P       1.00      1.00      1.00       205
           Q       1.00      1.00      1.00       247
           R       1.00      1.00      1.00       230
           S       1.00      1.00      1.00       249
           T       1.00      1.00      1.00       258
           U       0.99      0.99      0.99       271
           V       1.00      0.99      0.99       250
           W       1.00      1.00      1.00       224
           X       1.00      1.00      1.00       252
           Y       1.00      1.00      1.00       254
           Z       1.00      1.00      1.00       254

    accuracy                           1.00      6240
   macro avg       1.00      1.00      1.00      6240
weighted avg       1.00      1.00      1.00      6240
```

Fig. 14. Classification accuracy and precision

Table 1. Output with predicted gesture and accuracy

OUTPUT	PREDICTED GESTURE	ACCURACY (%)
	1	96.7%
	2	99.3%
	9	99.6%
	7	100%

(*continued*)

Table 1. (*continued*)

	8	93.4%
	6	100%
	3	91.8%
	4	99.9%
	5	97.2%

5 Conclusion

In this presented paper, a low cost, fast, appearance-based method for Indian sign gesture recognition using ORB feature extraction has been used. These approaches has been successfully skilled through prominent Binary classifier-Label Binarizer. The proposed technique outperform all the data preprocessing steps for hand detection like CANNY edge detection, Dilation and Thresholding. Although the approach gives substantially high accuracy for recognition of gestures. These properties combined with the provided simplicity during its development and its flexibility to be expanded by introducing and training the system with any hand gestures that the user prefers, justify its ability to be exploited on educational applications. The system is tested against static gesture images and may be further extended to acknowledge dynamic gestures in videos in real-time. The model are often trained on physical hand models to supply extra data which may be studied for improving the accuracy. This designed system is extremely much helpful and supply a standard mode of communication between deafdumb and normal people. Hence our system reduces the barrier of communication for deaf and dumb people. The proposed system is effective in recognizing the alphabets of Indian Sign Language. This system can definitely help millions of deaf people to communicate with other normal people.

The longer-term work of the proposed system is to extend a greater number of gesture images for gesture to speech recognition and using different sign languages. Moreover, We also aim to extend in the future new set of features could also be added as a future work to the features utilized in this paper to reinforce the performance of the system.

6 Future Work

The research work can be enhanced by taking the dataset captured in cluttered backgrounds and various illumination conditions. Indian Sign Language is still a very least explored field than the other sign language. A great achievement will be designing a real time Indian Sign Language recognition system which will be considering the facial expressions and the various contexts. As a future work 3D gestures and non-manual signs will also be included to make the system much more beneficial to the hearing impaired people.

References

1. Guangming, Z., Liang, Z., Peiyi, S., Juan, S., Shah, S.A.A., Bennamoun, M.: Continuous gesture segmentation and recognition using 3DCNN and convolutional LSTM. In: IEEE Transactions on Multimedia, Vol. XX, No. XX, XX (2018)
2. Wu, D., Pigou, L., Kindermans, P-J., Le, N., Shao, L., Dambre, J., Odobez, J.M.: Deep dynamic neural networks for multimodal gesture segmentation and recognition. In: IEEE Transactions on Pattern Analysis and Machine Intelligence (2016)
3. Neethu, P.S., Suguna, R., Sathish, D.: An efficient method for human hand gesture detection and recognition using deep learning convolutional neural networks. Soft Comput. **24**(20), 15239–15248 (2020). https://doi.org/10.1007/s00500-020-04860-5

4. Verma, P., Sah, A., Srivastava, R.: Deep learning-based multi-modal approach using RGB and skeleton sequences for human activity recognition. Multimedia Syst. **26**(6), 671–685 (2020). https://doi.org/10.1007/s00530-020-00677-2
5. Sharmaa, A., Mittala, A., Singha, S., Awatramania, V.: Hand gesture recognition using image processing and feature extraction techniques. In: International Conference on Smart Sustainable Intelligent Computing and Applications Under ICITETM2020 (2020)
6. Barbhuiya, A.A., Karsh, R.K., Jain, R.: CNN based feature extraction and classification for sign language. Multimedia Tools Appl. **80**(2), 3051–3069 (2020)
7. Adithya, V., Rajesh, R.: Hand gestures for emergency situations: a video dataset based on words from Indian sign language. Data Brief **31**, 106016 (2020)
8. Lai, K., Yanushkevich, S.N.: CNN+RNN depth and skeleton based dynamic hand gesture recognition. arXiv:2007.11983v1 (2020)
9. Yong, L., Zihang, H., Xiang, Y., Zuguo, H., Kangrong, H.: Spatial temporal graph convolutional networks for skeleton-based dynamic hand gesture recognition. EURASIP J. Image Video Proces (2019)
10. Athira, P.K., Sruthi, C.J., Lijiya, A.: A signer independent sign language recognition with co-articulation elimination from live videos: an indian scenario. J. King Saud Univ. Comput. Inf. Sci. (2019). https://doi.org/10.1016/j.jksuci.2019.05.002
11. Xu, S., Liang, L., Ji, C.: Gesture recognition for human-machine interaction in table tennis video based on deep semantic understanding. In: Signal Processing: Image Communication (2019)
12. Bhuvaneshwari, C., Manjunathan, A.: Advanced gesture recognition system using long-term recurrent convolution network. Mater. Today Proc. **21**, 731–733 (2020)
13. Wentao, C., Ying, S., Gongfa, L., Guozhang, J., Honghai, L.: Jointly network: a network based on CNN and RBM for gesture recognition. In: The Natural Computing Applications Forum 2018, Neural Computing and Applications (2019)
14. Yifan, Z., Congqi, C., Jian, C., Hanqing, L.: EgoGesture: a new dataset and benchmark for egocentric hand gesture recognition. IEEE Trans. Multimedia **20**(5), 1038–1050 (2018)
15. Harish Kumar, A., Kavin Kumar, S., Geetha, N., Kishore Kumar, J., Shri Hari, S.: Handwritten digit recognition. J. Xi'an Univ. Archit. Technol. **XII**(IV) (2020). ISSN No 1006–7930
16. Geetha, N., Sandhya, M., Sruthi, M., Subha Sri, S., Karthikeyan, G.: Fingerprint authentication system using convolution neural networks with data augmentation. Int. J. Adv. Res. Basic Eng. Sci. Technol. **5**(7) (2019), ISSN (ONLINE):2456–5717

An Application of Transfer Learning: Fine-Tuning BERT for Spam Email Classification

Amol P. Bhopale$^{(\boxtimes)}$ ⓘ and Ashish Tiwari

Department of Computer Science and Engineering, Visvesvaraya National Institute of Technology, Nagpur 440010, India
at@cse.vnit.ac.in

Abstract. The increased use of cyberspace and social media has resulted in a rise in the number of unsolicited bulk e-mails, necessitating the implementation of a reliable system for filtering out such anomalies. In recent years several deep learning based word representation techniques are devised. These advances in the field of word representation can provide a robust solution to such problems. In this paper, we applied a transfer learning technique, i.e., a pre-trained Bidirectional Encoder Representations from Transformer (BERT) model is fine-tuned on the required datasets for spam email classification. The classification results are compared with other state-of-the-art classification techniques such as logistic regression, SVM, Naïve Bayes, Random Forest, and LSTM. To evaluate the performance of a proposed technique, experiments are carried out on two well-known datasets viz. Enron spam dataset with 33,716 email messages and Kaggle's SMSS-pamcollection dataset containing 5574 messages. Significant improvements are observed in results generated by the proposed model over other models.

Keywords: BERT · Spam classification · Transfer learning · Contextual representation

1 Introduction

With the advancements in internet technology; communication through emails has become the fastest and most convenient mode of exchanging information. In last few years, use of social media has been tremendously increased which causes more internet traffic. The marketing industry has begun to take advantage of this low-cost mode of communication and thus mailboxes are getting filled with thousands of unwanted emails. These unwanted emails are commonly referred to as spam and are extremely annoying to users. It affects the system in terms of memory space on the email server, the bandwidth of a communication network, and more consumption of CPU power. More time is consumed in removing out this content from the system, making it difficult for users and internet providers. Spams are also a potential tool for virus attacks.

Several keyword-based approaches i.e. knowledge engineering approaches are documented in the literature for spam filtering. Knowledge Engineering Approach involves

© The Author(s), under exclusive license to Springer Nature Switzerland AG 2022
R. Misra et al. (Eds.): ICMLBDA 2021, LNNS 256, pp. 67–77, 2022.
https://doi.org/10.1007/978-3-030-82469-3_6

a collection of rules specified either by the user or by some authority to restrict unsolicited emails; however, spammers have overcome these strategies using tricky methods. Thus, in later years, automated self-adapting spam detection techniques are devised using prominent machine learning algorithms that do not specify any rules, but rather require training samples to learn the classification model. However, the performance of these algorithms is strongly influenced by the selection of a feature extraction method. Although ML-based methods have documented improvements over rule-based techniques, they are not considered to be sufficient. In recent years, context-based spam email detection techniques are being studied to assess the effect of a word in the presence of another word. This allows classification models to learn semantically from training samples. While several neural network-based techniques have been documented in the past for finding relevance between words, there is still scope with the advancement of recent word representation techniques. The BERT model [1] has popularly emerged in recent years as an important technique for natural language processing (NLP) tasks. It has done well in dealing with text classification, question-answering, grammar parsing, and natural language inference to other state-of-the-art techniques. The key contribution of this work is to use transfer learning for spam email classification, i.e. the pre-trained BERT model is fine-tuned on the appropriate dataset for spam email classification, and the impact of contextual word representations produced using it is analyzed on the classification results. Transfer learning is a technique that focuses on storing information acquired while solving one task and reusing it to solve another yet related domain tasks.

The outline of the paper is as follows. The related work is discussed in Sect. 2, followed by the BERT model architecture in Sect. 3. Section 4 presents the context-based spam email classification model architecture. Section 5 briefly outlines the state-of-the-art classification techniques. The experimental setup and the results generated are presented in Sect. 6, followed by the conclusion and potential future scope in Sect. 7.

2   Literature Survey

Over the past few years, the global research community has shown continued interest in spam email filtering techniques. The classification techniques can be briefly categorized as Rule-based knowledge engineering techniques, Machine Learning techniques, and Deep Learning based spam filtering techniques. In recent years, the deep learning based BERT has generated outperformed results in the downstream NLP tasks with its contextual word representation approach. In this literature, we briefly discuss some of the spam email filtering studies and the recent work done in classification using the BERT model.

2.1   Spam Email Classification

Researchers have presented many machine learning-based spam email classification techniques in the past, which typically include probabilistic approaches, decision tree [2], artificial immune system [3], support vector machine (SVM) [4, 5], artificial neural networks (ANN) [6], case-based techniques [7]. In the paper [8], authors studied the

content-based filtering techniques using machine learning approaches to produce automated filtering rules. Generally, rules examine the frequency and distribution of words
in the content of an email. Studies such as [9], focuses on developing heuristic rules to
analyze patterns, usually regular expressions to compute scores based on the occurrences
of patterns to prepare the spam filter model, however, the email with new spam patterns
can easily evade through such filters without being noticed.

Since the rules are continuously expected to be revised for naive users, knowledge
engineering-based approaches do not guarantee the reliability of the performance. On
the other hand, machine learning based approaches are proven to be more effective
because they use a collection of pre-classified emails to train the model. However, these
techniques find it difficult to mine adequately-represented features; these techniques
require expertise from domain experts to improve their low-learning capabilities and
also suffer from a curse of dimensionality and high computational costs.

Deep learning approaches are more effective with a high dimensional data for classification as a consequence of the fact that the features learned are automated and not
human engineered. BERT is a pre-trained, deep learning natural language system that
delivers state-of-the-art results on a wide range of downstream NLP tasks.

2.2   Fine-Tuning BERT for Classification

In several downstream NLP applications, the contextual word representation produced
using BERT has shown good results. In many classification activities, it has demonstrated
greater accuracy and is constantly being used in various domain-specific fields. In the
paper [10], the authors fine-tuned the BERT model for patent classification. The paper
[11] has employed BERT to predict future patents in electronic health records. To perform
document classification, Adhikari [12] applied BERT on four open-sourced datasets
and reported improved results over the baseline techniques. They fine-tuned BERT and
used the learned knowledge to improve the effectiveness of LSTM models through a
knowledge distillation process. A similar approach is followed in [13] where the BERT
model is customized with the LSTM technique to classify tweets in the field of disaster
management.

3   BERT Model Architecture

BERT stands for Bidirectional Encoder Representation from Transformers. As the name
implies, it is a combination of multi-layer bidirectional transformer blocks consisting
of a stack of 12 transformer encoders in the basic model and increases for larger pre-
trained models. Bidirectional representation of unlabeled text is used to obtain the pre-
trained model by governing right and left contexts. Input encoder maps the sequence x
$= (x_1,..., x_n)$ of symbol representations to the sequence $z = (z_1,..., z_n)$ of continuous
representations, which is a combination of different embeddings, i.e. a position segment
and token embeddings, and the sequence $y = (y_1,..., y_n)$ is produced in the output for
one element at a time.

Model Pre-Training

Pre-training increases the accuracy and reliability of the designed model. The model weights are initialised in the pre-training phase in such a way that it learns with experience from large datasets for every new task. The BERT model is pre-trained on two large-size corpora viz. BookCorpus (800 M Words) and English Wikipedia (2500M words) using two unsupervised prediction tasks as below.

- Masked Language Model (MLM): In this task, some random percentage say 15% of tokens are masked to train the deep bi-direction representation model and then predicts only those masked tokens. Thus, in the masked word prediction process, the model considers bi-directional contextual information.
- Next Sentence Prediction: Natural language processing tasks such as Question Answering (QA), Natural Language Inference (NLI), need a deeper understanding of the relationships amongst sentences that the language modeling does not adequately capture. Authors trained the model on the next sentence prediction task in order to understand the relationship amongst sentences produced by the monolingual corpus. If sentence B is the true next sentence of A, the output is labeled as IsNext otherwise the label NotNext is generated.

4   The Proposed Approach

The pre-trained BERT model is fine-tuned on the collected datasets. The fine-tuning is done using the BERT base uncased model. It has 12 transformer layers, 12 self-attention heads, and a hidden node size of 768. This study has used the python Flair library which has a framework for state-of-the-art NLP tasks developed on top of the PyTorch framework. Figure 1 demonstrates the flow of the classification approach followed in this study.

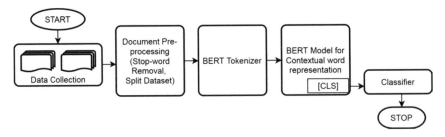

Fig. 1. The flow diagram.

4.1   Data Collection

This study used two publicly accessible spam datasets to test the proposed classification method and to make rational comparisons with other state of the art models.

- Enron Dataset: It is a corpus of emails exchanged between senior management of Enron Corporation. It consists of total 33716 emails, where 16545 ham-labeled legitimate emails and 17171 spam-labeled emails.
- SMSSpamCollection Dataset: It is obtained from the UCI machine learning repository, consisting of 5574 short massages named spam (747) and ham (4827). Table 1 gives the statistics of the datasets used in this study.

Table 1. Dataset summary

Dataset	HAM	SPAM	Total
Enron	16545	17171z	33716
SMSSpamCollection	4827	747	5574

4.2   Document Pre-processing

The selection of suitable combinations of pre-processing techniques significantly improves the classifier performance. Both datasets used for experimentation have separate email text files and are free from Html tags. Before applying pre-processing techniques, non-english characters and special characters are removed; and a single text file for each corpus is prepared to contain an email subject followed by a body massage per line. Stop word removal technique is applied to remove common words and articles that do not contribute to deciding classification labels. Also, stemming is applied to reduce the word to its basic form. For all experiments, 80% of the total instances are randomly sampled for training and 10% each for testing and validation.

4.3   BERT Tokenizer

For word-vector representations, BERT has used Wordpiece Embedding [14]. The tokenizer first checks the entire word in the vocabulary and, if not found, splits each token into the largest possible subwords present in the vocabulary. Thus, it always represents a word and does not suffer from the out of word vocabulary problem.

Eg:- The term **"Embeddings"** is tokenized as below using the WordPiece embeddings - ['em', '##bed', '##ding', '##s'].

In the above example, subword embedding vectors can be summed up to generate an approximate vector. With the help of BertTokenizer of the Pytorch library, each word from the text file is tokenized and only 512 tokens are considered per line to meet the maximum sentence length requirement for the BERT model. Since the SMSSpamCol-lection corpus comprises of short messages with a maximum length of 78, the original dataset is therefore considered for experimentation.

BERT uses the following special tokens for pre-training and fine tunning:

- **[CLS]**: It is the first token in any sequence, and it stands for classification. It is used for classification tasks in conjunction with the softmax layer.
- **[SEP]**: It is the sequence delimiter token used in the pre-training of sequence pair tasks i.e. the next sentence prediction task and in classification tasks it is appended to the end of the sequence.
- **[MASK]**: It is used to represent masked tokens in the pre-training phase.

4.4   Fine-Tuning BERT and Hyper-parameter Setting

The transformer's self-attention mechanism enables the BERT model to fine-tune by switching out the necessary inputs and outputs for several downstream tasks. Most of the hyper-parameters are same in the fine-tuning process as in pre-training, with some exceptions for batch size, learning rate, and the number of epochs of training.

Hyper-Parameter Setting
The original BERT-Base model has 12 transformer layers, 768 hidden nodes i.e. dimen-sion size, and 110 million parameters trained on lowercase English text. In the fine-tuning phase, parameters such as a number of epochs, learning rate, batch size are optimized and the maximum sequence length (MSL) is set at 512. The quality of the results produced by any model is highly sensitive to the number of epochs. Due to resource constraints, the model is trained at 2,3,4 epochs with learning rate 0.05 and 0.1 for Enron dataset. For SMSSpamCollection corpus, 10 epochs with a learning rate of 0.1 is used to train the model. The trade-off between the different hyperparameters and the results generated by the classifier is shown in the table below (Table 2).

Table 2. Trade-off between hyper parameters and BERT classifier results for Enron dataset.

Epoch	Learning rate	Accuracy	Precision	Recall	F1 measure
2	0.1	0.9582	0.9750	0.9753	0.9787
3	0.1	0.9570	0.9821	0.9667	0.9780
4	0.05	0.9639	0.9791	0.9780	0.9816

4.5   BERT as Classifier

For a classification task, final hidden state of the token [CLS] is provided as an input to the Recurrent Neural Network (RNN) classifier to obtain a text representation into a single vector. Then Softmax function is applied on the vector to get the linear layer and generate the class labels. In this work, with the help of Flair[1] Framework, the classifier

---

[1] https://github.com/flairNLP/flair.

model takes word embeddings and inputs them into the Recurrent Neural Network to obtain a text representation, and then puts the text representation in a linear layer at the end to get the actual class label.

5 State-Of-The Art Classification Techniques

Several classification techniques are studied in the past. In this study, some state-of-the-art classification techniques are discussed and classification results are compared against the results generated with BERT based classification model.

5.1 Logistic Regression

It is a machine learning algorithm used for predictive analysis based on the probability concept. It is seen as analogous to linear regression and an alternative to linear discriminatory analysis. The Bernoulli distribution is used to derive the conditional distribution and by means of a logistic function, the dependent variable values are measured. It explains the relationship between one dependent and one or more nominal, ordinal independent variables. This study has used the default constraints adjusted according to L2 Regularization.

5.2 Support Vector Machine (SVM)

It belongs to the set of supervised learning techniques which are designed to solve discrimination and regression problems. It is a non-probabilistic binary classifier, projects data points into a high dimensional feature space and a learned hyperplane is used to differentiate the points between two classes. The hyperplane is adjusted such that it maximizes the margin between points of two classes thus; achieve better generalization over test data. The discriminant function it uses is as below:

$$g(x) = w^T f(x) + b \tag{1}$$

Here, w is weighted vector, b is bias term; f(x) represents a non-linear mapping from input space to high dimensional feature space.

This study has used the linear SVM model for binary classification. Constraints are adjusted according to L2 regularization and other parameters are initialized to their default values.

5.3 Naïve Bayes (NB)

Naive Bayes is a probabilistic model based on a Bayesian theorem with strong independence of the hypothesis. In general NB, a conditional probability model determines the probability of class C in presence an event X.

$$P(C_k|X) = \frac{P(C_k)P(X|C_k)}{P(X)} \tag{2}$$

Out of several variants available, this study has used Gaussian NB and initialized all parameters to their default values.

5.4   Random Forest

It is an ensemble learning method based on bagging technique, suitable for solving problems related to the classification of data into groups. It learns parallelly on randomly constructed decision trees using different data subsets. Classification is performed by voting of multiple decision trees on training sub-samples. In this study, a random value 10 is used to generate decision trees, and all leaves are explored to the end; other parameters are initialized to their default values.

5.5   LSTM

LSTM is a deep learning methodology focused on an artificial recurrent neural network approach specifically designed to solve the issue of gradient explosion and disappearance. The LSTM layer and hidden layer are set to 1, and 15 epochs are used to learn the model at the rate of 0.05.

For all machine learning algorithms, features are extracted using the bag-of-words approach and each email is represented by considering only top 3000 words. In this work, the Scikit Learn Python library is used to implement machine learning algorithms.

6   Experiment and Results

This section describes the measures used for performance evaluation followed by results and discussion.

6.1   Performance Evaluation Measures

To assess the efficiency of the BERT based classifier and compare the results with other state-of-the-art classifiers; this study considers measures such as accuracy, precision, recall and F1 measure scores. Commonly, accuracy is used in classification techniques to measure the efficiency of the model. It is defined as the fraction of the number of samples correctly classified to the total number of samples in the dataset. However, in case of an imbalanced or uneven dataset, this measure produces an incorrect interpretation of the findings. For example, say the dataset consists of 95 legitimate records and only 5 spam records; although the classification model has incorrectly categorized all in one category, the accuracy score is still 95%. Thus to overcome such issues, Precision, Recall, and F1 Score are used along with Accuracy to measures and correctly identify samples. The following equations give a definition of each measure used in this study.

$$Accuracy(A) = \frac{number\ of\ correctly\ categorized\ samples}{total\ samples} \qquad (3)$$

$$Precision(P) = \frac{TP}{TP + FP} \qquad (4)$$

$$Recall(R) = \frac{TP}{TP + FN} \qquad (5)$$

$$F1 - Measure = \frac{2 * P * R}{P + R} \tag{6}$$

Where,

TP: True positive samples are defined as the correctly classified samples for the positive class; FP: False positive samples are defined as the incorrectly predicted samples for positive class; FN: False negative samples are defined as the incorrectly predicted samples for negative class.

All these measures focus on positive class than the negative class i.e. it provide more importance to identifying spams than the legitimate records.

6.2 Results and Discussion

As stated above, the accuracy, precision, recall, and F1 measure scores of the BERT-based classifier are recorded in this analysis. Experiments are conducted on two datasets, and the findings are compared with other state-of-the-art classification techniques. Tables 3 and 4 shows the classification results respectively, for the Enron and SMSSpamCollection Datasets. In the BERT architecture, words are represented by considering their context in both directions, this helps to understand the real meaning of a word in different scenarios and thus helps to improve the performance of classification models. In case of the Enron dataset, logistic regression generated higher accuracy than the proposed BERT-based classification model; however, for other performance evaluation measures, the proposed model performed better than the logistic regression technique. It is observed from Table 3 and 4 that for both datasets, the BERT-based classification model outperformed well compare to other state-of-the-art classification models across all evaluation measures.

Table 3. Classification results generated by each model on test set for Enron dataset.

Classification Model	Accuracy	Precision	Recall	F1 Measure
Naïve Bayes	0.9581	0.9578	0.9574	0.9576
Logistic Regression	0.9703	0.9702	0.9696	0.9699
SVM	0.9587	0.9576	0.959	0.9583
Random Forest	0.9347	0.933	0.9375	0.9343
LSTM	0.9470	0.9435	0.9440	0.9439
BERT	**0.9639**	**0.9791**	**0.978**	**0.9816**

Table 4. Classification results generated by each model on test set for SMSSpamCollection dataset.

Classification Model	Accuracy	Precision	Recall	F1 Measure
NB Multinomial	0.9767	0.9707	0.9393	0.9542
Logistic Regression	0.9713	0.9776	0.9127	0.9416
SVM	0.9767	0.9809	0.9299	0.9533
Random Forest	0.9677	0.9815	0.8965	0.9329
LSTM	0.9681	0.9672	0.9145	0.9416
BERT	**0.9846**	**0.9876**	**0.9726**	**0.9922**

7 Conclusion

In this paper, we presented a transfer learning application in which we attempted to exploit recent developments in contextual word representation techniques for one of the crucial downstream NLP tasks, namely spam email filtering or email classification. The study has enhanced the classification results over the state-of-the-art classification techniques by fine-tuning the pre-trained BERT model using the two labeled datasets, which contain both spam and legitimate records. The results show that the success of two-stage mechanism (pre-training and fine-tuning) in the area of deep learning is promising for more classification tasks. In future, it would be interesting to study the impact of ensembles of classifiers using different word representation techniques on large records.

References

1. Devlin, J., Chang, M-W., Lee, K., Toutanova, K.: BERT: Pre-training of deep bidirectional transformers for language understanding. In: NAACL, (2019)
2. Masud, K., Rashedur, M.R.: Decision tree and naïve Bayes algorithm for classification and generation of actionable knowledge for direct marketing. J. Soft Eng. Appl. **6**(4), 196–206 (2013)
3. Bahgat, E.M., Rady, S., Gad, W.: An e-mail filtering approach using classification techniques. In: Gaber, T., Hassanien, A.E., El-Bendary, N., Dey, N. (eds.) The 1st International Conference on Advanced Intelligent System and Informatics (AISI2015), November 28–30, 2015, Beni Suef, Egypt. AISC, vol. 407, pp. 321–331. Springer, Cham (2016). https://doi.org/10.1007/978-3-319-26690-9_29
4. Bouguila, N., Amayri, O.: A discrete mixture-based kernel for SVMs: application to spam and image categorization. Inf. Process. Manag. **45**(6), 631–642 (2009)
5. Torabi, Z.S., Nadimi-Shahraki, M.H., Nabiollahi, A.: Efficient support vector machines for spam detection: a survey. Int. J. Comput. Sci. Inf. Secur. **13**, 11–28 (2015)
6. Cao, Y., Liao, X., Li, Y.: An e-mail filtering approach using neural network. In: Yin, F.-L., Wang, J., Guo, C. (eds.) ISNN 2004. LNCS, vol. 3174, pp. 688–694. Springer, Heidelberg (2004). https://doi.org/10.1007/978-3-540-28648-6_110
7. Fdez-Riverola, F., Iglesias, E.L., Diaz, F., Mendez, J.R., Corchado, J.M.: SpamHunting: an instance-based reasoning system for spam labelling and filtering. Decis. Support Syst. **43**(3), 722–736 (2007)
8. Christina, V., Karpagavalli, S., Suganya, G.: Email spam filtering using supervised machine learning techniques. Int. J. Comput. Sci. Eng. **02**(09), 3126–3129 (2010)

9. Méndez, J.R., Fdez-Riverola, F., Díaz, F., Iglesias, E.L., Corchado, J.M.: A comparative performance study of feature selection methods for the anti-spam filtering domain. In: Perner, P. (ed.) ICDM 2006. LNCS (LNAI), vol. 4065, pp. 106–120. Springer, Heidelberg (2006). https://doi.org/10.1007/11790853_9

10. Lee, J.S, Hsiang, J.: PatentBERT: Patent classification with fine-tuning a pre-trained BERT Model. ArXiv abs/1906.02124 (2019)

11. Miotto, R., Li, L., Kidd, B.A., Dudley, J.T.: Deep patient: an unsupervised representation to predict the future of patients from the electronic health records. Sci. Rep. **6**(1), 26094 (2016)

12. Ashutosh, A., Ram, A., Tang, R., Lin, J.: DocBERT: Bert for document classification. arXiv preprint arXiv:1904.08398 (2019)

13. Guoqin, M.: Tweets classification with BERT in the field of disaster management. In: StudentReport@Stanford.edu (2019)

14. Wu, Y., et al.: Google's neural machine translation system: bridging the gap between human and machine translation. arXiv preprint arXiv:1609.08144 (2016)

Concurrent Vowel Identification Using the Deep Neural Network

Vandana Prasad$^{(\boxtimes)}$ and Anantha Krishna Chintanpalli

Department of Electrical and Electronics Engineering, Birla Institute of Technology and Science, Pilani Campus, Vidya Vihar, Rajasthan 333031, India
h20190092@pilani.bits-pilani.ac.in

Abstract. An alternative deep neural network model was developed to predict the effect of fundamental frequency (F0) difference on the identification of both vowels in concurrent vowel identification experiment. In the current study, the time-varying discharge rates, computed from the auditory-nerve model, to concurrent vowel were the inputs to a ten-layer perceptron. The perceptron was trained, using the gradient descent with momentum and adaptive learning rate backpropagation algorithm, to obtain a similar identification score observed in normal-hearing listeners for 1-semitone. The perceptron was then tested to predict the concurrent vowel scores for the other 5 F0 difference conditions. The proposed perceptron model was successful in qualitatively predicting the concurrent vowel scores across F0 differences, as observed in concurrent vowel data.

Keywords: Concurrent vowel identification · Deep neural network · Multi-layer perceptron · F0 difference

1 Introduction

The ability to attend to a particular sound in the presence of other competing sounds is an important aspect of the auditory (or hearing) system. Many acoustic cues are available for normal-hearing listeners to segregate the target sound. These cues are: (1) difference in fundamental frequency (F0) between the sounds, (2) onset and offset asynchrony (i.e., the difference in start and end times) between the sounds, (3) difference in the spectral characteristics between the sounds. Among these, the F0 difference cue is widely studied.

The concurrent vowel identification experiments are studied to understand the effect of the F0 difference cue. In this experiment, two simultaneously presented vowels, with equal levels and durations, are presented to one ear (through headphones) of the listeners. The listener's task is to identify two vowels (i.e., concurrent vowel) that were presented. Previous behavioral studies on normal-hearing listeners had shown that the percent correct of both vowels increases with F0 difference and then asymptotes at higher F0 difference (e.g., Assmann and Summerfield, 1990; Summers and Leek, 1998; Arehart et al., 2005; Chintanpalli and Heinz, 2013; Chintanpalli et al., 2016).

There are computational models that have successfully captured the effect of F0 difference on concurrent vowel identification. Meddis and Hewitt (1992) model was

© The Author(s), under exclusive license to Springer Nature Switzerland AG 2022
R. Misra et al. (Eds.): ICMLBDA 2021, LNNS 256, pp. 78–84, 2022.
https://doi.org/10.1007/978-3-030-82469-3_7

the first to capture these F0 differences on concurrent vowels. Later on, Chintanpalli and Heinz (2013) had used the neural responses from a more physiologically realistic auditory-nerve model with the same Meddis and Hewitt (1992) F0-guided segregation algorithm to predict the concurrent vowel scores across F0 differences. The predictions were qualitatively successful in capturing the effect of F0 difference on concurrent vowel scores. Settibhaktini and Chintanpalli (2020) had replicated this effect in their computational modeling study while analyzing across duration effects.

There are limited computational models on concurrent vowel identification using the deep neural network (DNN). Culling and Darwin (1994) had used a single layer perceptron using a simple model of the auditory-nerve fiber (only with gamma-tone filters) and computed the response probability for each concurrent vowel. However, it is an indirect approach to relate to the concurrent vowel scores. Recently, Joshi and Chintanpalli (2020) developed a multi-layer perceptron (with 8 hidden layers, 50 neurons in each layer) using the neural responses from a more advanced auditory-nerve model, to successfully capture the qualitative effect of level-dependent changes in concurrent vowel scores.

This study aims to develop a DNN model that could capture the F0 difference effect on concurrent vowel scores instead of using an indirect measure through the response probability. Additionally, a more advanced auditory-nerve model and a robust DNN will be employed in the current study than those used in Culling and Darwin (1994) to predict the scores.

2 Methods

2.1 Stimuli

The stimuli generations for this study were the same used in Chintanpalli et al (2016). Five synthetic English vowels (/i/, /ɑ/, /u/, /æ/, /ɝ/) were generated using a cascade formant synthesizer (Klatt, 1980). Table 1 shows the formant frequencies (F1 to F5) and their bandwidth for each vowel used in this study and these values were the same across previous studies on concurrent vowels (e.g., Summers and Leek, 1998; Arehart et al., 2005; Chintanpalli and Heinz; 2013). Each vowel contains the fundamental frequency (F0) and formant frequencies. The duration of an individual vowel was 400-ms, including 15-ms raised cosine rise and fall ramps.

A concurrent vowel pair was obtained by adding any two individual vowels. In each vowel pair, one vowel's F0 was always 100 Hz while the other vowel's F0 was either 100, 101.5, 103, 106, 112 and 126 Hz. These F0 difference conditions correspond to 0, 0.25, 0.5, 1, 2 and 4 semitones, respectively. Each non-zero semitones condition had 25 vowel pairs. To maintain the equal number of vowel pairs, the 0-semitone had 5 identical vowel pairs (e.g., /i, i/) and 10 different vowel pairs (e.g., /i, u/), but the latter was repeated twice. A total of 150 concurrent vowel pairs (25 vowel pairs x 6 F0 difference conditions) were used. Each vowel was presented at 65 dB sound pressure level (dBSPL). These vowel pairs were presented as an input to the auditory-nerve model to obtain the neural responses. A computational model was developed in this current study by cascading a physiologically realistic auditory-nerve model with a DNN to predict the concurrent vowel scores across F0 differences.

Table 1. Formants in Hz for five different vowels. Values in the parenthesis of the first column correspond to bandwidth around each formant (in Hz).

Vowel	/i/	/ɑ/	/u/	/æ/	/ɝ/
F1 (90)	250	750	250	750	450
F2 (110)	2250	1050	850	1450	1150
F3 (170)	3050	2950	2250	2450	1250
F4 (250)	3350	3350	3350	3350	3350
F5 (300)	3850	3850	3850	3850	3850

2.2 Neural Responses from the Auditory-Nerve Model

The auditory-nerve (AN) model developed by Zilany et al (2014) was used to obtain the neural responses to concurrent vowels. It is an extension of previous other models that have been successfully tested with the experimental data obtained from animals (e..g, Zhang et al., 2001; Zilany et al., 2009). The model captures the level-dependent changes in cochlear nonlinearities essential for the normal functioning of the human ear. The AN model's input was the concurrent vowel and the output was the time-varying discharge rate (or phase-locking cue) for a single characteristic frequency (CF) of the AN fiber. This study utilized 100 different logarithmically spaced CFs, ranging between 125 and 4000 Hz. The upper limit was 4000 Hz, as the vowel's formant is below 4000 Hz (see Table 1). The AN fibers are classified into high spontaneous rate (SR), medium SR, and low SR. These fiber types constitute the population response. Based on Liberman (1978), 61% HSR, 23% MSR, and 16% LSR fibers exist in the auditory-nerve. The population neural response at each CF was the weighted sum of discharge rates based on SR distributions (i.e., DR_HSR × 0.61 + DR_MSR × 0.23 + DR_LSR × 0.16; where DR stands for discharge rate). For each F0 difference, the AN population responses across 100 CFs were obtained to each 400-ms concurrent vowel.

2.3 Deep Neural Network to Predict the Concurrent Vowel Scores

The current study utilized a similar framework of the neural network used in Joshi and Chintanpalli (2020) to predict the effect of F0 difference on concurrent vowel scores. For each vowel pair, the discharge rates across CFs were passed as an input to a deep neural network to predict its identification score. Each vowel pair had a matrix feature space of 100 (i.e., number of CFs used) × 40,000 (i.e., number of samples = 100 kHz sampling frequency × 400-ms duration). The input layer's matrix dimensionality was 2,500 (100 CFs × 25 vowel pairs) × 40,000 for all 25 concurrent vowel pairs at each F0 difference. One hot encoding matrix with a dimension of 2,500 × 25 (i.e., the number of vowel pairs) was used as the target patterns for the neurons' outputs. The rows (specific set of 100 CFs) and single column corresponding to that vowel pair were set to 1; otherwise, 0. For instance, /i, i/ was allocated to the first column, and only the first 100 rows were set to 1 while the remaining 2400 rows were set to 0.

A multi-layer perceptron (MLP) was used to predict the concurrent vowel scores across F0 differences. More specifically, the ten-layer perceptron (i.e., with 9 hidden layers) was used in this study. Figure 1 shows the schematic diagram of the network architecture. Each of the hidden layers had 60 neurons, but the layers' activation functions were different. The first two hidden layers used the radial basis transfer functions ('radbas' in MATLAB) and while the rest of the hidden layers had used the log-sigmoid activation functions. The output layer utilized the linear activation functions. The neurons' output were ranged between 0 and 1 in the output layer. For each F0 difference, the discharge rates for a specific vowel pair were passed through a 10-layer perceptron to compute the neurons' output at the output layer. This procedure was repeated across 25 vowel pairs and 6 F0 difference conditions.

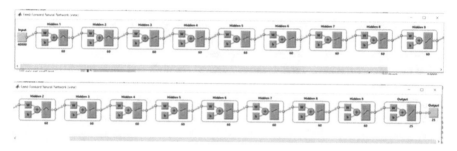

Fig. 1. Schematic of the 10-layer perceptron for concurrent vowels used in the current study.

The network architecture training was done using the gradient descent with momentum and adaptive learning rate backpropagation algorithm (Rumelhart et al., 1986). The weight (and bias) updation is the sum of the two terms. The first term is a product of learning rate, momentum and error gradient at the current iteration. In contrast, the second term is a product of momentum and the previous iteration's error gradient. The mean square error (MSE) metric was used to evaluate the performance of the neural network. The momentum term, along with the gradient descent, results in a faster convergence rate than the traditional gradient descent algorithm. The 'traingdx' command in MATLAB was used to train the network architecture. The initial values for learning rate and momentum were the default values (0.01 and 0.9, respectively) used in MATLAB. For each iteration, if the MSE error approaches closer to zero, then the learning rate was increased by a factor equal to 1.05; otherwise, the learning rate was decreased by 0.7. The network training halted when the validation error increased more than 6 times after it was last decreased. These are the default values in MATLAB.

3   Results

The MLP model (see Fig. 1) was trained using the time-varying discharge rates of concurrent vowels at 1-semitone. This F0 difference was selected for training, as the scores did not increase afterward in the concurrent vowel data (triangles in Fig. 2). Previous behavioral studies on concurrent vowels had also shown that the scores usually asymptote at 1-semitone (e.g., Assmann and Summerfield, 1990; Summers and Leek, 1998; Chintanpalli and Heinz, 2013). For each vowel pair, the outputs (2,500 × 1) of the MLP model were computed. Finally, the mean was calculated across those rows (or CFs), which corresponds to that specific vowel pair. For example, /i, i/ was assigned to the first column of the output dimensionality. The mean was computed only for those output values between the rows from 1 to 100 (i.e., corresponding to the CFs associated with /i, i/). If the individual output related to that vowel pair is greater than the mean, then the output was set to 1; otherwise, 0. The MLP model identified the correct vowel pair if more than 65% of the corresponding outputs had 1 after the mean thresholding. The number of neurons in each hidden layer, number of hidden layers and type of the activation functions of the MLP model were varied such that the percent correct score (across 25 vowel pairs) at 1-semitone was similar to the concurrent vowel score (Chintanpalli et al., 2016). With the network architecture shown in Figure 1, the training stage's accuracy rate was 80%, closer to the concurrent vowel score, which had 86.26%. Once the training was over, the MLP model was tested against other 5 F0 difference conditions using the same network architecture.

Figure 2 shows the MLP model scores (circles) for both vowels across F0 difference conditions. The identification score improved as the F0 difference increased from 0 to 1 semitones and then asymptoted from 1-semitone. This pattern of identification scores with an increasing F0 difference is qualitatively similar to that of listeners' identification scores (triangles). The F0 benefit is a widely used quantitative metric to determine the extent of benefit in identification score based on F0 difference (e.g., Assmann and Summerfield, 1990; Summers and Leek, 1998; Chintanpalli and Heinz, 2013) and is defined as the difference in the score between 4-semitones and 0-semitone. The predicted F0 benefit from the MLP was 24%, qualitatively similar to 32% in the concurrent vowel data (Chintanpalli et al., 2016).

4   General Discussions

The current study developed an alternative approach using the MLP model based on the AN fibers' discharge rates to predict the effect of F0 difference on concurrent vowel identification. The MLP model had nine hidden layers and each hidden layer had 60 neurons to achieve this effect. There was an improvement in identification scores as the F0 difference increased from 0-semitone to 1-semitone and then the score asymptoted (fairly remained the same) from 1-semitone to 4-semitone (Fig. 2, circles). These scores were qualitatively similar to the concurrent vowel data obtained from normal-hearing listeners (Fig. 2, triangles).

Even though the model captures the pattern of identification scores across F0 differences, there is still an absolute difference in scores between the MLP model and concurrent vowel data (compare circles with triangles in Fig. 2). This suggests that other cues apart from the phase-locking (using the discharge rate) of the AN fibers might be contributing to the identification scores. The other possible cue available at the auditory-nerve level for identification is the rate-place cue (e.g., Sachs and Young, 1979). A possible future work could be to incorporate both the rate-place cue and the phase-locking cue in the MLP model to minimize the absolute difference in scores.

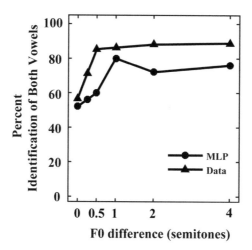

Fig. 2. Predicted effect of F0 difference on percent identification of both vowels, using the MLP model (circles). For comparison purposes, the actual concurrent vowel data (triangles) from Chintanpalli et al (2016) is used. The model captures the pattern of identification scores with an increasing F0 difference observed in the data.

The concurrent vowel data (Chintanpalli et al., 2016) and the previous modeling studies (Chintanpalli and Heinz, 2013; Settibhaktini and Chintanpalli, 2020) found that the one-vowel correct identification in the pair was 100% at each F0 difference. However, the MLP model had failed to predict this effect. A possible reason could be that the MLP model was trained to identify two vowels but not for one-vowel correct identification. As future work, the DNN model needs to incorporate both one-vowel and two-vowel scores. Nevertheless, the current study results had open up an alternative neural-network model to predict the concurrent vowel data for normal-aged listeners and hearing-impaired listeners that are published in the literature.

Acknowledgment. This work was supported by the second author's Outstanding Potential for Excellence in Research and Academics (OPERA) Grant and Research Initiation Grant (RIG), awarded by BITS Pilani, Pilani Campus, Rajasthan, India. The concurrent vowel data was used from Chintanpalli *et al.* (2016), which was collected at the Medical University of South Carolina, USA, under the supervision of Dr. Judy R. Dubno (NIH/NIDCD: R01DC000184 and P50 DC000422, and NIH/NCRR: UL1RR029882), Professor in Department of Otolaryngology-Head and Neck Surgery. Thanks to Mr. Harshavardhan Settibhaktini for generating the figures in a publishable format.

References

Arehart, K.H., Rossi-Katz, J., Swensson-Prutsman, J.: Double-vowel perception in listeners with cochlear hearing loss: differences in fundamental frequency, ear of presentation, and relative amplitude. J. Speech. Lang. Hear. Res. **48**(1), 236–252 (2005). https://doi.org/10.1044/1092-4388(2005/017)

Assmann, P.F., Summerfield, Q.: Modeling the perception of concurrent vowels: vowels with different fundamental frequencies. J. Acoust. Soc. Am. **88**(2), 680–697 (1990). https://doi.org/10.1121/1.399772

Chintanpalli, A., Ahlstrom, J.B., Dubno, J.R.: Effects of age and hearing loss on concurrent vowel identification. J. Acoust. Soc. Am. **140**(6), 4142–4153 (2016). https://doi.org/10.1121/1.4968781

Chintanpalli, A., Heinz, M.G.: The use of confusion patterns to evaluate the neural basis for concurrent vowel identification. J. Acoust. Soc. Am. **134**(4), 2988–3000 (2013). https://doi.org/10.1121/1.4820888

Culling, J.F., Darwin, C.J.: Perceptual and computational separation of simultaneous vowels: cues arising from low-frequency beating. J. Acoust. Soc. Am. **95**(3), 1559–1569 (1994). https://doi.org/10.1121/1.408543

Joshi, A., Chintanpalli, A.K.: Level-Dependent changes in concurrent vowel scores using the multi-layer perceptron. In: Goel, N., Hasan, S., Kalaichelvi, V. (eds.) MoSICom 2020. LNEE, vol. 659, pp. 393–400. Springer, Singapore (2020). https://doi.org/10.1007/978-981-15-4775-1_42

Klatt, D.H.: Software for a cascade/parallel formant synthesizer. J. Acoust. Soc. Am. **67**(3), 971–995 (1980). https://doi.org/10.1121/1.383940

Liberman, M.C.: Auditory-nerve response from cats raised in a low-noise chamber. J. Acoust. Soc. Am. **63**(2), 442–455 (1978). https://doi.org/10.1121/1.381736

Meddis, R., Hewitt, M.J.: Modeling the identification of concurrent vowels with different fundamental frequencies. J. Acoust. Soc. Am. **91**(1), 233–245 (1992). https://doi.org/10.1121/1.402767

Rumelhart, D.E., Hinton, G.E., Williams, R.J.: Learning internal representations by error propagation. In: Rumelhart, D.E., McClelland, J.L. (eds.) Parallel distributed processing, vol. 1, pp. 318–362. MIT Press, Cambridge, MA (1986)

Sachs, M.B., Young, E.D.: Encoding of steady-state vowels in the auditory nerve: representation in terms of discharge rate. J. Acoust. Soc. Am. **66**, 470–479 (1979). https://doi.org/10.1121/1.383098

Settibhaktini, H., Chintanpalli, A.: Modeling concurrent vowel identification for shorter durations. Speech Commun. **125**, 1–6 (2020). https://doi.org/10.1016/j.specom.2020.09.007

Summers, V., Leek, M.R.: F0 Processing and the seperation of competing speech signals by listeners with normal hearing and with hearing loss. J. Speech, Lang. Hear. Res. **41**(6), 1294–1306 (1998)

Zhang, X., Heinz, M.G., Bruce, I.C., Carney, L.H.: A phenomenological model for the responses of auditory-nerve fibers: i. Nonlinear tuning with compression and suppression. J. Acoust. Soc. Am. **109**(2), 648–670 (2001). https://doi.org/10.1121/1.1336503

Zilany, M.S.A., Bruce, I.C., Carney, L.H.: Updated parameters and expanded simulation options for a model of the auditory periphery. J. Acoust. Soc. Am. **135**(1), 283–286 (2014). https://doi.org/10.1121/1.4837815

Zilany, M.S.A., Bruce, I.C., Nelson, P.C., Carney, L.H.: A phenomenological model of the synapse between the inner hair cell and auditory nerve: long-term adaptation with power-law dynamics. J. Acoust. Soc. Am. **126**(5), 2390–2412 (2009). https://doi.org/10.1121/1.3238250

Application of Artificial Intelligence to Predict the Degradation of Potential mRNA Vaccines Developed to Treat SARS-CoV-2

Ankitha Giridhar[✉] and Niranjana Sampathila

Department of Biomedical Engineering, Manipal Institute of Technology, MAHE, Manipal
576104, Karnataka, India
niranjana.s@manipal.edu

Abstract. The Covid19 pandemic has impacted the entire world negatively, and scientists, healthcare professionals and engineers are all on the search for viable solutions. During the search for a vaccine for the virus, scientifically known as SARS-CoV-2, it was identified as an mRNA virus, which is why mRNA vaccines could be potential solutions. However, mRNA vaccines easily degrade, and the objective of this study was to predict the degradation rates of various potential mRNA strands to potentially select an ideal sequence for a vaccine. This paper details an approach that uses a Neural Network model with the LSTM (Long Short Term Memory) and GRU (Gated Recurrent Unit) architectures to predict the degradation of each sequence in the given data, which comprised of sequences of mRNA. The performance of the model was evaluated using the MCRMSE (Mean Columnwise Root Mean Squared Error) as the scoring metric.

Keywords: Artificial Intelligence · mRNA · SARS-COV-2 · Covid-19 · Vaccine · TensorFlow · GRU · LSTM

1 Introduction

The Covid-19 or Coronavirus pandemic has caught the world firmly in its grasp since its causative virus, SARS-CoV-2, short for Severe Acute Respiratory Syndrome Coronavirus 2, was first identified in Wuhan, China in December 2019 [1]. As of the 6th of April 2021, over 130 million positive cases have been confirmed worldwide, with more than 2.8 million people having died because of the virus [2].

The symptoms of infection primarily include fever, cough and fatigue along with breathing difficulties and the loss of olfaction and gustation. Further complications may include pneumonia and even acute respiratory distress syndrome, which may lead to death [3].

The transmission is generally via virus-containing aerosols or respiratory droplets from an infected person. Hence, the virus can enter the body via respiration or physical contact [4].

A case of infection can be managed with methods of supportive care such as fluid therapy and oxygen support [5]. Antiviral treatments are under investigation, however

© The Author(s), under exclusive license to Springer Nature Switzerland AG 2022
R. Misra et al. (Eds.): ICMLBDA 2021, LNNS 256, pp. 85–94, 2022.
https://doi.org/10.1007/978-3-030-82469-3_8

they have not been shown to affect mortality rates so far, though they can impact recovery time [6].

Ribonucleic Acid (RNA) is a nucleic acid in the form of a polymeric molecule. Messenger RNA (mRNA) is a type of RNA that is used by cellular organisms to convey genetic information, using Guanine(G), uracil (U), Adenine (A), and cytosine (C), which are nitrogenous bases [7]. Viruses often use RNA genomes to transmit their genetic information, as in the case of Covid-19 [8].

Coronavirus is a Baltimore class IV positive-sense single-stranded RNA virus, and thus, mRNA molecules are very plausible candidates for a potential vaccine [9]. The drawback is that these molecules can spontaneously degrade, rendering their vaccination capabilities. Since the vaccine would have to be transported to be made available to more populations, a refrigator-stable vaccine is essential.

The Eterna community at Stanford's School of Medicine collaborated with Kaggle to put together a database with which efforts could be made to apply Artificial Intelligence to predict the degradation rates of possible mRNA sequences in order to choose the vaccine. An Eterna dataset consisting of over 3000 RNA molecules was used for training. The models were then scored on a second set of RNA sequences that were devised by Eterna players specifically for COVID-19 mRNA vaccines [10]. This paper aims to detail one of the attempts made in this research procedure. The methodology is discussed in Sect. 3 and the working of the AI model is discussed in Sect. 4.

2 Current Work

Plenty of work has been done globally in the effort to unearth an effective vaccine. Work in the Artificial Intelligence domain, however, is more niche. A company called Inovio from San Diego attempted to use a gene optimization algorithm for the vaccine, which worked well on animals, but was benched for humans. Multiple efforts similar to this were undertaken by several research groups across companies and universities [11]. The National Institute of Allergy and Infectious Diseases in the USA conducted a trial on an AI-based flu vaccine developed by scientists at Flinders University using two AI programs; one that generated synthetic compounds, and another that determined which ones would be good candidates for vaccines [12]. Neither of these attempts revolved around mRNA-based vaccines, which is the target of this paper.

3 Methodology

3.1 Data Used

The data used was collected from the Competition Data of the OpenVaccine: COVID-19 mRNA Vaccine Degradation Prediction on Kaggle [10]. This data included the following:

i.   Training data in the form of a JavaScript Object Notation (.json) file – 2400 entries
ii. Test data in the form of a.json file – 3634 entries
iii. Base pairing probability matrices – 6267 matrices in the form of NumPy array files (.npy)

In the training data, 2400 RNA sequences were taken from a set of 3029 sequences of length 107. Experimental data is available for the first 68 of these bases in five conditions, as detailed below. The remaining 629 sequences were included in the test data (Public Test). The test data also includes 3005 new RNA sequences (Private Test) with 130 bases, making a total of 3634 sequences in the test data set. For these latter sequences, measurements for the first 91 bases are expected.

3.1.1 Data Features

The feature columns in the data provided include the following. Examples are given in Table 1.

- **id** - A unique identifier for each sequence.
- **seq_scored** - It refers to the number of positions used in scoring with predicted values in the target columns, i.e. 68 in the training and public test sets, and 91 in the private test set.
- **seq_length** - It refers to the length of the sequence, i.e. 107 in the training and public test sets, and 130 in the private test set.
- **structure** - These strings represent an array of the periods parantheses and periods, indicating paired or unpaired bases. These strings are of lengths 1×107 in the training and public test sets, and 1×130 in the private test set.
- **predicted_loop_type** - These strings describe the structural context or the loop type of each character in sequence. These strings are of lengths 1×107 in the training and public test sets, and 1×30 in the private test set.
- ***_error_*** – These columns exist for every deg_* column in the target columns (explained subsequently), and contain arrays of floating point numbers. They represent calculated errors in experimental values obtained in reactivity and the deg_* columns.

Table 1. Data features

Feature name	Example
Id	id_00073f8be_12
seq_scored	68 or 91
seq_length	107 or 130
sequence	GGAAAAGUAC... and so on
structure	......(((((((((.(((((.....)))))))) and so on
predicted_loop_type	EEEEESSS.... and so on
*_error_*	[0.2167, 0.34750000000000003, 0.188, 0.2124, 0... and so on for all target columns

3.1.2 Target Values

The target values in the data provided include the following. Examples are give in Table 2.

- **reactivity** - These are vectors representing arrays of floating point numbers, which denote reactivity values for the first few bases as denoted in sequence, and are used to determine the likely secondary structure of the RNA sample. The vectors are of lengths 1×68 in the training and public test sets, and 1×91 in the private test set.
- **deg pH10** - This consists of vectors of lengths 1×68 in the training and public test sets, and 1×91 in the private test set. These vectors represent arrays of floating point numbers which are reactivity values for the first 68/91 bases in case of icubation without magnesium at a pH of 10, and are used to determine the likelihood of degradation at the base/linkage in these conditions.
- **deg Mg pH10** - This consists of vectors of lengths 1×68 in the training and public test sets, and 1×91 in the private test set. These vectors represent arrays of floating point numbers which represent reactivity values for the first 68/91 bases in case of icubation with magnesium at a pH of 10, and are used to determine the likelihood of degradation at the base/linkage in these conditions.
- **deg 50C** - This consists of vectors of lengths 1×68 in the training and public test sets, and 1×91 in the private test set. These vectors represent arrays of floating point numbers which represent the reactivity values for the first 68/91 bases in case of icubation without magnesium at a temperature of 50 °C, and are used to determine the likelihood of degradation at the base/linkage in these conditions.
- **deg Mg 50C** - This consists of vectors of lengths 1×68 for the training and public test sets, and 1×91 for the private test set. These vectors represent arrays of floating point numbers which represent the base reactivity values for the first 68/91 bases in case of icubation with magnesium at temperature of 50 °C, and are used to determine the likelihood of degradation at the base/linkage in these conditions.

Table 2. Target values

Column name	Example
reactivity	[0.3297, 1.5693000000000001, 1.1227, 0.8686, 0.. and so on
deg_pH10	[2.3375, 3.5060000000000002, 0.3008, 1.0108, 0…and so on
deg_Mg_pH10	[0.7556, 2.983, 0.2526, 1.3789, 0.637600000000… and so on
deg_50C	[0.6382, 3.4773, 0.9988, 1.3228, 0.78770000000…and so on
deg_Mg_50C	[0.35810000000000003, 2.9683, 0.2589, 1.4552,… and so on

3.1.3 Other Data

The RNA base pairing probability matrices have been included as NumPy array files (.npy). There is little detectable sequence similarity between RNA molecular structures, and base pairing probability matrices create reliable multiple alignments taking into account shared structural features [13].

3.2 Algorithm and Procedure

The model was built using Python on a Jupyter kernel which used a Graphics Processing Unit (GPU). Python libraries that were used include:

- Pandas [14] – This library was used for data handling.
- NumPy [15] – This library helped with the numeric calculations done.
- TensorFlow (Keras) [16] – This was used to construct the model.
- Sklearn [17] – This was used for splitting the data for the model.
- Plotly [18] – The model performance was visualised with this library.

The workflow for the prediction of mRNA degradation to obtain a viable sequence for the Covid19 vaccine is described in Fig. 1:

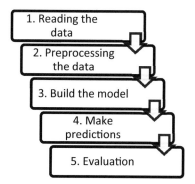

Fig. 1. Workflow for the prediction of mRNA degradation

3.2.1 Reading and Preprocessing the Data

The .json files were read into Pandas DataFrames. Each.npy file was read into a NumPy array.

Helper functions were written to obtain the sum, maximum and normalised values from the Base Pair Probability (BPP) arrays. The normalisation was done using the mean and standard deviation of these arrays.

In the columns with string sequences (sequence, structure and predicted_loop_type), each character was assigned a numeric value, and the sequences were thus converted into integers.

3.2.2 Building the Model

TensorFlow Keras was used to build the model. The architectures used were Gated Recurrent Unit (GRU) [19] and Long-Term Short-Term (LSTM) [20].

LSTM units have cells that remember values over time intervals (long-term) and the gates which regulate the data flow in and out of the cell (short-term).

A GRU is similar to LSTM, but it has fewer parameters and thus has better speed. The model is further explained in the next section.

3.2.3 Making Predictions

Predictions were made on the two given datasets – private and public, using the previous set's learning algorithm. The reactivity for each position on each strand was predicted.

3.2.4 Evaluation

The metric used for evaluation was the MCRMSE (mean columnwise root mean squared error) and given in the Eq. (1).

It is given by the formula given below, where N_t is the number of scored target columns, and $\hat{y}$ and y are the predicted and actual values, respectively [10].

$$MCRMSE = \frac{1}{N_t} \sum_{j=1}^{N_t} \sqrt{\frac{1}{n} \sum_{i=1}^{n} \left(y_{ij} - \hat{y}_{ij}\right)^2} \tag{1}$$

4   The LSTM and GRU Based AI Model

4.1   The Proposed Model

The very first hidden layer of the model was the embedding layer with a dimension of 100, where the integer embeddings of the sequences were used. Helper functions were written to add these layers to the model. The dimension of each hidden layer following the first one was 256.

As mentioned earlier, the model uses GRU and LSTM architectures, and these constituted the remaining hidden layers. These are both types of recurrent neural netwrok layers, where the internal memory is used to process inputs. These are useful here, as the input is sequential and unsegmented. The LSTM layer contains input, output and forget gates which regulate the flow of information inside and outside the layer. The GRU layer is similar, but without a forget gate.

The output layer was a dense layer with model a linear activation function.

To prevent overfitting, the model was 'diluted' with a dropout of 0.5 [21]. The function used to build the model also took in the parameters seq_len and pred_len, which were set to 107 and 68 to default, but could be changed to 130 and 91 for the private test data.

The batch size and epoch are hyperparameters that define the number of samples to work through before updating the internal model parameters and the number of times the learning algorithm will run through the entire training dataset. They were set to 64 and 60, respectively [22].

The model summary after building is displayed in Fig. 2.

```
Model: "functional_1"
_____
Layer (type)                    Output Shape        Param #    Connected to
================================================================================
input_1 (InputLayer)            [(None, 107, 6)]    0
_____
tf_op_layer_strided_slice (Tens [(None, 107, 3)]    0          input_1[0][0]
_____
embedding (Embedding)           (None, 107, 3, 100) 1400       tf_op_layer_strided_slice[0][0]
_____
tf_op_layer_Reshape (TensorFlow [(None, 107, 300)]  0          embedding[0][0]
_____
tf_op_layer_strided_slice_1 (Te [(None, 107, 3)]    0          input_1[0][0]
_____
concatenate (Concatenate)       (None, 107, 303)    0          tf_op_layer_Reshape[0][0]
                                                               tf_op_layer_strided_slice_1[0][0]
_____
bidirectional (Bidirectional)   (None, 107, 512)    861696     concatenate[0][0]
_____
bidirectional_1 (Bidirectional) (None, 107, 512)    1182720    bidirectional[0][0]
_____
tf_op_layer_strided_slice_2 (Te [(None, 68, 512)]   0          bidirectional_1[0][0]
_____
dense (Dense)                   (None, 68, 5)       2565       tf_op_layer_strided_slice_2[0][0]
================================================================================
Total params: 2,048,381
Trainable params: 2,048,381
Non-trainable params: 0
_____
```

Fig. 2. Model summary

4.2 Training, Validation and Prediction

The private and public test datasets were preprocessed separately due to separate sequence and target lengths.

A cross-validation technique called GroupKFold was used with the help of the Sklearn library. Cross-validation is a sampling technique that evaluates the model on a limited data sample. The parameter k refers to the number of folds that the data is split into, which was 5 in this case. In the GroupKFold method, the groups are ensured to be non-overlapping [23]. For each split, a training, validation and holdout set were considered, and the predictions on each split were concatenated together.

The loss, i.e. the penalty for bad predictions needs to be minimised. The training loss and validation loss were plotted. The loss depends on the evaluation metric, in this case, MCRMSE. The final loss curves are displayed in Fig. 3.

4.3 Evaluation

As mentioned earlier, the evaluation was done using MCRMSE after predicting reactivity values for each base on each sequence of mRNA. The public and private scores were evaluated separately. This model scored an MCRMSE of **0.37083** on the private data set, but fared better on the public data, with a score of **0.25467**.

Training History

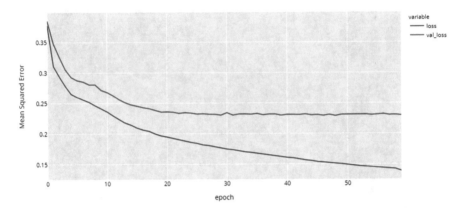

Fig. 3. Training and validation loss curves based on the MCRMSE metric

5 Results and Discussions

The public and private scores were evaluated separately. The MCRMSE scores were **0.37083** private data and **0.25467** on public data, perhaps indicating slight overfitting. This MCRMSE of roughly 0.37 is still a positive result, as the gap is that much smaller to fill. The highest score for this objective was around 0.34 [10]. Hyperparameter tuning of the model might help in enhancing the performance, enabling the proper selection of the vaccine based on the predictions.

At present, there are no mRNA vaccines in circulation in India [24], where this study was based, and even the existing vaccines are likely to become redundant owing to the fact that the SARS-CoV-2 virus mutates rapidly and has multiple strains [25], thus enhancing the potential uses of these methods.

6 Conclusion

The results obtained are a positive reflection on how AI can help in the Coronavirus situation, which is still very much at large. In this proposed work, we have managed to obtain a very low error (MCRMSE) in predicting the reactivity rates for various mRNA sequences that could be used as potential vaccines against SARS-CoV-2. While the results are not perfect, the MCRMSE scores lie in the 0.3–0.4 range, proving that with just a little tuning, such models could be used for a more dynamic vaccine manufacturing and selection process, especially as there are no mRNA vaccines in circulation in India as of the time at which this paper has been submitted. AI is a vast, formidable force which can be used for endless purposes in the future, as indicated by the work done in this paper.

Acknowledgements. The authors would like to thank the Department of Biomedical Engineering, Manipal Institute of Technology, Manipal Academy of Higher Education for the support in enabling this study.

References

1. Zhonghua, L.X.B.X.Z.Z.: The epidemiological characteristics of an outbreak of 2019 novel coronavirus diseases (COVID-19) in China. Epidemiology Working Group for NCIP Epidemic Response, Chinese Center for Disease Control and Prevention. Zhonghua Liu Xing Bing Xue Za Zhi **41**(2),145–151 (2020)
2. Dong, E., Hongru, D., Gardner, L.: An interactive web-based dashboard to track COVID-19 in real time. Lancet Inf. Dis. **20**(5), 533–534 (2020). https://doi.org/10.1016/S1473-3099(20)30120-1
3. Struyf, T., et al.: Signs and symptoms to determine if a patient presenting in primary care or hospital outpatient settings has COVID-19 disease. Cochr. Datab. Syst. Rev. (2020). https://doi.org/10.1002/14651858.CD013665
4. World Health Organization. Modes of transmission of virus causing COVID-19: implications for IPC precaution recommendations: scientific brief, (No. WHO/2019-nCoV/Sci_Brief/Transmission_modes/2020.1), 27 Mar 2020. World Health Organization (2020)
5. Whittle, J.S., Pavlov, I., Sacchetti, A.D., Atwood, C., Rosenberg, M.S.: Respiratory support for adult patients with COVID-19. J. Am. College Emer. Phys. Open **1**(2), 95–101 (2020)
6. Mitjà, O., Clotet, B.: Use of antiviral drugs to reduce COVID-19 transmission. Lancet Glob. Health **8**(5), e639–e640 (2020)
7. Higgs, P.G.: RNA secondary structure: physical and computational aspects. Q. Rev. Biophys. **33**(3), 199–253 (2000)
8. Gao, Y., Yan, L., Huang, Y., Liu, F., Zhao, Y., Cao, L., et al.: Structure of the RNA-dependent RNA polymerase from COVID-19 virus. Science **368**(6492), 779–782 (2020)
9. Machhi, J., Herskovitz, J., Senan, A.M., Dutta, D., Nath, B., Oleynikov, M.D., et al.: The natural history, pathobiology, and clinical manifestations of SARS-CoV-2 infections. J. Neuroimmune Pharmacol. **15**(3), 359–386 (2020). https://doi.org/10.1007/s11481-020-09944-5. PMC 7373339. PMID 32696264
10. Stanford University. OpenVaccine: COVID-19 mRNA Vaccine Degradation Prediction. (September 2020). https://www.kaggle.com/c/stanford-covid-vaccine/data. Accessed 14 Sept 2020
11. Waltz, E.: AI takes its best shot: what AI can—and can't—do in the race for a coronavirus vaccine - [Vaccine]. IEEE Spectr. **57**(10), 24–67 (2020). https://doi.org/10.1109/MSPEC.2020.9205545
12. Ahuja, A.S., Reddy, V.P., Marques, O.: Artificial intelligence and COVID-19: a multidisciplinary approach. Integrat. Med. Res. **9**(3), 100434 (2020)
13. Hofacker, I.L., Bernhart, S.H., Stadler, P.F.: Alignment of RNA base pairing probability matrices. Bioinformatics **20**(14), 2222–2227 (2004)
14. The Pandas Development Team: Pandas-dev/Pandas: Pandas (February 2020). https://doi.org/10.5281/zenodo.3509134
15. Harris, C.R., Millman, K.J., van der Walt, S.J., et al.: Array programming with NumPy. Nature **585**, 357–362 (2020)
16. Chollet, F.: keras. GitHub Repository (2015). https://github.com/fchollet/keras
17. Pedregosa, F., et al.: Scikit-learn: machine learning in python. J. Mach. Learn. Res. (2011)
18. Plotly Technologies Inc.: Collaborative data science. Montréal, QC (2015). https://plot.ly
19. Cho, K., et al.: Learning phrase representations using RNN encoder-decoder for statistical machine translation (2014). https://arxiv.org/abs/1406.1078
20. Hochreiter, S., Schmidhuber, J.: Long short-term memory. Neural Comput. **9**, 1735–1780 (1997)

21. Hinton, G.E., Srivastava, N., Krizhevsky, A., Sutskever, I., Salakhutdinov, R.R.: Improving neural networks by preventing co-adaptation of feature detectors (2012). https://arxiv.org/abs/1207.0580
22. Brownlee, J.: What is the difference between a batch and an epoch in a neural network? Mach. Learn. Mastery (2018)
23. Brownlee, J.: A gentle introduction to k-fold cross-validation. Mach. Learn. Mastery (2018)
24. Sahoo, J.P., Nath, S., Ghosh, L., Samal, K.C.: Concepts of immunity and recent immunization programme against COVID-19 in India. Biotica Res. Today **3**(2), 103–106 (2021)
25. van Oosterhout, C., Hall, N., Ly, H., Tyler, K.M.: COVID-19 evolution during the pandemic – Implications of new SARS-CoV-2 variants on disease control and public health policies. Virulence **12**(1), 507–508 (2021)

Explainable AI for Healthcare: A Study for Interpreting Diabetes Prediction

Neel Gandhi$^{(\boxtimes)}$ and Shakti Mishra$^{(\boxtimes)}$

School of Technology, Pandit Deendayal Energy University,
Gandhinagar, Gujarat, India
{neel.gict18,shakti.mishra}@sot.pdpu.ac.in

Abstract. With extensive use of machine learning in field of healthcare, explainable AI becomes a vital part of machine learning process by helping healthcare practitioners for understanding critical decision-making process. Machine learning has greatly impacted medical field by identification of factors responsible for vulnerable health risks and factors affecting them. Hence, interpretation of machine learning model is extremely critical in the healthcare sector. Healthcare practitioners must understand factors that affect decision-making process for diseases like diabetes made by machine learning model. Explainable artificial intelligence helps clinical practitioners in interpretation of black-box models and their decision-making process to verify why a particular decision was taken by machine learning model that is extremely important in medical field. Interpretability in machine learning methods can help healthcare practitioners to understand the features that affect the decision-making process for detection of diseases like diabetes. In the past few years, researchers have been successful in detecting diabetes using machine learning models. Explainable artificial intelligence is exceptionally useful in turning black-box into glass box machine learning model and explains their respective decision-making process in healthcare. The paper presents various interpretable machine learning methods for understanding factors affecting decision-making in case of diabetes prediction that could be explained using model agnostic methods.

Keywords: Explainable AI · Interpretable machine learning · Diabetes · Model agnostic methods

1   Introduction

Diabetes is a global health concern, according to International Diabetes Federation, about 8% of the people were likely to have diabetes in 2007 and it is estimated to increase 7.3 % by 2025. Presently, nearly 246 million people are suffering from diabetes [22]. Diabetes mellitus is a disease that leads to high level of blood sugar responsible for disturbing the metabolism in human body. Diabetes occurs usually due to less secretion of insulin from pancreas and also problems of utilization of insulin effectively. Diabetes [12] being a long-lasting disease, it

© The Author(s), under exclusive license to Springer Nature Switzerland AG 2022
R. Misra et al. (Eds.): ICMLBDA 2021, LNNS 256, pp. 95–105, 2022.
https://doi.org/10.1007/978-3-030-82469-3_9

is necessary for it to be diagnosed in the early stages to prevent any further damages to human body. Recently, machine learning techniques have been used for prediction of diabetes. Use of artificial intelligence in medicine has proved to be effective from earlier days. It is necessary for the clinician to understand the basis for the particular prediction taken by a machine learning model [14]. Hence, explainable artificial intelligence helps clinicians decide why a model had taken a particular decision and evaluate if a model is functioning properly based on the explanation provided by explainable ai methods. Also, machine learning model should be evaluated on the basis of reasons for selection of particular decision for given prediction problem. Clinicians should understand why a particular decision was taken by black-box machine learning models used in healthcare. Explainable AI ensures that machine learning models are not being biased and at the same time fair against most of selection criteria. In recent years, explainable AI has been mostly used in supervised machine learning models to understand decisions taken by healthcare systems.

2   Artificial Intelligence in Diabetes

2.1   Explainable Artificial Intelligence

Explainable Artificial Intelligence [14,18] has emerged as a crucial area in field of healthcare. Explainable AI (XAI) has major aim of reducing opacity of machine learning model. For explaining machine learning model majorly two types of methods are used namely model agnostic method applicable to all types of machine learning model and model-specific method being limited to few variants [17]. The emergence of explainable AI would help users as shown in Fig. 1(a) to understand the model's behavior detect errors in model predictions by providing deeper insight into the decision-making process of machine learning model.

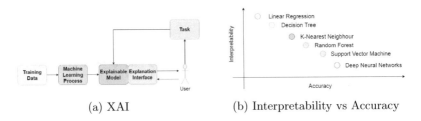

(a) XAI   (b) Interpretability vs Accuracy

Fig. 1. Explainable artificial intelligence

It might be used in various formats like text, tabular, and image format. Limitations of explainable ai is the trade-off between the accuracy and interpretability because when a model is simple like the linear regression it becomes very easy to explain but the complicated neural networks having multiple layer architectures are comparatively harder to explain as depicted by Fig. 1(b). Therefore, maintaining accuracy and interpretability trade-off is important in the sector of healthcare.

2.2 Diabetes Dataset Description

The dataset used for the purpose of studying explainable ai for diabetes prediction was provided by National Institute of Diabetes and Digestive and Kidney Diseases [1] named Pima Indian diabetes. Dataset was acquired considering women's above 21 years of age of Pima Indian origin, outcome of dataset was to predict whether patient has diabetes or not. Outcome of value equal to 1 was considered for patient having diabetes. 268 cases of diabetes out of 768 were tested positive for diabetes while others were not suffering from diabetes. The data set was used to derive factors that were most important for determination of diabetes with use of model agnostic methods in explainable ai. Here, Table 1 illustrates medical predictor variables used and feed as input to machine learning model.

Table 1. Dataset description

Sr no.	Attributes	Sr no.	Attributes
1	Pregnancies	6	BMI
2	Glucose	7	DiabetesPedigreeFunction
3	BloodPressure	8	Age
4	SkinThickness	9	Outcome
5	Insulin		

3 Model-Agnostic Methods

Model agnostic methods are used for interpreting any type of machine learning model. The properties that make model agnostic methods desirable include their flexibility in terms of model explanation as well as representation. They are better compared to model-specific methods that are limited to few machine learning models. In this paper, we present variety of model agnostic methods for problem of interpreting diabetic healthcare prediction [20].

3.1 Permutation Feature Importance and Feature Interaction

Permutation feature importance (Fig. 2(a)) help us to measure importance of a particular feature towards contribution of predicted output [2,5,6]. Permutation feature importance is measured by permuting a given data set for a particular feature and calculation of corresponding increase or decrease in prediction error for deriving strength of positive or negative relationship between output and corresponding feature permuted.

Merits of feature permutations are providing global interpretation by taking into account all interaction and it does not require retraining of the model at the same time provides a fine interpretation. Uncertainties including unclear representation of use of training or testing data, bias nature, correlated feature problem, and variation in result are observed due to error in some cases with use of permutation feature importance.

Feature interaction (Fig. 2(b)) helps in explaining interaction of the feature between two or more features after accounting individual feature effects mostly following Friedman's H-statistic rule. Friedman's H-statistic rule is generally applied to explain the interaction of features for a particular machine learning model, suppose we consider two features that are decomposed into following terms including a constant, term for first feature, term for second feature, and term for explaining interaction between two features that help us in getting the desired prediction [8, 10].

Friedman's H-statistic-

$$H_{jk}^2 = \sum\nolimits_{i=1}^{n} \left[PD_{jk}\left(x_j^{(i)}, x_k^{(i)}\right) - PD_j\left(x_j^{(i)}\right) - PD_k\left(x_k^{(i)}\right) \right]^2 / \sum\nolimits_{i=1}^{n} PD_{jk}^2\left(x_j^{(i)}, x_k^{(i)}\right)$$

$$(1)$$

where $PD_{jk}(x_j, x_k)$ represents 2-way partial dependence function for both features and $PD_j(x_j)$ and $PD_k(x_k)$ represents partial dependence functions for single features. In our feature interaction graph, we have illustrated feature interaction of different features of our model in accordance with H statistic rule as depicted by Fig. 2(b). Feature Interaction has several advantages like helping user understand meaningful interpretation from all kinds of interactions between features and get significantly higher interaction features. However, feature interaction method might be computationally expensive and sometimes results generated may be unstable.

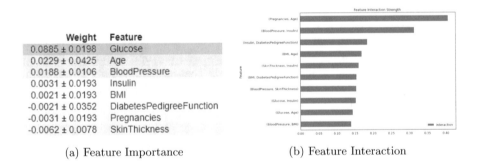

Weight	Feature
0.0885 ± 0.0198	Glucose
0.0229 ± 0.0425	Age
0.0188 ± 0.0106	BloodPressure
0.0031 ± 0.0193	Insulin
0.0021 ± 0.0193	BMI
-0.0021 ± 0.0352	DiabetesPedigreeFunction
-0.0031 ± 0.0193	Pregnancies
-0.0062 ± 0.0078	SkinThickness

(a) Feature Importance   (b) Feature Interaction

Fig. 2. Interpreting feature importance and interaction

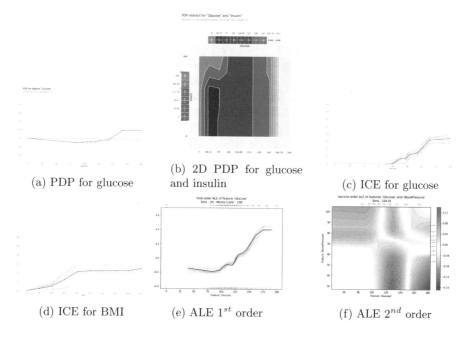

(a) PDP for glucose

(b) 2D PDP for glucose and insulin

(c) ICE for glucose

(d) ICE for BMI

(e) ALE 1st order

(f) ALE 2nd order

Fig. 3. PDP, ICE and ALE

3.2 Partial Dependence Plot (PDP), Individual Conditional Expectation (ICE) and Accumulated Local Effects (ALE) Plot

Partial dependency plot (PDP) helps understand relationship between predicted outcome and features to identify relationship between them that may be monotonic, linear, or polynomial for a more complex relationship. PDP explains effect of one or two features on overall machine learning models outcome by considering marginal effect. PDP can be used for regression as well as classification models. PDP is model agnostic global method that considers all the instances and generates corresponding prediction [7,23]. PDP for glucose shown in Fig. 3(a) and 2D-PDP glucose and insulin shown in Fig. 3(b). For PDP application, we have considered two features, glucose and insulin, glucose being most significant feature in detection of pima indian diabetes, PDP helps monitor glucose as well as insulin taking two parameters contributing to predict the outcome of diabetes in a particular patient. PDP method being intuitive, clearly understandable and representing casual interpretation for one or two features in given machine learning model. In the case of correlation, PDP might hide heterogeneous effect.

Individual conditional expectation (ICE) are useful for understanding individual instances separately by visualization of each line for instance that helps understand change in each instance's predictions for corresponding feature change. [9] Unlike PDP, ICE focuses on specific instances rather than average effect of a particular feature like partial dependency plot. We have taken an example of ICE plot for glucose (Fig. 3(c)) and BMI (Fig. 3(d)) represented by lines derived from each instance that clearly helps us understand change over each instance rather than an average of PDP. The advantage of implementing ICE curve compared to partial dependency plot helps us to uncover the heterogeneous relationships between features and give an even more intuitive understanding of feature instances of interest. Due to the representation of each individual feature instance separately, ICE sometimes might get very overcrowded and maybe it might be difficult to figure out average unlike a PDP plot.

Accumulated local effects (ALE) are alternative to partial dependence plot, they should be used when features are highly correlated. ALE plot [4] describe feature influence on the outcome of machine learning model on an average. ALE calculates the average of changes in predictions that are accumulated over grid/outcome unit. ALE approach calculates average marginal distribution by solving the issue of computing difference occurring and outcomes over conditional distribution in case of partial dependence plot. The advantage of using ALE is an unbiased approach to correlation in dataset. The interpretations also provide insights for 1D plot in case of one feature and 2D plots in case of two features as presented in Fig. 3(e) and Fig. 3(f) respectively. ALE is computational fast as well as unbiased in comparison to PDP. Due to varying stability across the feature space, it becomes difficult to interpret from ALE especially from second-order ALE plot.

3.3   Local Interpretable Model-Agnostic Explanations (LIME)

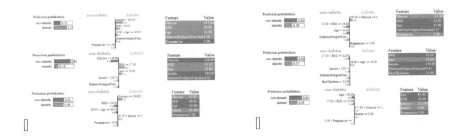

Fig. 4. LIME

Local interpretable model-agnostic explanations (LIME) are used for approximating predictions made by machine learning model assuming it to be a black box [3,13,19]. LIME falls in category of local surrogate model (applied on some rather than all). As LIME focuses on local surrogate models rather than global surrogate models as it is applied to individual predictions for the concerned machine learning black-box model. It is clearly visible as depicted by Fig. 4 that LIME prediction varies extensively due to nature of being local rather than global but being easy for human interpretation. LIME is used to perturb features individually from selected instances of interest that are used for drawing meaningful by weighting new samples and fitting on given surrogate model for approximating machine learning model for explaining the prediction.

The advantages of LIME include human-friendly explanations for underlying black-box model. Also, selective explanations and contrastive explanations are derived from black-box model. LIME could be applied to our tabular data with fidelity measure for interpreting effectively. LIME with different kernel settings might generate inconsistent explanation, may ignore correlation and trade-off must also be maintained between fidelity and sparsity are some of the limitations of LIME.

3.4 Shapley Additive Explanations (SHAP)

Shapley additive explanations [11] are used for individual predictions functioning on basis of Shapley values, they are mostly kernel SHAP were kernel-based estimation approach that works on conversion of instance collation to feature values and generate corresponding predictions on table or image data whereas tree SHAP is a faster variant using tree-based machine learning models for efficient estimation approach.

SHAP feature importance also helps us to calculate the importance of various features, it is an alternative to permutation feature importance as shown in Fig. 5(d). SHAP summary plot provides a summary of the importance of the overall features with their respective effects, they are denoted by custom colors to understand the effect more precisely depicted in Fig. 5(c). SHAP dependence plot are used in place of PDP and ALE to show the dependence of one or two features on each other in Fig. 5(b). Lastly, clustering SHAP values is used to

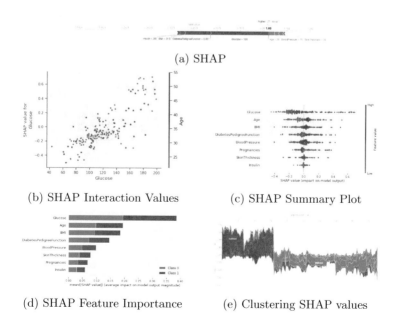

(a) SHAP

(b) SHAP Interaction Values   (c) SHAP Summary Plot

(d) SHAP Feature Importance   (e) Clustering SHAP values

Fig. 5. Shapley Additive Explanations(SHAP)

investigate groups of similar instances based on Shapley values in accordance to clustering similarity Fig. 5(e). Also, SHAP has strong foundations with connections to LIME and shapely values. SHAP [21] has global model interpretation yet the two variants have their own drawbacks like kernel shap is slow and ignores feature-dependence whereas the tree shap [15, 16] can produce unintuitive feature attribution.

4   Model Evaluation

The study on various model agnostic methods on variety of machine learning models were used to obtain interpretation and respective results that are recorded to analyze the functioning of black box model. Major factors were consider for developing a comparative analysis of various black box machine learning models. Machine learning algorithm used for interpretation include Logistic regression, Random Forest, Naive Bayes and K Nearest Neighbour provided by Table 2. Table also illustrates F1 score,precision ,accuracy,recall,monotone,interaction and task performed by various machine learning models as listed.

Table 2. Comparative analysis of various machine learning models for interpretability

	Accuracy	AUC	Recall	Precision	F1 score	Monotone	Interaction	Task
Logistic regression	0.76	0.75	0.72	0.67	0.70	Yes	No	Classification
Random forest	0.77	0.76	0.68	0.71	0.70	Some	Yes	Classification, regression
Naive Bayes	0.75	0.73	0.63	0.69	0.66	Yes	No	Classification
K nearest neighbour	0.81	0.79	0.72	0.77	0.75	No	No	Classification, regression

Correlation Matrix shows the relationship between various features illustrated by the Fig. 6(a). Also, analysis was developed considering area under curve receiver operating curve as depicted by Fig. 6(b).

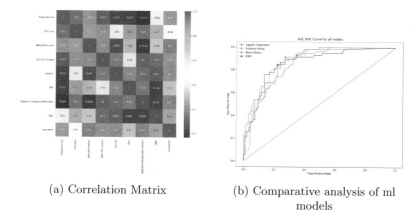

(a) Correlation Matrix   (b) Comparative analysis of ml
                                       models

Fig. 6. Model evaluation

5   Conclusion

In this paper, we have presented a variety of model agnostic methods that are applicable to all types of machine learning models for explaining the reason behind particular prediction. The paper takes into consideration various machine learning models and evaluate their performance based on various performance parameters by presenting a comparative analysis of models based of their extent of interpretability. The paper majorly focuses on model agnostic methods consisting of permutation feature importance, individual conditional expectation, partial dependency plot, accumulated local effects, LIME, and SHAP model agnostic method. The paper presents results obtained from each method for the given problem of Pima Indian diabetes. The paper shows that none of methods can completely be used for interpretation of outcome but the combination of described model agnostic methods would surely be sufficient for providing an insight into black-box decision-making process and reducing opacity of the model.

For future work, we will focus on applying explainable AI to gain insights on more complicated and deep neural networks and understand reason for their predictions. In the current scenario, explainable AI is limited to some extent for deep neural network-based machine learning models. Also, we can consider other model-specific methods for understanding interpretability of predictions made by the machine learning model and example-based explanations would also be helpful in some cases of text, image, and other areas of machine learning interpretation. Lastly, paper proves to be effective in examining various types of model agnostic methods helpful for healthcare practitioners for interpretation of predictions made by machine learning model. The paper contributes to field of machine learning and healthcare by improving the process of interpretation of black-box model that is critical in healthcare by understanding the reason for a particular decision with use of model agnostic methods.

References

1. Pima Indians Diabetes Database. https://www.kaggle.com/uciml/pima-indians-diabetes-database
2. Altmann, A., Toloşi, L., Sander, O., Lengauer, T.: Permutation importance: a corrected feature importance measure. Bioinformatics **26**(10), 1340–1347 (2010). https://doi.org/10.1093/bioinformatics/btq134
3. Alvarez-Melis, D., Jaakkola, T.S.: On the robustness of interpretability methods (Whi). arXiv:1806.08049 (2018)
4. Apley, D.W., Zhu, J.: Visualizing the effects of predictor variables in black box supervised learning models. J. R. Stat. Soc.: Ser. B (Stat. Methodol.) **82**(4), 1059–1086 (2020)
5. Fan, A., Jernite, Y., Perez, E., Grangier, D., Weston, J., Auli, M.: ELI5: long form question answering. In: ACL 2019 - 57th Annual Meeting of the Association for Computational Linguistics, Proceedings of the Conference, pp. 3558–3567 (2020)
6. Fisher, A., Rudin, C., Dominici, F.: All models are wrong, but many are useful: learning a variable's importance by studying an entire class of prediction models simultaneously. J. Mach. Learn. Res. **20**(177), 1–81 (2019)
7. Friedman, J.H.: Greedy function approximation: a gradient boosting machine. Ann. Stat. **29**(5), 1189–1232 (2001). https://doi.org/10.1214/aos/1013203451
8. Friedman, J.H., Popescu, B.E.: Others: predictive learning via rule ensembles. Ann. Appl. Stat. **2**(3), 916–954 (2008)
9. Goldstein, A., Kapelner, A., Bleich, J., Kapelner, M.A.: Package 'ICEbox' (2017)
10. Greenwell, B.M., Boehmke, B.C., McCarthy, A.J.: A simple and effective model-based variable importance measure. arXiv preprint arXiv:1805.04755 (2018)
11. Janzing, D., Minorics, L., Blöbaum, P.: Feature relevance quantification in explainable AI: a causality problem. arXiv preprint arXiv:1910.13413 (2019)
12. Kalyankar, G.D., Poojara, S.R., Dharwadkar, N.V.: Predictive analysis of diabetic patient data using machine learning and Hadoop. In: Proceedings of the International Conference on IoT in Social, Mobile, Analytics and Cloud, I-SMAC 2017 (DM), pp. 619–624 (2017). https://doi.org/10.1109/I-SMAC.2017.8058253
13. Kumari, P.K., Haddela, P.S.: Use of LIME for human interpretability in Sinhala document classification. In: Proceedings - IEEE International Research Conference on Smart Computing and Systems Engineering, SCSE 2019, pp. 97–102 (2019). https://doi.org/10.23919/SCSE.2019.8842767
14. Lauritsen, S.M., et al.: Explainable artificial intelligence model to predict acute critical illness from electronic health records. Nat. Commun. **11**(1), 1–11 (2020). https://doi.org/10.1038/s41467-020-17431-x
15. Lundberg, S.M., Erion, G.G., Lee, S.I.: Consistent individualized feature attribution for tree ensembles. arXiv preprint arXiv:1802.03888 (2018)
16. Lundberg, S.M., Lee, S.I.: A unified approach to interpreting model predictions. In: Advances in Neural Information Processing Systems, pp. 4765–4774 (2017)
17. Molnar, C.: Interpretable Machine Learning. Lulu. com (2020)
18. Pawar, U., O'Shea, D., Rea, S., O'Reilly, R.: Explainable AI in Healthcare. In: 2020 International Conference on Cyber Situational Awareness, Data Analytics and Assessment, Cyber SA 2020 (2020). https://doi.org/10.1109/CyberSA49311.2020.9139655
19. Ribeiro, M.T., Singh, S., Guestrin, C.: "Why should I trust you?" Explaining the predictions of any classifier. In: Proceedings of the 22nd ACM SIGKDD International Conference on Knowledge Discovery and Data Mining, pp. 1135–1144 (2016)

20. Ribeiro, M.T., Singh, S., Guestrin, C.: Model-agnostic interpretability of machine learning (Whi). arXiv:1606.05386 (2016)
21. Sundararajan, M., Najmi, A.: The many Shapley values for model explanation. In: International Conference on Machine Learning, pp. 9269–9278. PMLR (2020)
22. Tabish, S.A.: Is diabetes becoming the biggest epidemic of the twenty-first century? Int. J. Health Sci. **1**(2), V (2007)
23. Zhao, Q., Hastie, T., et al.: Causal interpretations of black-box models. J. Bus. Econ. Stat. **39**, 272–281 (2019)

QR Based Paperless Out-Patient Health and Consultation Records Sharing System

S. Thiruchadai Pandeeswari[✉], S. Padmavathi, and S. S. Srilakshmi

Department of Information Technology, Thiagarajar College of Engineering, Madurai, Tamil Nadu, India
`eshwarimsp@tce.edu`

Abstract. Clinics and hospitals always have this exhausting task of handling patients' health records. Especially in this time of Covid-19 pandemic, number of outpatients visiting hospitals surge and it is also paramount to maintain the outpatient records to retrieve patient's medical history at times of emergency. However, there has been no standard, easy to use, cost-effective and secure applications available for the maintenance of the out-patient records that contain their medical history, medication details and diet suggestions of the outpatients who walk in and out of clinics that are present across the country. Also, there is no system of cross-sharing patients' records among the hospitals securely. It is noteworthy that these health records are of value at critical times. The proposed system employs a lightweight, holistic approach to maintain outpatient records and facilitate sharing of the same among clinics at times of need. The proposed system aims at recording vital information (Audio) such as Doctor's advice, medications and dietary suggestions during consulting and converts the same to text and subsequently to QR Code. The QR codes along with medical tests data, if any are stored chronologically. The application also facilitates sharing of this systematically stored data among clinics, as and when required. In this paper, implementation of the above said system is presented with techniques used for voice capturing, speech to text conversion, bilingual translation and QR code generation. Notable benefits of the proposed application, especially during pandemic times include paperless medical consultation and prescription related records and a framework for sharing the same across clinics as needed.

Keywords: Voice capture · Speech to text conversion · Bilingual translation · QR Codes · Fog · YANDEX API

1 Introduction

With the growing population and advancement in medical technology and increasing expectation of the people especially for quality curative care, it has now become imperative to provide quality health care services through online applications like mobile, web, etc. In the recent times of the Covid-19 pandemic, digital solutions for out-patients medical consultation and maintenance of records gain importance as they promote paperless and contactless storage

© The Author(s), under exclusive license to Springer Nature Switzerland AG 2022
R. Misra et al. (Eds.): ICMLBDA 2021, LNNS 256, pp. 106–116, 2022.
https://doi.org/10.1007/978-3-030-82469-3_10

and retrieval technique for medical records. In the Indian medical scenario, both government and private hospitals are present in large number. In addition to multispecialty and super speciality hospitals present in metro cities, a large number of ad-hoc clinics that deal with a small-time illness like fever, diabetes etc., are present. As per the numbers mentioned in [1], 717860 registered medical practitioners are practising in 13550 hospitals and 27,400 dispensaries scattered across the country. Millions of patients all around the country visit these clinics every day. Most of these transactions happen on paper and are not stored for follow up. When a system is evolved to capture these large volumes of data, it gives rise to scope for performing data analytics on the medical data so that demographic-based knowledge on diseases may be obtained. Also, such systems will help to build medical datasets that may be leveraged for developing prediction and recommender systems. However, in a country like India where a larger percentage of the population is economically weak, it is very difficult to include all the hospitals in a hi-tech, expensive framework for storing and sharing electronic health records. Though there are high-end cloud-based cryptosystems available for storing and retrieving electronic health records of in-patients in some hospitals, such systems are local. No large-scale deployment of such systems has been experimented with so far. In this paper, we propose a simple framework that leverages the widespread, ramping up usage of mobile phones for establishing a large-scale framework for storing and sharing electronic medical records especially for the out-patients that walk into Adhoc clinics. The proposed solution integrates the voice capturing module, speech to text conversion module and bilingual translation module into a full-stack mobile application that works on a fog-to-cloud backbone. The proposed solution leverages QR codes to store the medical details of the out-patients and their oral and written prescription which is given by doctors.

2   Related Works

The use of electronic health records as a decision support component is highlighted in [2]. It is said that any recommender system is as good as the training set is used. Since real-time electronic health records are used as a decision support component, the proposed classifier was observed to produce phenomenal results in [2]. Many other works that deal with the storage of health records leverage cloud infrastructure for the storage of records. However, to ensure security, the health records are encrypted using a number of cryptographic techniques as suggested in [3,4]. It has been highlighted in [5] that when patients are transferred from one healthcare facility to some other healthcare facility, the patients' medical records must be carried physically. [5] proposes a solution that leverages the Quick Response (QR) codes to transmit sensitive medical information from one hospital to another securely. Using QR codes is an affordable solution as Mobile phone usage has increased exponentially. Also, QR codes are flexible as they can be easily integrated with applications. Ease of use and data integrity are some other benefits of using QR codes. The solution proposed in [5] has also been evaluated in terms of the capability of the QR codes to maintain security during transit and ease of availability at the intended location. The advantages mentioned in [2] include a reduction in mortality rates. This is mainly because with QR codes medical information are documented accurately without any errors. So wrong diagnosis or treatment due to misinterpretations are reduced to a greater extent. The use of QR-codes for disseminating patient information at different locations within the hospital is emphasized in [6]. The QR-code is used wherever the identification of the patient is required. The paper [6] also indicates the creation of medical information network in Turkey where medical information is accessed using a QR code identity tag. The QR code identity tag allows the members of the hospital to access patients medical information at times of emergency. The use of QR codes for transmitting sensitive information is highlighted in [7]. It is stated that QR codes are chosen instead of other cryptographic techniques due to the wide use of mobile devices among the end-users. It is also highlighted that use of QR codes exhibit efficient storage, flexible usage and maintenance of data integrity as well An overview of speech to the text conversion process and various stages of the process is explained in [8]. The method employed in [8] classifies the speech as Voiced, unvoiced and silent acoustic segments and then proceeds to convert into text. The study carried out on the existing works focused on the projects implemented in Zimbabwe and Turkey that leveraged QR codes for storing and retrieving medical records. Both the projects leveraged QR codes for the smart and secure transmission of patient records. Both the projects have evaluated the use of QR codes in terms of flexibility, availability and feasibility and the QR code-based medical records sharing system was found quite advantageous. However, these works are limited to storing and retrieving static health records. In this paper, ideas presented in the above mentioned past projects have been improved with the integration of speech to text conversion module to facilitate paperless medical consultation in this work. Also, the QR based medical records system is designed as a mobile/web application that can be provisioned using a Fog-Cloud framework.

3 Motivations

From the survey carried out on the related works and observing the Healthcare systems in our country, the following problems were identified and considered as our motivations for carrying out the work proposed in this paper.

- There is no well-evolved mechanism for documenting and maintaining Out-patient health records, sparing some high-end proprietary software.
- There is no framework for sharing paperless and contactless health records among the hospitals.
- The widespread usage of mobile phones in our country is untapped in the context of medical records storage and access.

4   System Description

In this paper, a lightweight holistic mobile and web-based application that maintains out-patient health records and medical information using QR codes have been proposed. The application is hosted in fog-to -cloud continuum utilizing the fog nodes to store and process data locally and the cloud to store records on a long-term basis. The Indian healthcare system is multilayered as stated in [1]. The layers of the Indian healthcare system is shown in Fig. 1 below. In addition to the hospitals involved in the pyramid given in Fig. 1, there are also other private clinics of large to small scale. Fog to cloud continuum-based architecture is leveraged to realize a framework for sharing electronic health records among these hospitals. A representative framework for sharing the health records is shown in Fig. 2 below. The networked computers and mobile devices present in the edge layer use the web console and mobile app of the proposed system for storing and retrieving health records from their fog nodes. The edge layer is considered as the primary source of data generation where the health records are created and stored on to the database in the fog nodes. The district hospitals and other large clinics that have the infrastructure to host a server may act as fog nodes. Fog nodes perform the functions listed below:

Fig. 1. Multilayered healthcare system in India

– host the application logic and render service to the subset of devices present in the edge layer
– temporarily store the data
– communicate with the peer fog nodes for data sharing
– communicate with the cloud for upward processing and storage

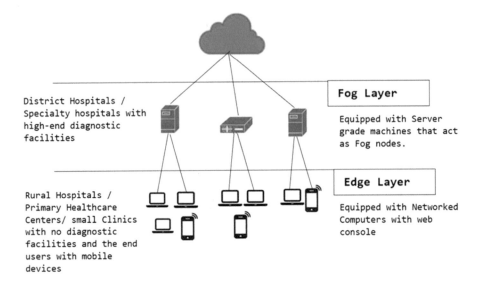

Fig. 2. Fog based architecture for Health records sharing

The application logic hosted in the fog nodes contains the following modules.

1. **Appointment booking module** – This module allows users (patients) to view the list of hospitals around them and book a visit to the hospital/clinic for consultation. This helps to create a list of patients waiting for consultation for the physician from his login dashboard. The patient may visit the clinic without prior booking of an appointment too. In that case, the physician should make an entry for the patient in the system.

2. **Voice Capture Module** – This module allows the physician to record his medical advice during the consultation. This module captures the doctor's medical advice and removes unwanted noise.

3. **Bilingual Translation Module** – More often than not, the physician records his advice bilingually (English and vernacular language combined). So the audio is translated from bilingual format to English by this module

4. **Speech to Text Conversion Module** – This module converts the physician's recorded audio to text.

5. **QR Code Generation Module** – This module trims the text generated by the speech to text converter and generates a QR code for the trimmed text. The QR code generated would contain medical prescription and other important details related to the consultation.

The QR codes are stored chronologically against the patient's history in the database in the fog nodes. This facilitates less storage. Other relevant medical records that occupy more storage space are periodically moved to the cloud and stored there. The application logic is made available to the end-users viz patients and physicians through a mobile app and web console. Thus the proposed system simplifies the maintenance and sharing of out-patient records. Major challenges in developing such a system involve translating the bilingual audio format into the text format and accommodating the medical data within QR Code.

5   Application Description

The application has been designed with three user roles viz Patients, Doctors and Clinic admins. The application allows the new users to register with the desired role. A registered user may book appointments at any clinic listed in the application. Appointments may also be booked based on consulting a particular physician. Based on the appointments booked, the patients' list is loaded for the Physician. At the time of consultation, the physician may select the patient from the list or create a new entry for the patient and start recording the medical advice. Once the audio recording is done, the bilingual translation and speech to text conversion take place. Once the text becomes available, a corresponding QR code is generated. The QR code is generated is stored in the database against the given patient's record date-wise. The application also provides an option to attach files to the given patient history so that other medical records of the patient like blood report, ultrasound report etc., may be stored in the database. The block diagram in Fig. 3 below indicates the flow of functions in the application. Further, the application allows the users to retrieve the QR codes and other records at times of need and also provisions sharing of the same using file-sharing mechanism. Patients may share the QR codes over mails and messengers through the application. When sharing patient records is needed between clinics, the files are shared using FTP protocol.

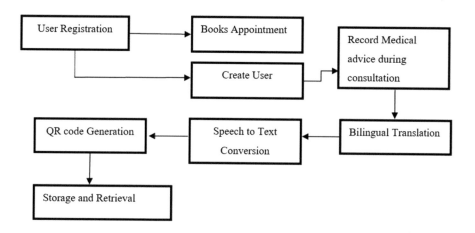

Fig. 3. Block diagram of the flow of actions in the application

6   Implementation

The implementation of the proposed system has been carried out using several APIs. The implementation also involves the design and development of a full-stack mobile app and a corresponding web console that allows the users to feed in the information. The APIs and tools that are used as part of the implementation are listed below:

1. **Port Audio API** – is used for capturing medical advice audio of the physicians. It is an open-source, cross-platform API for recording and playing back the audio.
2. **IBM Watson Speech to Text API** – is integrated into the application for translation of bilingual speech into text. This API performs speech to text conversion in the preferred language and returns the output as a JSON file. Yandex API is also integrated into the application for translation into several vernacular languages.
3. **Zxing library** - is a barcode image library implemented in Java to generate the QR code. It is an open-source library. Zxing can easily be embedded in an application. Zxing allows easy scanning of barcode images
4. **SQLite** SQLite database is a relational database. It is available as open-source. It is used to perform database operations on android devices. This is used to store the patient records in the proposed application

The mobile application was developed using android studio. The development majorly involved the frontend design of the mobile app and the integration of various required APIs into the application. To realize the fog nodes, the application instances were deployed on two server machines that have Intel Xeon Processor with 32 GB RAM independently. The sharing of medical records in the form of QR codes between these fog nodes is implemented through file transfer operations over FTP. The communication between fog to cloud has been left out of the scope of the experimentation carried out.

7   Result Screenshots

The web console and mobile app UI for User registrations in the application developed is shown below in Fig. 4a and b The patient list as seen from the Mobile app and web console is shown in Fig. 5a and 5b. Voice capture and translation screenshots are shown in figure Fig. 6a and 6b QR code generation is shown in Fig. 7 given below.

Fig. 4. (a) Registration page in mobile app (b) patient dashboard in mobile app

Fig. 5. (a) Physician dashboard showing list of patients in mobile app (b) patient dashboard in Mobile App

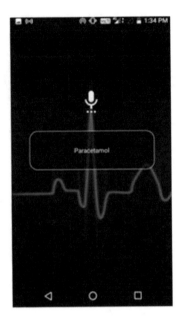

Fig. 6. (a) Screenshot showing Voice capture feature in mobile app (b) screenshot showing translation feature using Yandex API

Fig. 7. QR Code Generation in the app

8   Performance Evaluation and Future Work

While storing the records, the amount of data that can be embedded on a QR code proved to be major limitation of the system. With the lowest error correction level, a QR code can at most hold around 4200 alphanumeric characters which would be sufficient to hold a single prescription made out from a consultation session and other text based records. However, other multimedia medical records of high resolution cannot be stored using a QR code. The efficiency of speech capture, speech to text and translation modules depends on the APIs used for implementation. Port Audio API, IBM watson Speech to Text API and Yandex API have been experimented in the proposed work. For the evaluation of the proposed system, 30 different users bilingual speech was captured, converted to text and translated to English. The overall efficiency was found to be around 80%. Both the speech to text conversion and translation can be made more efficient by implementing suitable machine learning models which is considered for the future work.

9   Conclusion

Thus the system developed provisions for easy storage and retrieval of out patients' medical data with a help of a mobile application and QR codes. The entire application is deployed on a fog based application framework for fast and secured transmission of medical data among clinics. The major challenges in developing the system include bilingual translation and voice to text conversion, given that the data is medical and there is no scope for errors. The system involves a lot of scope for future development which includes a specific mobile device for carrying out the mentioned transactions viz recording medical advice, speech to text conversion and QR generation. The application may also be further improved on security aspects.

References

1. Mehta, P.: Framework of Indian healthcare system and its challenges: an insight. In: Health Economics and Healthcare Reform: Breakthroughs in Research and Practice, pp. 405–429. IGI Global (2018)
2. Wang, Y., et al.: A shared decision-making system for diabetes medication choice utilizing electronic health record data. IEEE J. Biomed. Health Inform. **21**(5), 1280–1287 (2016)
3. Maganti, P.K., Chouragade, P.M.: Secure application for sharing health records using identity and attribute based cryptosystems in cloud environment. In: 2019 3rd International Conference on Trends in Electronics and Informatics (ICOEI). IEEE (2019)
4. Buchanan, W.J., et al.: Secret shares to protect health records in Cloud-based infrastructures. In: 2015 17th International Conference on E-health Networking, Application and Services (HealthCom). IEEE (2015)

5. Czuszynski, K., Ruminski, J.: Interaction with medical data using QRcodes. In: 2014 7th International Conference on Human System Interactions (HSI). IEEE (2014)
6. Uzun, V.: QR-code based hospital systems for healthcare in Turkey. In: 2016 IEEE 40th Annual Computer Software and Applications (COMPSAC), vol. 2. IEEE (2016)
7. Dube, S., et al.: QR code-based patient medical health records transmission: Zimbabwean case. In: Proceedings of Informing Science & IT Education Conference (InSITE) (2015)
8. Khilari, P., Bhope, V.P.: A review on speech to text conversion methods. Int. J. Adv. Res. Comput. Eng. Technol. (IJARCET) **4**(7), 3067–3072 (2015)

Searching Pattern in DNA Sequence Using ECC-Diffie-Hellman Exchange Based Hash Function: An Efficient Approach

M. Ravikumar, M. C. Prashanth$^{(\boxtimes)}$, and B. J. Shivaprasad

Department of Computer Science, Kuvempu University, Shimoga, Karnataka, India

Abstract. In this paper, the efficient Elliptic Curve Cryptography with Diffie-Hellman exchange is implemented in order to perform the searching pattern effectively from DNA sequences. The elliptic curve cryptography is defined as public key cryptography which utilizes the elliptic curve properties in terms of finite field. For key exchange process, ECC is combined with Diffie-Hellman algorithm (EC-DH) which is defined as technique for exchanging cryptography keys securely through the public channel. It ensures the security and it highly resistive against the brute force attack. Lesser computational time is obtained and effective security with power consumption achieved. Experimentation results are carried out by using the publicly available dataset. The effectiveness is proved from the comparison results of the proposed and existing study.

Keywords: Searching · Patterns · DNA sequence · Elliptic curve cryptography · Diffie-Hellman exchange · Multiple hash

1 Introduction

A DNA pattern identifies or accepts one or more sequences with respect to given search pattern and it provides the positions and number of the matching pattern. In computational biology, searching pattern from sequence of DNA shows the broader applications. Searching a pattern is significant in information processing in computer science for identifying the structural and functional genes behavior. The microbiologists frequently searched the significant information in databases. To retrieve the pattern, DNA database is highly complex and huge and it considered as difficult process. To maintain the ever growing demands searching process is required in comparison of DNA sequence. In case of DNA sequences of multiple detections, these kind of algorithms are easier to implement and shows effectiveness [1].

In similar with the large DNA sequences the DNA sequence size is grown which is critical in recognize the genomes biological features. The genetic information is presented in the DNA sequences illustrates the cellular forms with respect to biological development. In differentiating the living objects features while processing the biological data representing the genetic form for living object identification and it emphasizes on the genomes personality with respect to protein cells collected with modified segment

© The Author(s), under exclusive license to Springer Nature Switzerland AG 2022
R. Misra et al. (Eds.): ICMLBDA 2021, LNNS 256, pp. 117–127, 2022.
https://doi.org/10.1007/978-3-030-82469-3_11

such as compounds, amino acids and so on. For DNA formation these are transformed in to the molecule form. For handling the DNA groups, the pattern searching and sequence processing is required. Searching process is the essential one and the precision values are based on the repository, search query and integrated system [2].

Number of occurrences of provided pattern is examined between the large sequences of DNA. However the existing studies revealed that it is concentrated only on the particular search type which differentiates in same functional processes and optimal processing time is obtained [3]. Efficient algorithm is developed based on the specific sequences from larger database which search the number of repetitions with respect to initial and end index to know the intensity of pattern occurs the most. Computational complexity and security are considered as the major issue. Efficient encryption and key exchange algorithms has to be focused to perform the pattern recognition in an effective way. The elliptic curve cryptography is defined as public key cryptography which utilized the elliptic curve properties in terms of finite field. For key exchange process, ECC is combined with Diffie-Hellman algorithm which is defined as technique for exchanging cryptography keys securely through the public channel. It ensures the security and it highly resistive against the brute force attack [4–6].

The remaining part of the paper is organized as follows. Section 2 discusses the related work on pattern searching. In Sect. 3, proposed methodology is given, results and discussion illustrated in the Sect. 4, finally conclusion in given in Sect. 5.

2   Related Work

This section gives, some of the approaches which are related to searching a pattern in DNA sequence.

This study is based on the DNA coding with respect to the encryption methods related with images. this research work focused on the five categories of DNA coding such as DNA dynamic coding, various DNA coding types [7], various base complement operation, DNA fixed coding and several DNA operations. These kinds of techniques have been illustrated and compared further. Optimal DNA coding mechanism has been utilized in this study for implementing the new encryption mechanism to illustrate the security and effectiveness. This study highlighted the drawbacks of the several image encryption techniques. Dynamic DNA operations and DNA coding have been studied and the various techniques influences have been discussed [8]. Further, asymmetric image encryption technique has been focused in this study to focusing on the symmetric image encryption distribution and potential security issue related with key management. The asymmetric encryption focused is the chaotic theory and elliptic curve ElGamal cryptography. The initial values are generated by the SHA 512 hash values and the chaotic index sequence with cross over permutation has been utilized for the plain image scrambling. Moreover, the utilization of El Gama encryption can able to solve the key management issues and also the security improvement, with respect to the scrambled image generation. To obtain the cipher image, chaos game combined with diffusion in terms of DNA sequence has been implemented. The results show that this study exhibited high robustness towards plain text attack and thus provides better security mechanisms.

On the other hand, this study focused on image encryption mechanism merging the DNA sequences, SHA 256 hash and chaotic system [9]. The plain image has been encoded into DNA matrixes which have been then performing the row by row diffusions and wave based permutations. The plain image is determined the key matrix by the DNA encoding and decoding which are resulted in close dependency. The security evaluation results exhibited the better secure key space, effect of better encryption, secret key with respect to high sensitivity. Various attacks are also resisted by this algorithm. This algorithm has been appropriate to the digital image encryption and provided higher security level.

The DNA sequence operation and nested chaotic maps have been focused using the rapid secure mechanism based on image encryption process. The chaotic attractor initial conditions have been generated by SHA-256 algorithm [10]. Diffusion layer and confusion layer have been focused based on the private key encryption mechanism. High sensitivity is proved against the various differential and secret attacks. Finally, correlation analyses have been performed based on the plain image pixels. Encrypted and plain images have been verified later. These proposed methods are resistive against data loss attacks and noise. This method shows efficiency compared with various other schemes [11]. For security and privacy, the organizational data to be stored in cloud requires the effective algorithm. Thus this study focused on the elliptic curve based Diffie-Hellman algorithm. It provides encryption and decryption to protect the data and enables confidentiality. The sensitive information is leaked in some cases. Data security has been enhanced and the computational complexity is reduced. The parameter focused is the key generation time, encryption and decryption time and computational overhead. This algorithm provides 70% better performance related with the existing encryption algorithms.

The paper presented an innovative method for encryption in accordance with DNA sequence operations comprising of three processes [12]. The suggested method could be easily processed and found to be computationally simple for achieving high speed, security and sensitivity. Further this system could be implemented for encrypting color images. The suggested method employed a 256-bit length secret key for increasing the security. Similar to the existing work the proposed work also employed ECC- Diffee Hellman based encryption system for finding the sequence pattern of DNA.

We can also find more information on DNA sequence analysis in [13–18].

3   Proposed Method

In this study, the large sets of DNA sequences are focused and the pattern searching has been performed and recognised. The DNA sequences is an input sequences, in which the patterns have to be searched. Before searching those patterns initially, DNA sequences have been pre-processed for collecting the data points such as sequences length, space among the sequences and sequence index. For numeric value generation the hash value function will be operated for every sequence and it is expressed as specific hash value of every sequence and followed by patterns searching which are query patterns leads to input sequences. This can be pre-processed to evaluate the searching pattern length. It can be functioned with hash function for identifying the pattern length to generate hash values. After finding the hash values of both input sequences and the searching pattern, the searching pattern will be matched with the input DNA sequences till the end of the sequences to find the number of occurrences of the patterns found and the taken time to search all patterns will be calculated as the computational time in milliseconds. By comparing this computational time, the efficiency of the algorithm will be known. If the pattern is not found in the input DNA sequences, it will show as the pattern does not found.

For secure searching and time complexity, ECC encryption method is combined with Diffie-Hellman algorithm for large DNA sequential patterns which is shown in following proposed flow Fig. 1.

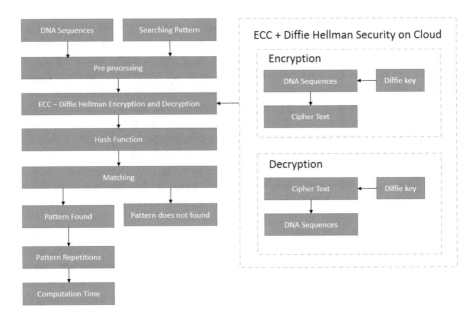

Fig. 1. Block diagram of the proposed method

ECC-Diffie-Hellman Mechanism (EC-DH):

An elliptic curve E with respect to GF (p) comprised with (x, y) solutions expressed in the equation given below

$$(y^2 = x^3 + ax + p) mod p \tag{1}$$

The general elliptic curve functions are the point doubling and point addition. Scalar multiplication is required by the ECC primitives. The series of addition and doubling of p point is obtained. The notations are defined as,

Generator point G with m prime order.

The pair of key is (P, n) in which p is the public key $p = n * G$ and private key is n which is lesser than m.

Alice user is Us_{alc} and Bob user is Us_{bob}.

X-OR operation is $\oplus$.

The significant EC-DH algorithm is tabulated below,

Algorithm
1.key generation
Start
Initiate the connection between user U_S and U_{S_i}
$Us_{alc} = (P_{alc}, n_a)$ is the key pair for $Us_{alc}.Us_{\bar{i}} = (P_{alc}, n_a)$ is the key pair for U_{S_i}
U_S send the point P_i to U_S similarly U_S send the P to U_{S_i}
2. Encryption: hiding
U_S computes the point $n_{alc}(P_{bol}$ let it be $Sp_1 = (x_{sp_1}, y_{sp_1})n_{alc}(P_{bol}$
and let it be $Sp_1 = (x_{sp_1}, y_{s_1}$
U_S calculate the distance between S and S let it be
U_S calculate the $R_1 = x$ and $R_2 = x$ cipher$= R_1 \oplus R_2$ message
U_S send the chipper to U_{S_i}
For the next step for Alice change the value to $R_1 = R_1 +$
3. Decryption
U_{S_i} compute the point and S
U_S calculates $R_1 = x$ and $R_2 = x$
In the end U_{S_i} decrypt the message using Message $= R_1 \oplus R_2$ Cipher
For the next session bob change the value to $R_1 = R_1 +, \quad R_2 = R_2 +$

The elliptic curve cryptography is defined as public key cryptography which utilized the elliptic curve properties in terms of finite field. Generally smaller keys are required by ECC which is then related with non ECC cryptography which exhibits the security. For key exchange process, ECC is combined with Diffie-Hellman algorithm which is defined as technique for exchanging cryptography keys securely through the public channel. The keys are not exchanged generally and they are combined. For example, Alice selects secret integer as private key and measures the Eq. (1) as the public key.

The above described algorithm is the key exchange, encryption and decryption process which consumes lesser power compared with simple hiding encryption algorithm. It ensures the security and it highly resistive against the brute force attack. Lesser computational time is obtained and effective security with power consumption achieved due to the applicable XOR operations $\oplus$ in encryption algorithm.

Hash Function

The hash function will convert each string into a numeric value, it will be called as the hash value or string value, hash function gives the values for both the input pattern and the pattern which had to be searched, and then it will compare the values of the searching pattern with the input pattern. If the values get matched it exhibits as the pattern has been found, otherwise the pattern does not exist. For this condition the hash is that the sequence s and t are equal (s = t) and also their hash values should be equal hash(s) = hash (t). Otherwise, it won't be able to compare the sequence.

4   Results and Discussion

This section explains the results and discussion of the DNA sequence searching pattern with the employed ECC- Diffee Hellman approach. This section deliberated the performance of the proposed system in reducing the searching time of a particular DNA pattern. The generated computationally efficient algorithm utilized ECC- Diffee Hellman based hash function. This experiment that has been performed on a publically available dataset was also compared with various methods in order to prove its efficacy. To conduct experimentation, we have collected two publically available datasets [NCBI] one of length 2311 and another is 5301 and the sample of the datasets is as shown below

AGCCCAGCTCTTAAGCTGCGCTAGAAAAGCTAGCCCAGCTCTTAAGCTGC
TCCGGAAAAGCTAGAGCCCAGCTCTTAAGCTGCTCCGGAAAAGCTAGAGC
CCAGCTCTTAAGCTGCTCCGGAAAAGCTAGCCCAGCTCTTAAGCTGCTCC
GGAAAAGCTAGCCCAGCTCTTAAGCTGCTCCGGAAAGCTAAGCCCGCTAC
CTCCGGAAAAGCTAGAGCCCAGCTCTTAAGCTGCTCCGGAAAAGCTAGCC
CAGCTCTTAAGCTGCTCCGGAAAAGCTAGCCCAGCTCTTAAGCTGCTCCG
GAAAGCTAAGCCCGCTACTTGCAGCTAAGCCCGCTACTTAG............

We carryout experimentation to prove the efficiency of the proposed method; the experimentation is conducted on publically available two DNA sequences datasets containing the length of 2311 and 5103 sequence, and the time is calculated in milliseconds. In this, we focus on searching the total occurrences of the given pattern and also the index of the given pattern in the dataset. Here, the DNA sequence searching methods like Naive search, Z search, KMP search, RP search, Boyre-Moore search, Optimized Naïve search, BMHorspool search, Shift Based search, Finite Automata search are used. The proposed method performs better with respect to computational time when compared with all the searching algorithms and then the results are tabulated and plotted in

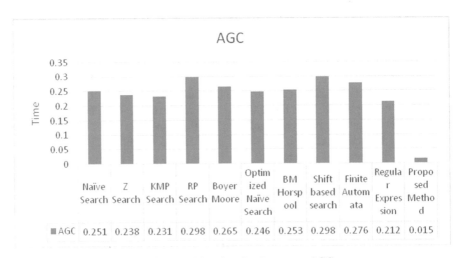

Fig. 2. Searching time for the pattern AGC

the graphs. In this, experimentation is conducted for different cases i.e., the sequence of three, sequence of four, sequence of five and intentionally we have taken the pattern which is not appeared in the entire input sequence and the algorithm shows the given pattern does not exist.

Case 1: AGC Pattern

Case 1 gives the results for searching methods for the length of three pattern sequence AGC in total 2311 length of sequence, which has repeated forty-two times and the computational time has measured in milliseconds which is given in the Fig. 2 and also graph is plotted in Fig. 2.

The above figure gives the computational time for all the methods to search the pattern AGC in the 2311 length of database. Where the proposed method gives the best computational time when compared with the other methods.

Case 2: CTGA Pattern

In Fig. 3, Case 2 gives the results of searching methods and their computational time is measured in milliseconds for the length of given four CTGA sequence pattern, where it repeats for thirteen times in the sequence. From the Fig. 3, it is clear that the proposed system provides the better computational time for all the methods to search the given pattern CTGA in the 2311 length of database.

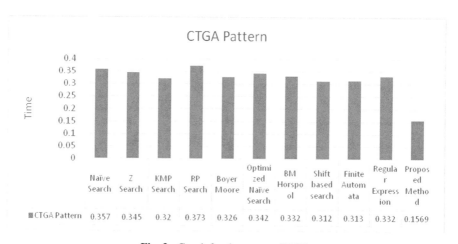

Fig. 3. Graph for the pattern CTGA

Case 3: TGCAG Pattern

Case 3 gives the results of searching methods and their computational time is measured in milliseconds for the length of given five TGCAG sequence pattern in total of 2311 sequence length, where it repeats for nine times in the sequence and the graph is plotted in Fig. 4.

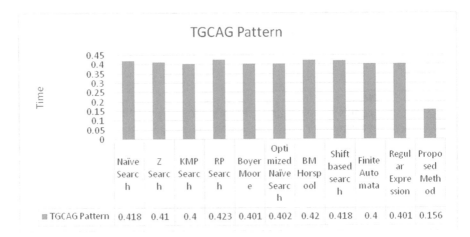

Fig. 4. Graph for the pattern TGCAG

The above table in the Fig. 4 gives the computational time for all the methods to search the given pattern TGCAG in the 2311 length of database. Where the proposed method gives the best computational time when compared with the other methods.

Case 4: For GCTGCT Pattern
Case 4 gives the results of searching methods and their computational time is measured in milliseconds which is given in the Fig. 5 for the length of given six GCTGCT sequence pattern in total of 2311 sequence length, where it repeats for Eight times in the sequence and also the graph is plotted in Fig. 5.

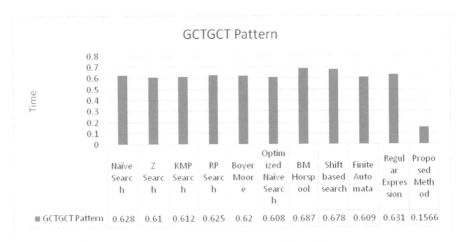

Fig. 5. Graph for the pattern GCTGCT

The above table in the Fig. 4 gives the computational time for all the methods to search the given pattern GCTCGT in the 2311 length of database. The proposed method gives the best computational time when compared with the other methods.

Case 5: Overall Comparative Analysis for 2311 Sequences

Case 5 gives the overall results of searching methods in the table in the Fig. 6 and its computational time is measured in milliseconds for all given pattern in total of 2311 sequence length, also the graph is plotted in Fig. 6.

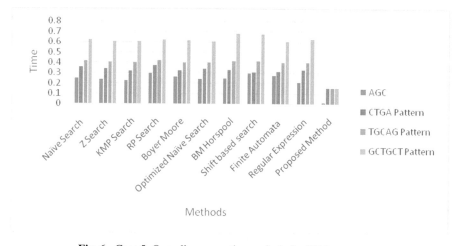

Fig. 6. Case 5: Overall comparative analysis for 2311 sequences

Case 6: Overall Comparative Analysis for 5103 Sequences

Case 6 gives the overall results of searching methods in the Fig. 7 and its computational time is measured in milliseconds for all given pattern in total of 5103 sequence length, also the graph is plotted in Fig. 7. The above table in the figure gives the overall analysis of the computational time for all the methods to search the given pattern in the 5103 length of database. Where the proposed method gives the best computational time when compared with the other methods.

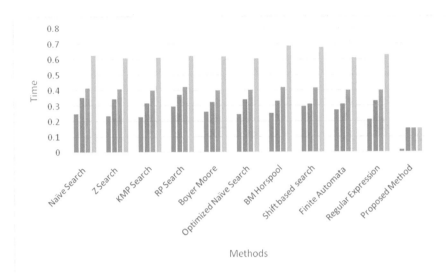

Fig. 7. Graph on overall values of 5103 total sequences

5 Conclusion

In this work, we have presented an efficient approach for searching a pattern in DNA sequence, which employ ECC- Diffie-Hellmann based on multiple hash function. Experiment is con-ducted on two different datasets length of sequence which is shown in different cases. From the experimentation, it is observed that the proposed system gives the best result with respect to the computational time when compared with the ten other methods. By considering the orders of computational time performance, the proposed technique can be an alternative for multiple searching with large set of patterns of biological sequences.

References

1. Busia, A., et al.: A deep learning approach to pattern recognition for short DNA sequences. BioRxiv 353474 (2019)
2. Cherry, K.M., Qian, L.: Scaling up molecular pattern recognition with DNA-based winner take all neural networks. Nature **559**, 370–376 (2018)
3. Kalsi, S., Kaur, H., Chang, V.: DNA cryptography and deep learning using genetic algorithm with NW algorithm for key generation. J. Med. Syst. **42**, 1–12 (2018)
4. Jalali, A., Azarderakhsh, R., Kermani, M.M., Jao, D.: Supersingular isogeny Diffie-Hellman key exchange on 64-bit ARM. IEEE Trans. Depend. Secure Comput. **16**, 902–912 (2019)
5. Mehibel, N., Hamadouche, M.H.: A new approach of elliptic curve Diffie-Hellman key exchange. In: 2017 5th International Conference on Electrical Engineering-Boumerdes (ICEE-B), pp. 1–6 (2017)
6. Bodur, H., Kara, R.: Implementing Diffie-Hellman key exchange method on logical key hierarchy for secure broadcast transmission. In: 2017 9th International Conference on Computational Intelligence and Communication Networks (CICN), pp. 144–147 (2017)

7. Xue, X., Zhou, D., Zhou, C.: New insights into the existing image encryption algorithms based on DNA coding. PLoS ONE **15**, e0241184 (2020)
8. Luo, Y., Ouyang, X., Liu, J., Cao, L.: An image encryption method based on elliptic curve elgamal encryption and chaotic systems. IEEE Access **7**, 38507–38522 (2019)
9. Chai, X., Chen, Y., Broyde, L.: A novel chaos-based image encryption algorithm using DNA sequence operations. Opt. Lasers Eng. **88**, 197–213 (2017)
10. Slimane, N.B., Aouf, N., Bouallegue, K., Machhout, M.: An efficient nested chaotic image encryption algorithm based on DNA sequence. Int. J. Mod. Phys. C **29**, 1850058 (2018)
11. Subramanian, E.K., Tamilselvan, L.: Elliptic curve Diffie–Hellman cryptosystem in big data cloud security. Cluster Comput. **23**, 3057–3067 (2020)
12. Norouzi, B., Mirzakuchaki, S.: An image encryption algorithm based on DNA sequence operations and cellular neural network. Multimedia Tools Appl. **76**, 13681–13701 (2017)
13. Neamatollahi, P., Hadi, M., Naghibzadeh, M.: Efficient pattern matching algorithms for DNA sequences. In: 25th International Computer Conference, Computer Society of Iran (CSICC), 2020, pp. 1–6 (2020)
14. Tahir, M., Sardaraz, M., Ikram, A.A.: EPMA: efficient pattern matching algorithm for DNA sequences. Exp. Syst. Appl. **80**, 162–170 (2017)
15. Fostier, J.: BLAMM: BLAS-based algorithm for finding position weight matrix occurrences in DNA sequences on CPUs and GPUs. BMC Bioinformatics **21**, 1–13 (2020)
16. Ryu, C., Lecroq, T., Park, K.: Fast string matching for DNA sequences. Theoret. Comput. Sci. **812**, 137–148 (2020)
17. Munirathinam, T., Ganapathy, S., Kannan, A.: Cloud and IoT based privacy preserved e-Healthcare system using secured storage algorithm and deep learning. J. Intell. Fuzzy Syst. **39**, 3011–3023 (2020)
18. Najam, M., Rasool, R.U., Ahmad, H.F., Ashraf, U., Malik, A.W.: Pattern matching for DNA sequencing data using multiple bloom filters. BioMed Res. Int. **2019**, 1–9 (2019)

Integrated Micro-Video Recommender Based on Hadoop and Web-Scrapper

Jyoti Raj$^{(\boxtimes)}$, Amirul Hoque, and Ashim Saha

Department of Computer Science and Engineering,
National Institute of Technology Agartala, Agartala, India

Abstract. Social-media is the biggest modern platform for sharing and gaining knowledge without geographical restrictions. Videos have became the key of social-media. Short length videos considered here as Micro-videos. Some topics can be effectively elaborated in short time-span. Trending videos are generally short length videos. But every platform like Facebook, YouTube, twitter is focusing on own platform. Cross-platform is the major issue in micro-video recommendation. Users can't look at every platform for their required videos. It is quite hectic and time-consuming to surf every social-media websites. In this paper an approach is taken to remove cross-platform issue using web-scrapers. Algorithms used for recommendation generation are Content based and Collaborative based algorithms. And the content of the micro-videos are the keywords searched by users. Generally most algorithms work by taking into account the rating given by users. But some platforms don't have rating options like YouTube. User-based recommendation is a traditional approach. It requires time and space for calculation. Hadoop is the best solution for it. User-based recommendation algorithm based on Map-Reduce without the use of rating has been elaborated in detail. Combination of recommendations generated from different algorithms gives the best result.

Keywords: Social-media · Web-scrapper · Map-reduce · Micro-video recommendation engine · Platform-independent · Big-data

1 Introduction

Videos give a better and effective understanding compared to textual information. People watches videos for stress reduction, entertainment, news, tutorial and so on. They also watches for the troubleshoot of issues related to several fields like automobile, installation of software. Every social platform is increasing its space for videos. Like Twitter, Facebook have a separate tab for uploaded, shared or trending videos. Users have lots of platforms to watch videos like YouTube, Facebook, Twitter, Vimeo. More-over these, various countries also have their own platforms for video sharing. And new apps and websites are launching day-by-day. Due to numerous options, users are not able to get/choose the best

© The Author(s), under exclusive license to Springer Nature Switzerland AG 2022
R. Misra et al. (Eds.): ICMLBDA 2021, LNNS 256, pp. 128–140, 2022.
https://doi.org/10.1007/978-3-030-82469-3_12

out of all. So one can easily visualize about the future and complexity of video-market. The whole complexity reduces little when all social-media's micro-videos is available at one platform. Video producer will have to focus more on content to produce short length video by cutting irrelevant matters. So good micro-videos will come out. Users can get what they want in short time-span. Micro-videos also provide a way to advertisers to attain users attention in short time span. It can't be denied that some field or topics will certainly need long-length videos, but generally popular videos are of short length. Users even loose attention after a certain period of watch-time. So, shift from video to micro-video is a good solution for reduction of complexity in video-market.

Recommendations help users and producers in their choice and to get more target users. A recommender directly affect the companies growth-factor. A good recommender engine enhances the profile of websites. Various authors take different parameters of videos for considering them as micro-videos. Some parameters are frame per second (fps), quality, watch-time by users, actual duration of videos. In this paper, micro-videos means videos of duration less than 300 s. Micro-video Recommendation System (MRS) is a engine which here recommends the URLs of micro-videos. These URLs further embedded on a built Web-interface. Lots of works are going on to improve the quality of recommendation. Various Web scrapers built for cross-platform recommendation. Web Scrapers [1] are data harvesting mechanism which extracts data from websites. This paper has taken an approach towards dynamic recommendation. Dynamic in the sense that scrapers extract data from several websites and recommends on timely-basis. This paper has also elaborated the working steps of recommendation generation based on user similarity on Hadoop platform.

2   Related Work

There are numerous algorithms for micro-video recommendation generation [10]. Some of the implemented algorithms by various authors are:- Shang et al. [2] has implemented the Slope-one algorithm using Map-Reduce and has also proposed web-crawlers for cross-platform recommendation. But the generated recommendation is static in nature. Jiang et al. [3] has worked on Wechats and Weibos public API data and has used k-means clustering algorithm for categorization. They have used Bias classification for classification purpose. Hot topic is recommended in each category such as trailers, comedy, records, spoof. Balachandra et al. [4] has acquired various parameters of a video like ranking rundown, brief outline, click rate. And they have used Neural Association algorithm for recommendation generation. Ramakrishnan et al. [5] have used feature vector approach for recommendation. One approach for recommendation is also through tripartite graph propagation. Chen et al. [6], has used a graph with Query node, User Node, Video nodes and has done normalization in loops where trade off factor is high. Brbic et al. [7], has generated recommendation for YouTube videos. They have defined one formula for rating calculation based on view count, like count, dislike count, user factor and appearance number. And they have taken up-loader of the videos as related users. Taking up-loader as related user is a plus point of this method. Next uploaded video by particular up-loader as related user is most likely to be watched by user. Jyoti et al. [10] has shown in a survey that recommendation using improved slope one algorithm gives better outcome.

Web Scrapers are very useful in various ways especially when data-science field is on apex. It is also a way for the conversion of unstructured data of world wide web into structured one. There are lots of techniques for web-scraping such as Hypertext Transfer Protocol Programming, Hyper Text Markup Language Parsing, Computer vision web-page analysers, Document Object Model Parsing, Web Scraping Software [8]. Certain web scraping software tools reduce the complexity of scrapers by making it abstract in nature. Some of the tools are Mozenda, Visual Web Ripper, Web Content Extractor, Import.io, Scrapy. Few of the mentioned scraper tools are complete GUI based while few needs code to run. It needs browser extension for implementation. Web scraping provides a base for boosting web-advertisement. It gives a way to know the customers, websites and market trend in a better way. Eloisa, et al. [9] have shown how collaborative filtering can be used in Web Advertisement using Web Scrapers.

3   Web-Scraper Using Selenium and Beautifulsoup

Web scraper is data harvester used for extracting data from websites. While it can be done manually by a user, the term typically refers to automated implementation performed by a bot. Though there are lots of tool and software available for scraping, we found Selenium and Beautifulsoup most appropriate and easy to work around as per required need.

3.1   Selenium Web-Driver

Selenium supports web-scraping by providing a web-driver named as Selenium Web-Driver. It is mainly useful in different types of testing. This driver acts as an interface between built scraper and specific browser. Web-Driver gives the full automated access control to web by providing an instance of a browser. It supports various browsers such as Firefox, Internet Explorer, Chrome, Microsoft-Edge or Safari. And the driver is browser-specific, like geckodriver is for Firefox. Selenium 2.0 is fully implementable in Python, C#, Java and Ruby.

3.2   Beautifulsoup

Beautiful Soup is a Python package for parsing HTML and XML documents (including having malformed markup, i.e. non-closed tags, so named after tag soup). It creates a parse tree for parsed pages that can be used to extract data from HTML, for web scraping. It supports various parsers like Python's html parser, lxml's HTML parser, lxml's XML parser, html5lib. And "html5lib" is pure python-parser , which parses HTML in the way a web browser does. In this paper, html5lib parser is used.

3.3   Implementation

"driver = webdriver.Firefox()" gets the handle of the driver. Then any url is opened by driver.get (url). Any element of the web-page is found by "driver.find_element_by_". There are different ways for passing keys to textbox or clicking on button or finding any tag. Elements can be found by css.selector, xpath, name, id. "time.sleep()" is used to make bot behave like humans. So that id doesnot get blocked by the server.All these driver work is supported by *Selenium*. After scrolling the window to the lowest point whole content of the webpage is passed to *Beautifulsoup*. From here soup's parsing work starts. Elements like tag, class-name, href, etc. can be found in bs4 by using "soup.find"/"soup.findAll". "soup.find" finds the first element only and "soup.findAll" finds all the element which can be handled using iterative loop. Extracted data are stored in csv format in file. Software needed for running Scraper-code are:

1. selenium WebDriver 3.141.0
2. geckodriver
3. beautifulsoap/bs4
4. python3

Different types of scraper has been built for data extraction from YouTube and Facebook. They are as follows:-

1. Scraper for search-history extraction
2. Scraper for keyword-based search and extraction of result
3. Scraper for watch-history extraction

3.4   Scraper for Search-History Extraction

Keywords of search-history are extracted by this scraper. These keywords act as input for content-based recommendation. Scraper first login the particular website (here YouTube) and then clicks on search history option. From there it extracts all keywords of search-history. Figure 1 shows the search history of a user from YouTube extracted by scraper for a particular user.

```
mahout tutorial for beginners
big data tutorial
ubuntu
cricket
comedy
```

Fig. 1. Sample of extracted data (search history) from YouTube of a particular user

3.5   Scraper for Keyword-Based Search and Extraction of Result

Scraper first opens the particular website(here YouTube and Facebook). It passes the particular keyword(here comedy) to search text box and then clicks the search option. Window is scrolled down to the lowest point to get all the search result. In Facebook search-result has lots of tabs, so video tab is clicked by scraper for videos only. It extracts link, time-span, views-count. Data pre-processing is done for conversion of data extracted into a particular format. Like extracted time-span is in "HH:MM:SS" text-format. Whole string is broken-down to get hours, minutes and seconds in integer format and then mathematics is applied to convert time-span in seconds. And then compared to 300 s. So only those videos are extracted whose time-span is less than 5 min. From here we get micro-videos URL, which will be the result of search by the user. Figure 2 displays the extracted data (URL,views) in csv format based on search-history keyword(comedy) from Facebook.

```
https://www.facebook.com/DryBarComedy/videos/1835323220101824/    578224
https://www.facebook.com/DryBarComedy/videos/532123057298201/     579797
https://www.facebook.com/Likekomsan/videos/2042233249218023/      33253614
https://www.facebook.com/THECOSMOQUEENS/videos/369400863847666/   2060593
https://www.facebook.com/awwmygoshlol/videos/2924075760951730/    52608
https://www.facebook.com/DryBarComedy/videos/271715270151672/     1323158
https://www.facebook.com/DryBarComedy/videos/748902582147631/     883825
https://www.facebook.com/DryBarComedy/videos/534919767001459/     544637
```

Fig. 2. Sample of extracted data (URL, views) for keyword (comedy)-based search from Facebook

3.6   Scraper for Watch-History Extraction

```
Game of thrones 2018,https://www.youtube.com/watch?v=Tgi-Yj0xrig
Game of Thrones : Crypts of Winterfell,https://www.youtube.com/watch?v=wA38GCX4Tb0
Game of Thrones : Aftermath (HBO),https://www.youtube.com/watch?v=vwmAWOE5F9o
Marvel Studios' Avengers: Endgame,https://www.youtube.com/watch?v=KCSNFZKbhZE
Marvel Anthem | A.R. Rahman | Hindi,https://www.youtube.com/watch?v=w5LGdqQAChs
BHARAT | Official Trailer | Salman Khan,https://www.youtube.com/watch?v=Ea_GKoe81GY
Taarak Mehta Ka Ooltah Chashmah,https://www.youtube.com/watch?v=B4i3caVq53w&t=2s
Tom y Jerry en Español ,https://www.youtube.com/watch?v=PxrnoGyBw4E&t=52s
Teri Mitti - Kesari | Akshay Kumar ,https://www.youtube.com/watch?v=wF_B_aagLfI
```

Fig. 3. Sample of extracted watch-history data in csv format

Videos titles and links which are watched by users are extracted in csv format. Scraper first opens the YouTube watch history page. It passes username and password for login. Scrolls-down to the lowest point of window, so that full his-tory can be extracted. Here all-length videos data are retrieved for user-based recommendation. So that good Similarity-value come into figure for better rec-ommendation. Figure 3 shows the sample of collected watch-history data of a particular user from YouTube.

4   Proposed Model

Aim of the proposed model is to fetch the micro-videos from different social-media platforms and recommend them to the users on single platform. This model has devised a way to bring several websites' micro-videos on a single platform and to generate a good recommendation. Model can be revised as per requirement. Like user based recommender can be replaced with item-based, deep-learning, etc. cetera by keeping scrapers model same. Here URLs of micro-videos acts as recommendation for users. In this paper, Web Scrapers may violate certain websites terms and conditions. For better scientific results, we have not taken into account the concerned legal issues. This section defines the whole working in brief. Furthermore all are defined in detail in following sections.

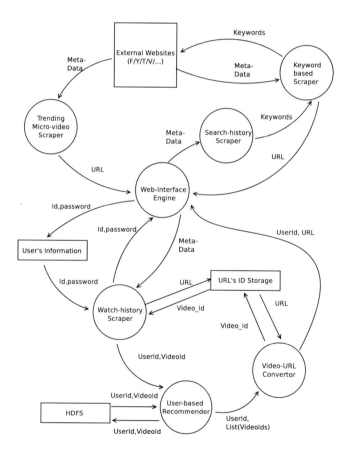

Fig. 4. Data-flow diagram of proposed-model for micro-video recommendation system (F-Facebook, Y-YouTube, T-Twitter, V-Vimeo)

Figure 4 is the DFD diagram of the proposed model. Web-Interface is the interface where all the recommended URLs embedded for users. This interface is a website. But it can be a website, an app or a browser extension. App like SHARE-it is recommending video from YouTube only to users. Any video-sharing website can use this model for recommendation generation from multi-platform social-media. External websites here refer to popular-video sharing websites like Facebook, YouTube, Twitter, Vimeo and so on. All scrapers used in the model built using Beautiful-soup and Selenium.

First using Trending Micro-video Scraper scraped all external websites trending micro-videos URL and provided to Web-Interface Engine. All these URLs is embedded in sub-containers of website. Search history scraper scrapes the Web-interface for search-history keywords for a particular user. This scraper will not be needed if user's search history is stored in database by Web-Interface Engine. These keywords act as content of a particular user. Based on these keywords Content-based recommendation is generated for users. These keywords are passed to Keyword-based Scraper which scraped all external websites and gave output as URLs. Required number of URLs recommended to users by Web-Interface Engine.

Watch history scraper takes the user-id and password of a particular user from user information database and scrapes the watch history of a user. It gives URL as output and these URLs are passed to URL'S Id file storage database. From here scraper gets the video-id and passes the user-id and video-id to User-based Recommender. User-based Recommender stores all user-ids and video ids in hadoop distributed file system(hdfs). And generates User-Id and list of recommended Video Ids. These data are passed to Video-URL converter. Video-URL converter converts the video-id to URL by accessing URL'S Id file storage database. After conversion of video-id into URL, (UserId, URL) is passed to Web-interface Engine, which recommend these URLs to user.

5   Keyword-Based Recommendation

In proposed Model, Web-interface engine is considered here as YouTube. It needs user's Web-interface Engine's id and password for search-history extraction. Search-history scraper scrapes the keywords searched by user. And passes recent keywords to Keyword-based Scraper. KBS first parses the YouTube videos by searching those keywords one-by-one and extracts the videos whose time-duration is less than 5 min. Same is done for Facebook also. Here for simplicity only top-5 search result is extracted for each keywords from both Facebook and YouTube. After completion of all-websites scraping, URLs of all extracted micro-videos are stored in CSV format file for recommendation. So, recommendation is generated from more than one platforms to one platform dynamically. These URLs is further passed to Web-interface Engine, which embedded these URLs to sub-containers of Web-page as recommendation for users. Recommendation is generated on timely-basis. As scraper takes time for recommendation generation based on keywords passed for searches.

6   Map-Reduce Programming Paradigm

Map-Reduce is a programming paradigm which provides an algorithm for implementation of any logic in parallel, distributed computing. Hadoop Map-Reduce is
a software framework for easily writing applications which process vast amounts
of data (multi-terabyte data-sets) in-parallel on large clusters (thousands of
nodes) of commodity hardware in a reliable, fault-tolerant manner. It is the
model/algorithm used by Hadoop for processing large data-sets. Map-reduce
works on (key, value) pair only. There are can be any number of mappers in one
job but there will be only one reducer for each job.

It composed of mainly two functions mapper and reducer. Figure 5 depicts the
working model of map-reduce. Whatever we want to implement on Hadoop, we
have to logically break the whole work into the working model steps. Input can
be command-line Text, text-file, group of text files. Whole input file is chunked
into various splits based on HDFS block size (generally 64 MB). Each Inputsplit
is assigned a Record-reader which reads one line at a time and splits them into
(key, value) pair based on defined Input-Format. TextInputFormat is the default
input format of MapReduce. It is mainly used for unformatted data or line-
based records. Its key is the byte Offset of each line of different splits. And
value is the contents of the line, excluding line-terminators. Different box classes
are defined for input types like Text for string, IntWritable for integers. Each
split is assigned a single mapper. And different mappers will run in parallel on
different data-nodes decided by yarn(resource manager) and name-node. Mapper
has map function where we can write our logic for each line. All Mappers output
get collected as intermediate data. Inbuilt shuffling and sorting get applied on
intermediate data. Shuffling is a phase on intermediate data to combine all values
into a collection associated to same key. Sorted and shuffled data is passed to
reducer. Reducer also read each line at a time. Reducer input is (key, list(value)).
RecordWriter writes the output of reducer into Output File.

7   User Based Collaborative Recommendation Using Map Reduce on Hadoop Platform

User-Based Collaborative filtering algorithm says that if two users User1, User2
are similar then User2's watched videos can be recommended to User1 and vice-
versa. Similarity is calculated by the number of common watched videos by
both users. Two users are similar if they have highest similarity number when
calculated over all users. In the Figure 6 similarity value among users are as
follows:-

- sim(user1,user2) = 1
- sim(user1,user3) = 2
- sim(user2,user3) = 1

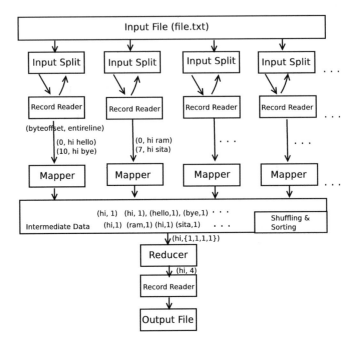

Fig. 5. Map-reduce working model

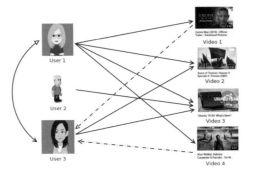

Fig. 6. User-based collaborative filtering

So user1 and user2 are most similar. User1's watched history is recommended to User2.

This algorithm can be applied in different platforms but hadoop is best fit for this algorithm. This algorithm comes under traditional way of recommendation. Lots of new recommendation has already been discovered and implemented for recommendation like slope-one, deep learning, associative clustering. But all are based on ratings given by users. User-Based recommendation does not need rating given by users. YouTube videos does not having rating option for videos. User-Based has disadvantage of computational-complexity. Distributed-mode of computing can easily handle computational-complexity. So, User-based collaborative filtering algorithm using Map-Reduce is the best for micro-videos recommendation which does not having rating option. Also 80% of users does not give rating to videos/items where rating options are available.

User-based collaborative filtering can be implemented on Map-Reduce by diving the whole task into five different map-reduce jobs. Whole algorithm is tasted by taking known output recommendation. All jobs are discussed briefly below:-

Input to the first map-reduce job is csv or tsv format of data which contains user-ids and corresponding videos ids watched by the user. Key is the byte-Offset of the beginning of the line within the file(not whole file just one split). We don't need to worry about byte Offset value because each Record-Readers assigns its value after every Input Splits. Here mapper's output is reverse of each line of input data. The use of applying mapper is that after all combined mapper's output during intermediate stage all users id who watched a particular video get in one list. This list acts as an input value to the reducer stage. Reducer's output key is videoid and value is comma-separated all userids who watched that video.

First job's output acts as input to the second job's mapper. And its mapper-output is $userid_i : userid_j$, all userids pairs who have at least one common video-ids in watch history. Reducer's input value is the list of 1's value whose sum shows the similarity-count between two pairs. This job is similar to basic word-count problem in map-reduce. Similarity-count refers to maximum number of common videos in watch history. It basically generates the similarity-matrix of all users.

Third job's mapper just passes the value of second job's reducer output. During reducer phase highest similarity value is calculated for every pair of user-id. So reducer output is pair having highest similarity-count.

Fouth job generates recommendation for every user based on similarity-value calculated in third job. But it has replication of data due to the fact that two users whose watched history is recommended may have common video-ids.

Fifth job removes all duplicate recommended video-ids and also removes the video-ids which have been already watched by the user. And generates final recommendation in tsv format. Fourth and fifth job both have two mappers and one reducer.

Library file "org.apache.hadoop.mapreduce.lib.chain.Chain Mapper" supports chaining of mappers. Mapper stage 2 takes original data set as input and Mapper Stage 1 takes previous job's reducer's output as input. Reducer work on both mappers output in combined way."o:" or "r:" is used to differentiate between original data-set and calculated data set by previous job. ChainMapper library file is included in the "hadoop-mapreduce-client-core-2.0.0-cdh4.3.1 .jar" jar package. ArrayList of Collection framework is used for storing data in reducer phase for computation. Whole Algorithm can be broken down in terms of Mapper-Reducer Input and Output of all jobs.

Job 1:

Mapper Stage:

Mapper Input: $< byteOffset, < UserId, VideoId >>$,

Mapper Output:$< VideoId, UserId >$

Reducer Stage:

Reducer Input: $< VideoId, List < UserId >>$

Reducer Output: $< VideoId, < UserId_a, UserId_b, \cdots >>$

Job 2:

Mapper Stage:

Mapper Input: $< byteOffset, < VideoId, UserId_a, UserId_b, \cdots >>$

Mapper Output: $< Pair < User_i : User_j >, 1 >$

Reducer Stage:

Reducer Input: $< Pair < User_i : User_j >, List < 1, 1, 1, \cdots, 1, 1 >>$

Reducer Output: $< UserId_i, < Similarity\_value : UserId_j >>$

Job 3:

Mapper Stage:

Mapper Input: $< byteOffset, < UserId_i, Similarity\_value : UserId_j >>$

Mapper Output: $< UserId_i, < Similarity\_value : UserId_j >>$

Reducer Stage:

Reducer Input: $< UserId_i, List < Similarity\_value : UserId_j >>$

Reducer Output: $< UserId_i, UserId_k >$

Job 4:

Mapper Stage 1:

Mapper Input: $< byteOffset, < UserId_i, UserId_k >>$

Mapper Output: $< UserId_k, < r : UserId_i >>$

Mapper Stage 2:

Mapper Input: $< byteOffset, < UserId, VideoId >>$

Mapper Output: $< UserId, VideoId >$

Reducer Stage:

Reducer Input: $< UserId_k, List < r : UserId_i, VideoId_k >>$

Reducer Output: $< UserId_i, VideoId_k >$

Job 5:

Mapper Stage 1:

Mapper Input: $< byteOffset, < UserId_i, VideoId_k >>$

Mapper Output: $< UserId_i, VideoId_k >$

Mapper Stage 2:

Mapper Input: $< byteOffset, < UserId_i, VideoId_i >>$

Mapper Output: $< UserId_i, < o : VideoId_i >$

Reducer Stage:

Reducer Input: $< UserId_i, List < VideoId_k, o : VideoId_i >>$

Reducer Output: $< UserId_i, VideoId_r >$

7.1   Implementation on Hadoop

Following steps has been taken to run User-based recommendation on Hadoop-cluster with 5-nodes.

1. "start-dfs.sh" for data-nodes activation. It establishes connection between all nodes.
2. "start-yarn.sh" for resource-manager activation(name-nodes).
3. Uploading of standard-dataset on Hadoop. Command for uploading dataset on HDFS-"hadoop fs -put Dataset/videos.csv /user/hduser/userbcf/input/".
4. Compilation of java classes and geneartion of jar file.Command for compilation and jar file conversion - "javac -classpath hadoop-core-1.2.1.jar : hadoop-mapreduce-client-core-2.0.0-cdh4.3.1.jar -Xdiags:verbose -d recommender *.java | jar -cvf recommender.jar -C recommender/."
5. Run generated jar file on Hadoop with specification of input and output of all jobs directory in HDFS. Command for running jar file on Hadoop - " hadoop jar recommender.jar Driver /path-to-input-file /path-to-output-file1 /path-to-output-file2 /path-to-output-file3 /path-to-output-file4 /path-to-output-file5 ".
6. Get the generated recommendation on HDFS to local-file-system. Command for copy from HDFS to local file system is " hadoop fs -get /path-to-output-file5/part-r-00000 recommender.csv " Fig. 7 shows the result of recommendation generated as list of video-id for each user-id.

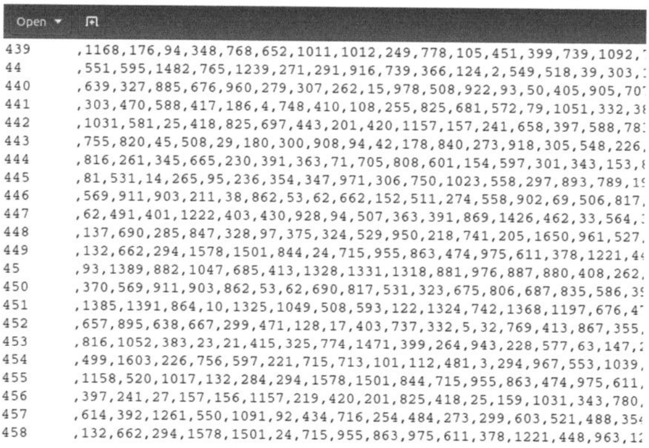

Fig. 7. Recommendation generated by user-based collaborative filtering on Hadoop

8 Result

Combination of scrapping and collaborative filtering algorithm gave the better clusters of recommendation. Scrapping enabled the platform independence. And scrapping is implemented by having the user's consent for their account at different platforms like Facebook, YouTube. By taking the advantage of distribute computing using hadoop, we were able to work on large data-set and fast processing. We divided the whole bunch of recommendations mainly into micro-videos:

1. watched by similar users
2. based on user's past behaviour
3. trending on different platforms

9   Conclusion

Micro-Video Recommendation System is a demanded recommender in the market. It is a newly emerging field nowadays. As these systems help users to choose easily their favourite ones. Data is also increasing tremendously day by day in all types of business. So choosing favourites ones is becoming quite a hectic or time-consuming for users, which increases its demand.

In this paper, different types of scrapers like Trending Micro-video scraper, Keyword-based Scraper, Watch-history Scraper, Search history Scraper has been implemented for bringing dynamic nature of the project. Content-based recommendation, Collaborative-filtering recommendation (User-based and Slope one recommendation) using map-reduce has been implemented for built Web-interface and external websites. User-based recommendation using map-reduce algorithm is implemented and discussed in brief. Hadoop single-node and the multi-node cluster has been set-up for recommendation processing. In brief Map-Reduce programming model has also been discussed. It has been also discussed how multi-platform recommendation is implemented.

References

1. Mahto, D.K., Singh, L.: A dive into Web Scraper world. In: 2016 3rd International Conference on Computing for Sustainable Global Development (INDIACom), New Delhi, pp. 689–693 (2016)
2. Shang, S., Shi, M., Shang, W., Hong, Z.: A micro video recommendation system based on big data. In: ICIS, Okayama. IEEE, Japan (2016)
3. Jiang, D., Shang, W.: Design and implementation of recommendation system of micro video's topic. In: ICIS 2017, Wuhan, China, pp. 483–485. IEEE (2017)
4. Balachandra, K., Chethan, R.: The video recommendation based on machine learning. Int. J. Innov. Res. Comput. Commun. Eng. 6(6), 6434–6439 (2018)
5. Ramakrishnan, R.: HADOOP based recommendation algorithm for micro-video URL. IDL(International Digital Library) Technol. Res. 1(7), 1–9 (2017)
6. Chen, B., Wang, J., Huang, Q., Mei, T.: Personalized video recommendation through tripartite graph propogation. In: Proceedings of the 20th ACM international conference on Multimedia, Nara, Japan, pp. 1133–1136. ACM, October 2012
7. Brbic, M., Rozic, E., Podnar Zarko, I.: Recommendation of YouTube videos. In: MIPRO 2012, 21–25 May 2012, Opatija, Croatia, pp. 1775-1779 (2012)
8. Saurkar, A.V., Pathare, K.G., Gode, S.A.: An overview on web scraping techniques and tools. Int. J. Future Revol. Comput. Sci. Commun. Eng. 4(4), 363–367 (2018)
9. Vargiu, E., Urru, M.: Exploiting web scraping in a collaborative filtering- based approach to web advertising. Artif. Intell. Res. 2(1), 44–54 (2013)
10. Raj, J., Hoque, A., Saha, A.: Various methodologies for micro-video recommendation system: a survey. In: International Conference on Computational Intelligence & IoT (ICCIIoT) 2018, 12 May 2020. https://ssrn.com/abstract=3598885 or https://doi.org/10.2139/ssrn.3598885 (2018)

Comparison of Machine Learning Techniques to Predict Academic Performance of Students

Bhavesh Patel[✉]

MCA Department, Ganpat University, Gujarat, India

Abstract. Many organizations use machine learning to analyze data and find significant hidden patterns in the data including healthcare, finance, online service provider, education institute, software companies etc. and based on getting rules from the pattern take the appropriate decisions in the favor of organization. This research paper has used four machine learning techniques to generate the models. This models are compared by various accuracy measured parameters to find the best suited model for the student's dataset. This paper has used various academic and demographic parameters of students to create the dataset. This research article includes logistic regression, decision tree, artificial neural network and Naïve Bayes machine learning techniques. As accuracy measurement parameters on model this research article has used ROC index, Error Rate, F-measure, and accuracy. The data set is collected from sharing the drive sheet among students of various institute. As a result, found that ANN model is best suited and highest accurate model for this dataset so by applying this model institute got the highest accurate result and take the wise decision to improve the performance of students in academic.

Keywords: Performance · Classification · Machine learning · Accuracy · F-measure

1 Introduction

Machine learning techniques are used by many researchers in education field to get the pattern and take the beneficial decisions. So many researchers have used it to find the solution of the problems like student's retention, selection of the course, find the week and average students, help in placement activities and so many same type of problems. The objective of article is to predict student's performance using various academic and demographic parameters. Student performance can be predicted using machine learning techniques to identify students at risk, so appropriate actions can be taken for these types of students to improve performance. Therefore, this research treatise contains a variety of machine learning techniques for building models and predicting student performance. This research treatise used a small data set of students to investigate the results.

This research paper has used logistic regression, decision tree, artificial neural network and naïve bayes classification techniques to build the model. This research article used ROC index, Error Rate, F-measure, and accuracy parameters to find the highly accurate mode.

© The Author(s), under exclusive license to Springer Nature Switzerland AG 2022
R. Misra et al. (Eds.): ICMLBDA 2021, LNNS 256, pp. 141–149, 2022.
https://doi.org/10.1007/978-3-030-82469-3_13

There are so many methodologies to collect the dataset like questionnaire's, interview, review, survey and many more. This research article has used survey methodology to collect the student's dataset. At the end of collection from various institute we have collected total 1467 instances. This research article follows the preprocessing steps to normalize and clean the dataset.

This research paper is divided into four sections: first section is Introduction, second section literature Survey, third section research model and its implementation and fourth section experiment and result analysis.

This research article is written to fulfill the objective "find out the highest accurate model by comparing machine learning techniques and applied this model on dataset to predict the performance of students into the academic".

2   Literature Review

This research is made by reviewing the following literature.

Zaffar M. et al. used various prediction models for students they have taken admission in programming courses. Filter Feature selection algorithms have been used by them in the pre-processing stage for generating the result. Based on the result they found that student's attendance, mathematics result, physics result plays an important role in programming course [1]. Hari S. et al. learn several Data Mining applications in their research work. This research work surveyed different techniques of data mining and its algorithms in several regions of EDM. They discover that, EDM can also be used to find out the knowledge based process for the problems of primary students [2]. Jacob J. et al. used different mining methods to predict performance of students. They have utilized Linear regression and decision tree techniques to identify the poor students in academic. Clustering techniques also used by them to cluster the students based on performance in the academic [3]. Xu J. et al., have used various machine learning models to forecast student's performance. They have used kNN, RF, Logistic Regression, Linear Regression, and Proposed Progressive Prediction algorithm in their research work. As a result, they found that proposed progressive prediction algorithm is the best algorithm among all the algorithms [4]. Barrak M.A. and Razgan A. describe that ANN and Decision Tree both are well known classification methods for classifying the data and prediction. This research paper has used decision trees for predicting performance of students that can be helpful to students who need special attention. In this research paper, researchers have predicted students drop out rate using academic, socio-demographic and institutional data to forecast the final GPA of students [5]. Mishra T. et al. surveyed student performance and employment prediction using data mining. The focused primarily on the traditional educational establishment into his research work. They have demographic and socio-economic factors into the prediction model. He also said that some research work also has been done on employment forecasting [6]. Shahiria A.M. et al. has used several mining techniques to evaluate the performance of students. They evaluated how these predictive algorithms could be used to find out the highly significant attributes in the student database [7]. Kavipriya P. used various data mining methods for predicting, analyzing, early warning, and evaluating student performance. He reviewed various classification methods such as decision trees, inexperienced Bayes algorithms, etc.

He suggested that because it is difficult to predict student performance due to many challenges such as statistical imbalances, there is a need to install a support vector machine that offers the best accuracy in his study [8]. Asiahs M. et al. has reviewed prediction modelling technique for academic performance of students. They have nursing various learning tasks for making the predictive models. Finally, they have used this models for course recommendation and career path planning [9].

3   Research Model

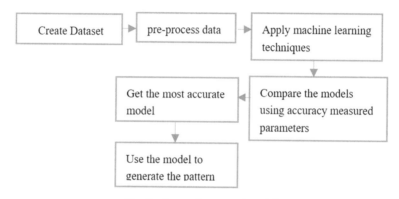

Fig. 1. Steps of proposed model

In this research paper followed the steps as per described in the Fig. 1. Initially collect the student's data set using the survey methodology and generate a dataset of student. After that, apply preprocess steps to remove the noisy and missing data from the dataset. After the preprocess, applied machine learning algorithms using weka tools and generate the models. Further, compare models with accuracy measured parameters and find the highest accurate model and applied it to predict academic performance of students.

4   Data Collection and Parameters

This research paper has been collected total 1467 instances from the various computer institute. The data has been collected by preparing a drive sheet and shared it among the students. After collecting the student's data, we have transformed it.csv file to processed the data using WEKA tools. Further experiment is done using the WEKA tools.

In this research article prepared a datasheet by following the parameters like gender, department, attendance ratio, family_income, family_education, siblings, attendance, internal_result, regularity_submission, interent_uses, social_media_uses, pratical_performance, theory_performance, End_sem_result. These parameters are categorized into appropriate class labels to classify the result.

5   Research Methodology

This research paper has been used following algorithms. Their description is as per the following:

5.1   Logistic Regression (LR)

LR is a one type of mathematical modeling technique that defines association among different independent variables X1...XK and a dependent variable. LR uses logistic function as a mathematical equation and it has value between 0 and 1 for specified input. This logistic model shows the probability of any event that is always a value between 0 and 1. The following equation shows the logistic model.

$$P(D = 1|X1, X2, \ldots, XK) = \frac{1}{1} + e^{-\left(\alpha + \sum_1^k \beta i Xi\right)} \tag{1}$$

Here α and β shows the parameters of model [11].

5.2   Decision Tree (DT)

DT model shows tree structure that resembles flowchart. In this structure, internal node shows attributes of test, branch shows the result of the test. Further, leaf node shows the target object label, and the first node shows the root node. The decision tree is either binary or non-binary tree. Decision tree has not required previous knowledge of any problem. So it is commonly used classification technique. It can also be easily converted into classification rules and generated rules are easy to understand. They are used in several real-world applications like molecular biology, manufacturing, medicine and financial analysis etc. The most common decision tree algorithms include CART, C4.5 and ID3 [10, 14, 15].

5.3   Artificial Neural Networks (ANN)

ANN is group of input and output units those are linked together using weighted connections. ANN absorbs by altering the weights to predict correct target. Backpropagation algorithm is highly used to train ANN. ANN has several benefits to use, like high resistance against noisy data and it gives good performance for classifying patterns using untrained dataset. ANN is used in real world applications also like speech recognition, handwriting and image identification etc. RNA can be recognize using their architecture. The architecture of fully connected multi-layer front feeder ANN is: an input layer, one or more hidden layers, and an output layer. Here, connections not going rear to previous layer. In addition, every element in the L layer gives inputs to every element in the L + 1 layer [10].

5.4 Naïve Bayes (NB)

It is one type of classification model. It is simplest disparity of Bayesian network. It describes each instance is independent than other instance. The following formula is used in Naïve Bayes model.

$$V_{max} = \max_{v_j \in V} P(v_j) \prod_i P(a_i|v_j) \tag{2}$$

Here v shows target of the model, $P(a_i|v_j)$ and $P(v_j)$ both can be find out by counting the frequencies in training dataset [12, 14, 15].

5.5 Validation Methods and Performance Measure Parameters

This research paper used 3-fold cross validation technique. Here, the database will be distributed in three equal sets. Due to three fold's cross validation the testing and learning sets are executed three times. In this method, the machine learning algorithm picks one set for testing purpose and other two sets for training purpose. Finally aggregate all folds or execution to count performance and accuracy of model. Classification models performance can also be evaluated using the ROC index. It is also a one type of most useful performance measure under the curve. ROC index is calculated using predicted score. Following equation three is used to count ROC index [21]. Furthermore, except ROC index, other important measures also used to find the accuracy of model like F-Measure and Error rate of classification error. Here, the following formula four is used for calculating the F-Measure. Generally, F-Measure use to find out the misclassification rate [13].

$$ROC\ index = \sum_{i=2}^{|T|} (FPR(T[i] - FPR(T[i-1])) \times (TPR(T[i] + TPR(T[i-1]))/2 \tag{3}$$

Here $|T|$ shows thresholds those used in research. $FPR(T[i])$ shows false positive rate. $TPR(T[i])$ shows true positive rate. Model with greater ROC-index describes good classification model. A model is strong model if value above 0.7 and a model is week if value below 0.6 [13].

$$F - measure = 2 * \frac{(precision * recall)}{(precision + recall)} \tag{4}$$

$$precision = \frac{TP}{TP + FP} \tag{5}$$

$$Recall = \frac{TP}{(TP + FN)} \tag{6}$$

Here TP is a True Positives. It shows the data rows in test sets having positive target and also those are predicted as positive target. Here TN is a True Negatives. It shows the data rows in test sets having negative target and also those are predicted as negative target. Here FP is False Positive. It shows the data rows in test set having negative target but those are predicted as positive target. Here FN is False Negatives. It describes the number of data rows in the test set positive target but those are predicted as negative target [13].

6   Experiment and Result Analysis

In this research paper, Decision Tree (DT), Naïve Bayes (NB), Artificial Neural Network (ANN) and Logistic Regression (LR) models are used. Each model's accuracy and performance measures are described into the Table 1.

To perform the experiment, data set into two folds, one for training model and other for testing the models. Here 80% dataset is used to train the model and 20% dataset is used to test the model.

Table 1. Result of machine learning models using accuracy measure parameters

MODEL	ANN	DT	LR	NB
TRUE POSITIVE	81.00	79.01	77.02	75.03
F-MEASURE	81.22	80.2	79.23	76.45
ACCURACY	81.52	79.15	78.32	75.32
ERROR RATE	18.48	20.85	21.68	24.68
ROC INDEX	0.831	0.792	0.762	0.752

Used Abbreviations:
LR – Logistic Regression
DT – Decision Tree
NB – Naïve Bayes
ANN – Artificial Neural Network

The following Figures 2, 3, 4, 5 show the result analysis into the form of charts.

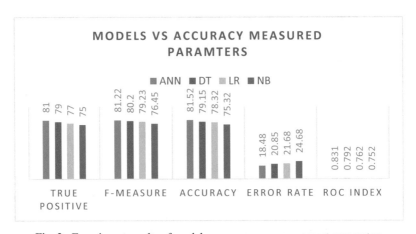

Fig. 2. Experiment results of models vs accuracy measurement parameters

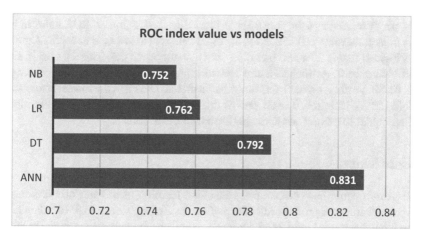

Fig. 3. Experiment result of ROC Index of different models of research

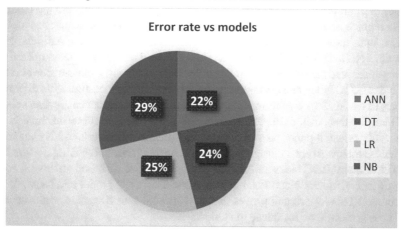

Fig. 4. Experiment result of error rate of machine learning techniques

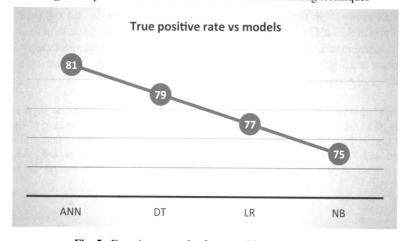

Fig. 5. Experiment result of true positive rate of models

As per described into the above Table 1 and Figs. 3, 4, 5, lowest ROC index in Naïve Bayes model. Its value is 0.752. Even accuracy rate is also lowest that is 75.32 and error rate is highest that is 29%. So based on above result prove that Naïve Bayes is a week model. At contrast, Artificial Neural Network (ANN) model having highest accuracy that is 81.52. Further, higher ROC index value that is 0.831 and the lowest error rate that is 18.48 (22%). This result itself proves that ANN model is highly accurate machine learning model to predict academic performance of students.

7 Conclusion

This research paper has compared the machine learning techniques and find the best suited model among them to predict the performance of students into the academic. This research has been used total 1467 student's data from the various institutes to address the problem. This dataset is prepared using academic achievement and demographics parameters of students. The objective this research article is to compare the model and find the highest accurate model. To achieve this goal, this research paper has used four machine learning techniques Decision tree (DT), Naïve Bayes (NB), Artificial Neural Network (ANN) and Logistic Regression (LR). These used models are compared using accuracy measured parameters like true positive (TP), accuracy, f-measure, error rate and ROC Index. ROC Index and error rates are highly acceptable measurements to prove the accuracy of model. So, considering these two accuracy measurement parameters checked the model, find that Artificial Neural Network (ANN) is highly accurate machine learning model. This research paper proved that, Artificial Neural Network (ANN) has highest ROC index and lowest error rate among all the compared model. At contrast, Naïve Bayes model has Lowest ROC index and highest error rate model. So based on this experiment it proves that, Artificial Neural Network is highest accurate model and Naïve bayes is week model. This experiment proves that ANN model is best model for this dataset to predict the academic performance of students.

Future Work
This research paper has used only four machine learning techniques and applied in Weka for analysis and result but as a future work we can elaborate this work with many other machine learning techniques with python programming for getting the analysis and result of machine learning techniques. Further, this work also elaborates by creating our own algorithm and compare its accuracy with other existing algorithm.

Acknowledgement. I would like thanks to my colleague those have supported me to prepare this research paper on machine learning techniques. I also would like thank to journal of computer science to give me the chance to present my research paper into your valuable journal. How can I forget to my research guide? I would like thanks to my guide to support me and guide me to prepare this article. Thanks to all direct and indirect hand those have helps me into this research article.

References

1. Zaffar, M., Hashmani, M.A., Savita, K.S.: A study of prediction models for students enrolled in programming subjects. In: 4th International Conference on Computer and Information Sciences (ICCOINS), Malaysia. IEEE (2018)
2. Hari, S., Ganesh, A., Joy, C.: Applications of educational data mining: a survey. In: IEEE Sponsored 2nd International Conference ICIIECS. IEEE (2015)
3. Jacob, J., Kavya, J., Paarth, K., Shubha, P.: Educational data mining techniques and their applications. In: IEEE International Conference on Green Computing and Internet of Things (ICGCIoT), Great Noida, India (2015). https://doi.org/10.1109/ICGCIoT.2015.7380675
4. Xu, J., Moon, K.H., Schaar, M.V.: A machine learning approach for tracking and predicting student performance in degree programs. IEEE J. Sel. Top. Signal Process. **11**(5), 742–753 (2017)
5. Barrak, M.A., Razgan, A.: Predicting students final GPA using decision trees: a case study. Int. J. Inf. Educ. Technol. **6**(7), 528–533 (2016). https://doi.org/10.7763/IJIET.2016.V6.745
6. Mishra, T., Kumar, D., Gupta, S.: Students' performance and employability prediction through data mining: a survey. Indian J. Sci. Technol. **10**(24), 1–6 (2017). https://doi.org/10.17485/ijst/2017/v10i24/110791
7. Shahiria, A.M., et al.: A review on predicting student's performance using data mining techniques. Pro. Comput. Sci. **72**, 414–422 (2015)
8. Kavipriya, P.: A review on predicting students' academic performance earlier, using data mining techniques. Int. J. Adv. Res. Comput. Sci. Softw. Eng. **6**(12), 101–105 (2016)
9. Asiah, M., et al.: A review on predictive modeling technique for student academic performance monitoring. MATEC Web Conf. **255**, 03004 (2019)
10. Han, J., Kamber, M., Pei, J.: Data Mining: Concepts and Techniques, 3rd edn., pp. 327–383. Elsevier, Amsterdam (2012)
11. Kleinbaum, D.G., Klein, M.: Logistic Regression. A Self-Learning Text, 3rd edn. Springer, New York (2010).https://doi.org/10.1007/978-1-4419-1742-3
12. Kotsiantis, S., Pierrakeas, C., Pintelas, P.: Predicting students' performance in distance learning using machine learning techniques. Appl. Artif. Intell. **18**, 411–426 (2010). https://doi.org/10.1080/08839510490442058
13. Kelleher, J.D., Namee, B.M., Arcy, A.D.: Fundamentals of machine learning for predictive data analytics. In: Algorithms, Worked Examples, and Case Studies, Kindle, 2nd edn (2015)
14. Golino, H.F., Gomes, C.M.: Four machine learning methods to predict academic achievement of college students: a comparison study. Rev. e-psi-Res. **1**, 68–101 (2014)
15. Kumari, J., Venkatesan, R., Jemima Jebaseeli, T., Abisha Felsit, V., Salai Selvanayaki, K., Jeena Sarah, T.: A Comparison of machine learning techniques for the prediction of the student's academic performance. In: Hemanth, D.J., Kumar, V.D.A., Malathi, S., Castillo, O., Patrut, B. (eds.) COMET 2019. LNDECT, vol. 35, pp. 1052–1062. Springer, Cham (2020). https://doi.org/10.1007/978-3-030-32150-5_107

Misinformation–A Challenge to Medical Sciences: A Systematic Review

Arpita Sharma[✉] and Yasha Hasija

Delhi Technological University, Shahbad Daulatpur, Main Bawana Road, Delhi 110042, India

Abstract. Misinformation is proving to be a big problem for health care professionals and various research studies are going on finding the way to spread prevent of misinformation and how to correct the misinformation found on internet and social media. The purpose of writing this review is to prospect why social media and internet being a boon to the society is proving to be a cause of distress in the field of medical sciences. The review also explores the various research that has been done on combating misinformation spread and what are the future scope of research which can be done to prevent spread of misinformation.

Keywords: Misinformation · Social media · Internet · Correction

1 Introduction

In recent years, the internet has changed the scenario how we were before internet. It has changed our lives in the positive ways and has made our life easier. Before internet we had to go to libraries to get any information on any topic and had to go to markets for shopping but now, we can get anything from clothes to groceries on a click and can also get information about each and everything on internet on a single click but everything has two face despite of being a boon to the society internet has been a cause of distress to the medical sciences. The variety of misinformation related to different diseases going viral on different search engines and social medias is proving to be very harmful to the society. The false myths about different type of medical conditions like eating this can cure that disease, doing this can cure this and many more, these types of myths we see on a daily basis on social media and internet which do not have any credible source leads to misperception among people and can cause health issues to them.

Many studies have been done how one can correct the misinformation that is flowing on internet but every study has one or the other limitations which is a major roadblock in improving internet in the field of medical sciences.

The reason for writing this review is what we know about the subject, what are the different studies done on the subject, what are the research questions that still exists and what we can do in further research studies to improve the existing condition. All the things will be discussed one by one further.

© The Author(s), under exclusive license to Springer Nature Switzerland AG 2022
R. Misra et al. (Eds.): ICMLBDA 2021, LNNS 256, pp. 150–159, 2022.
https://doi.org/10.1007/978-3-030-82469-3_14

2 What is Social Media and Why It is Proving to Be a Cause of Distress in the Field of Medical Sciences?

Social media also known as Web 2.0 is generally defined as the internet-based tool where individuals or the different communities are allowed to interact and gather to receive or share information, ideas or any other content like related to different diseases [1]. Social media sites may include different applications or websites depending upon the role they have, it may include twitter, Facebook, Instagram, WhatsApp, Wikipedia, blogs or many more sites. According to their function social media can be categorized into different groups [1]:

- Social Media for social interactions like Facebook, WhatsApp
- Social Media for professional interactions like LinkedIn
- Social Media for video sharing like YouTube
- Social Media for producing content like Twitter or Blogger
- Social Media for getting Information like Wikipedia
- Social Media for experiencing reality in the virtual world and for gaming experience like Second Life

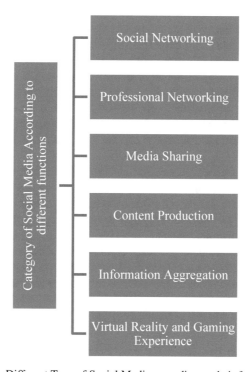

Fig. 1. Different Type of Social Media according to their functions

The main problem about social media is misinformation found on it. Misinformation can be defined as the information that deviates from the reality, the information which is factually incorrect and has no validation or verified source from where it has come which often leads to misperceptions [2]. One can get misinformation from fake news links, fake messages being spread on WhatsApp, Facebook or any other social media platforms (Fig. 1).

The information about medical sciences which spreads on the social media are of poor quality and has limited, unreferenced, informal or incomplete information [1]. Due to spread of poor quality of information anxiety and panic situation arises among people, many people without checking the source spreads the false information which spreads to more people and situation gets worsened.

According to the study about the spread of right and wrong news about 126.000 fallacious news were outspread by almost 3 million people on twitter between the year 2006–2017 [3]. Instead of truth, lies spreads more quickly on twitter according to this study, lies spread to between 1000–10,000 people whereas truth only reaches hardly to more than 1000 people [3]. This data shows how quickly the false information spread on social media and why it is a need to correct misinformation or prevent spread of misinformation. Social media while promoting many awareness campaigns has also at the same time have provided the platform for spreading of many misinformation regarding various diseases.

In one study it showed that the twitter bots and unidentified accounts were the ones which post most about the vaccines and mostly those were the misinformation that they were spreading which can result in increase of vaccine preventable diseases [4]. Due to these unidentified accounts misinformation about vaccines and diseases spread among people and they start various campaigns about vaccine which is mostly anti vaccine campaign and they refuse to take vaccines and these myths spread to other people also who without checking the credible source starts believing the false myths about vaccines and start refusing to take vaccines which leads to the serious concern among healthcare professionals and government on how to combat this problem and eradicate the vaccine preventable diseases.

According to Pew Research center survey around 72% adults in past 12 months had searched online about the medical condition from which they are suffering or others are suffering [5]. This survey indicates how people depends on internet to know about their health condition and if they counter any misinformation, that may lead to misperceptions which can be harmful for their health.

A study was conducted between 2019–2020 and from the study it was intimidated that the people who relied on social media for getting news about Covid-19 pandemic has the minimal knowledge about the real facts and truth during the outbreak of Covid-19 [6]. In another study it was revealed about 74% of public posts regarding Covid-19 outbreak had a source from news organization and only 1% were associated with health and science sites [7].

In February 2020 also, WHO had warned everyone regarding the infodemic that is happening with covid-19 outbreak where lot of information is available making difficult for everyone to distinguish between the information and the misinformation [8].

These studies are important in understanding why social media is detrimental to medical sciences and why there is requirement to combat the issue regarding misinformation (Fig. 2).

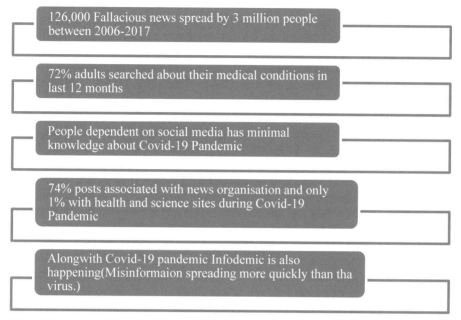

Fig. 2. Data extracted from Various Surveys showing Social media is causing distress to medical sciences

3   Various Methods to Correct Misinformation Related to Medical Sciences

John Cook et al. [9] conducted a study and laid the platform how misinformation can be corrected. They conducted a study how to effectively correct the information found on internet without having any "backfire effects". They concluded that how misinformation with the help of combination of information and computer science can helps in removing the misinformation from the internet. Two important findings were concluded from the study that various backfire effects can arise when correcting misinformation can ironically reinforce misinformation and how the view of world can play a part in the persistence of misinformation.

Ullrich K.H. Ecker et al. [10] conducted a research that can corrections made on social media spread misinformation to newer audiences and they also tested the familiarity backfire effect. The findings from the studies concluded that after corrections were not able to create misconceptions strongly as compared to the person who were not exposed to the false claims or corrections and it was safe to repeat corrected misinformation.

Leticia Bode et al. [11] performed the experiment of correcting misinformation by using Facebook new function which provides a link within Facebook which provides related link when people click on the former link. By using social media function misinformation and misperception were significantly reduced. The inference can be drawn from this study that if social media platforms come together and work on building new features which can help in preventing spread of misinformation will be beneficial to the society as people mostly come across false information on different social media.

In another study by Leticia Bode et al. [12] it was found that expert organization Centers for Disease Control and Prevention (CDC) can stop the spread of misinformation and misperception when they themselves correct the information on Twitter in comparison to the misinformation corrected by another user. This study provides a way how an expert organization if actively involved can significantly help in reducing spread of misinformation. Many organizations along with CDC like WHO and many more organizations can be formed which can work together actively by forming committees especially for the correction and preventing spread of misinformation. In their another study Leticia Bode et al. [13] also concluded that social media can be a great help in reducing the spread of misinformation and people using it should be encouraged to correct any misinformation they found in any tone they feel comfortable as their study provided the evidence that misperceptions were reduced when people correct the misinformation in any tone be it uncivil, affirmational or neutral. From the study it can be inferred that people should be motivated through various campaigns be it online campaigns on social media or offline campaigns through print media, electronics media to correct any misinformation they came across on any social media platform and should also be motivated to first know the source about the information before forwarding any information to other people.

In the study by Patrick R. Rich et al. [14] it was investigated that whether correcting information immediately or the correction is delayed does it have an effect on efficacy of correction and also if the correction is made for how many days it stops the spread of misinformation. In their investigation they found that there was not any evidence which can conclude that time matters in the efficacy of the correction. In another experiment also even if the people know or remember the correct knowledge and their misperceptions are reduced initially but after some time when misinformation spreads again people start believing in that despite having knowledge about the correct information. This study raises concerns even after correction of misinformation by experts like journalists, healthcare experts and others does not show the lasting durability in the beliefs in spite of people have lasting durability in the knowledge.

In a study by Heena Sahni et al. [15] it was found that how at the time of Covid-19 pandemic social media had created so much anxiety and panic due to spread of misinformation and how misinformation has spread more quickly than the virus. The study had opened the research because correcting misinformation is the need of time so that in future if any pandemic come there is no issue of misinformation as misinformation do more harm than the disease itself.

In the study by John Robert Bautista et al. [16] study was conducted how healthcare experts can rectify the misinformation present on social media like Facebook and Twitter. The study can lay the platform for health professionals how experts can rectify the misinformation on social media (Fig. 3).

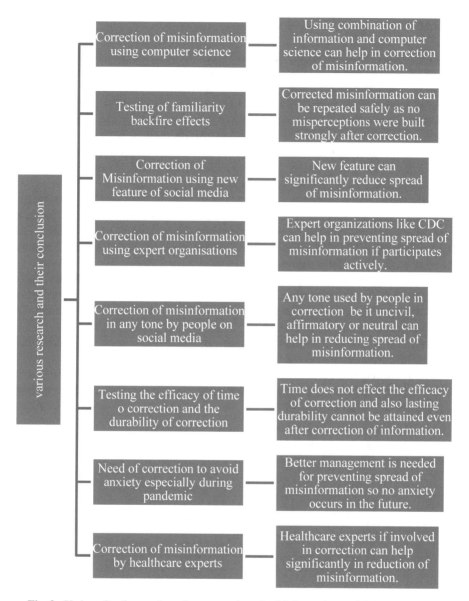

Fig. 3. Various Studies conducted on correction of misinformation and their conclusions

4 Limitations of the Research

Despite various studies being conducted on correction of misinformation there is still a need of more research and opting other ways including present ones to combat the spread of misinformation as misinformation still spreads on internet at a rapid speed. The studies which have been done till now has not come into practical. The studies done till now has been done in a limited area and on limited social media platforms [16]. There are many other social media and websites other than Facebook and Twitter for instance, Instagram, Tik Tok and others where people engagement is increasing at a rapid rate and chances of spreading of misinformation is also increasing and there is scope of research on these platforms too to prevent the spread of misinformation.

In the studies it has been concluded how expert organizations like CDC can help in correcting misinformation on social media but there is a limitation that CDC is not as active as it is a need for them to be active in correcting misinformation [12]. Another limitation is that there are only few credible sources for correcting so much misinformation available on social media platform.

In other study where health professionals were included to correct the misinformation only registered nurses and medical doctors were employed to correct misinformation, other healthcare professionals like physiotherapists, nutritionists and others should also be included in the study [16].

There is also need of checking the durability of the corrected information along with creating different measures to correct the misinformation [14].

By considering these limitations as a base, further research can be done on how to prevent spread of misinformation successfully and how corrected information can have lasting durability (Fig. 4).

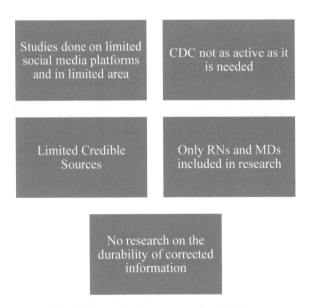

Fig. 4. Scope for future research and studies

5   Recommendations for the Prevent of Misinformation

There are many ways by which there can be significantly reduction in the spread of misinformation:

- Expert organizations can come together and help in preventing the spread of false information available on the social media and should be given the responsibility of correcting the misinformation. Many committees should be formed which should include various healthcare professionals, research scholars, journalists and others and should be given a sole responsibility of correcting misinformation. These committees should be active in correcting misinformation as their activeness is the key for preventing and correcting misinformation successfully.
- The study had shown that how using Facebook new feature helped in reducing the misperceptions, other social media platforms can also develop a feature which before letting user share the information should authenticate and validate the information the user is posting on social media which can help in preventing the misinformation to be posted on social media.
- We have seen on Twitter and other platforms how one can differentiate a public figure or a celebrity from other fake accounts of the celebrity by checking the blue tick after the name of the celebrity. Just like how blue tick validates the real accounts of the public figure this feature should also come on the different websites that provide health related information. It can help people in finding correct website and information among various websites which can misled them.
- Disclaimers should be given in the form of videos that the information is general and can vary situation wise as people often ignore the disclaimer which is in the written form and starts panicking by reading health related stuffs present on the websites. The information available related to health conditions can also be in form of video and the presenter should aware the public that the information can vary from people to people and before taking any precautions or medications one should definitely seek an expert.
- Various health campaigns can be organised for people to spread awareness about the misinformation. We have seen till date many campaigns like campaigns for spreading awareness about breast cancer, HIV-AIDS and others is proven successful in spreading awareness about the different diseases among lot of people. Print media, electronic media can play a big role in spreading awareness about misinformation also.
- Different search engines can make sure only websites which are authenticate and provide correct information should come when searches are done related to health stuff. Search algorithm should be set in such a way that only authenticate information should come on searching and websites having no credible source about the information they are putting on their sites should be blocked from putting any information online.
- Algorithms can be made in such a way that it does not let the user put any information having no credible sources to be posted on social media platforms. No forwarding of fake messages should be done and people using social media should be aware time to time by broadcasting videos, messages spreading awareness about the misinformation and the harmful effects misinformation can cause.

6 Conclusion

Internet and Social Media is a boon when used in a right manner but when it does not happen it leads to havoc by spreading misinformation which can be harmful to the society. Misinformation leads to unnecessary panic and anxiety among people by spreading false news about the diseases, medications and precautions one should take. Misinformation spread also leads to misperceptions about the vaccines and various anti vaccine campaign starts rolling out on social media leading to unwanted panic and difficulties for healthcare professionals and government on how to combat this situation and aware people about the safety and side effects of vaccines. Various research studies have been done till now which provides a way of correcting misinformation, how healthcare professionals, expert organizations can help in reducing spread of misinformation and correction of misinformation. Various studies have been done for testing the familiarity backfire effect and checking the efficacy of time on correction of misinformation and the durability correct information possess. But still there are limitations of the studies done till now as the research is done in a limited geographical area and on the limited social media platforms and only limited organizations and credible sources are available which can reduce spread of misinformation and only RNs and MDs were included in correcting misinformation and one of the major limitation is that even after correction of misinformation, false beliefs among people again starts spreading raising concern about the durability of correct information and also there is no full proof plan which can prevent the spread of misinformation as studies have been done on correcting misinformation but there are no studies which focus on lasting durability of the correct information. There is a scope of research on finding ways to prevent misinformation and future studies should be done how misinformation spread can be prevented and also focus on the durability of the correct information. Many recommendations can be looked upon for preventing spread and correction of misinformation by forming committees having experts for correcting misinformation, by developing new features on social media which can help in reducing spread of misinformation, by organizing various awareness campaigns for spreading awareness about misinformation, by broadcasting disclaimers in the form of video on websites and by allowing only authenticate data to be posted on websites and social media. Keeping all this as base future research should be done in a way that finds a suitable way of not only correcting misinformation but also on increasing the durability of corrected information. These recommendations can also open a way for future studies and methodologies should focus on these recommendations and research should also be expanded to more geographical areas and more social media platforms.

References

1. Ventola, C.L.: Social media and health care professionals: benefits, risks, and best practices. P & T: a peer-reviewed. J. Form. Manag. **39**(7), 491–520 (2014)
2. Vraga, E.K., Bode, L.: Defining misinformation and understanding its bounded nature: using expertise and evidence for describing misinformation. Polit. Commun. **37**(1), 136–144 (2020)
3. Vosoughi, S., et al.: The spread of true and false news online. Science **359**(6380), 1146–1151 (2018)

4. Broniatowski, D.A., et al.: Weaponized health communication: Twitter bots and Russian trolls amplify the vaccine debate. Am. J. Publ. Health **108**, 1378–1384 (2018)
5. Fox, S.: The social life of health information. In: Pew Research Center: Fact Tank, 15 January 2014
6. Mitchel, A., Jurkowitz, M., Oliphant, J.B., Shearer, E.: Americans who mainly get their news on social media are less engaged, less knowledgeable. In: Pew Research Center: Journalism & Media, 30 July 2020
7. Stocking, G., Matsa, K.E., Khuzam, M.: As COVID-19 Emerged in U.S., Facebook posts about it appeared in a wide range of public pages, groups. In: Pew Research Center: Journalism & Media, 24 June 2020
8. Munich Security Conference. World Health Organization, 15 February 2020
9. Cook, J., Ecker, U., Lewandowsky, S.: Misinformation and how to correct it. In: Emerging Trends in the Behavioral and Social Sciences (2014)
10. Ecker, U.K.H., Lewandowsky, S., Chadwick, M.: Can corrections spread misinformation to new audiences? Testing for the elusive familiarity backfire effect. Cognit. Res. **5**, 41 (2020)
11. Bode, L., Vraga, E.K.: In related news, that was wrong: the correction of misinformation through related stories functionality in social media. J. Commun. **65**(4), 619–638 (2015)
12. Vraga, E.K., Bode, L.: Using expert sources to correct health misinformation in social media. Sci. Commun. **39**, 621–645 (2017)
13. Bode, L., Vraga, E.K., Tully, M.: Do the right thing: Tone may not affect correction of misinformation on social media. In: The Harvard Kennedy School (HKS): Misinformation Review, vol. 1, p. 4 (2020)
14. Rich, P.R., Zaragoza, M.S.: Correcting misinformation in news stories: an investigation of correction timing and correction durability. J. Appl. Res. Mem. Cognit. **9**(3), 310–322 (2020)
15. Sahni, H., Sharma, H.: "Role of social media during the COVID-19 pandemic: beneficial, destructive, or reconstructive"? Int. J. Acad. Med. **6**(2), 70–75 (2020)
16. Bautista, J.R., Zhang, Y., Gwizdka, J.: Healthcare professionals' acts of correcting health misinformation on social media. Int. J. Med. Inf. **148**, 104375 (2021)

Comparative Analysis Grey Wolf Optimization Technique & Its Diverse Applications in E-Commerce Market Prediction

Shital S. Borse[✉] and Vijayalaxmi Kadroli

Terna Engineering College (TEC), Mumbai University, Navi Mumbai, India
vijayalaxmikadroli@ternaengg.ac.in

Abstract. In the E-commerce industry, promotional tactics are used to drive traffic to the selling platform/online portal. This traffic is converted into customers who pay and efforts are made to retain those customers. This survey Paper is showing a detailed & thorough survey of different techniques & describes, compares many techniques of identification through tables used for prediction analysis in E-commerce sites & exploring different important terms & know-hows in the current market research fields. The predictive performance of many E-commerce sites is upgrading using algorithms like SVM, Particle Swarm Optimization, Neural Network, GWO, Machine learning (ML), Genetic algorithm, etc. Main aim is to utilize these algorithms which provide optimum results in various applications such as predicting the customer's behavior, generate product titles of E-commerce & predict the quality, also a prediction of relevant products and user ratings for those products, and many more existing E-commerce scenarios. The purpose is to perform a comparison analysis of various available techniques in E-commerce market predictions along with some diverse applications.

Keywords: E-commerce · Prediction analysis survey · Machine learning · Grey wolf optimization · Genetic algorithm · AI · Fuzzy C-means · ANN

1 Introduction

E-Commerce websites are buying and selling products on online platforms which are having a vast and diverse catalog of products. A catalog made up of a unique series of products and can be identified by their brand, model, and main features which vary according to the type of product (books, clothes & electronics). Online platforms expose the product information through product pages and use the title as the product's main summary [1]. Due to which most people are dependent on these websites for daily activities. Recommender System & prediction system both are a new approach which most of the people are dependent on this website for daily activities as suggested by the user for handling the vast quality of information [2].

Closeness, surprising elements, and phenomena inspires the development of Prediction techniques. The biological behaviour of fishes, elephants, wolves, bees, ants, etc. along with social interactions and lifestyles too inspire the researcher to develop

© The Author(s), under exclusive license to Springer Nature Switzerland AG 2022
R. Misra et al. (Eds.): ICMLBDA 2021, LNNS 256, pp. 160–174, 2022.
https://doi.org/10.1007/978-3-030-82469-3_15

nature-inspired optimization techniques or algorithms which will help to solve many real-time problems [3–6]. Genetic Algorithm (GA) is inspired by a group of animals or natural swarms and insects. (Davis) [4]; Ant Colony Algorithm (ACO) (Dorigo & Gambardella) [6]; Swarm Intelligence (SI) algorithms (Engelbrecht) [7]; Particle Swarm Optimization (PSO) (Eberhart & Kennedy) (Pant et al.) [11, 12]; Firefly Algorithm (FA) (Yang); Sine cosine Algorithm (SCA) (Mirjalili) [12]; Grey Wolf Optimization (GWO) (Mirjalili et al.) [13].

The main challenge is to solve the real-time problems which require a global optimum solution with an optimal convergence rate. As compared to existing methods, better results must be obtained, depending on the methodology been defined to solve the problem using a nature-inspired algorithm called Metaheuristics. This algorithm is dependent on population (PSO, ACO, GWO) [14] or is dependent on the trajectory (SA) (Sharman) it has been central in clearing E-commerce Prediction issues, its mechanism of being derivation free and avoiding getting stuck in local stagnation. Grey Wolf optimization algorithm (GWO) is a rapidly progressing algorithm using for optimizing that implements the intelligence of a swarm of animals, and also it attracts attention of the author for solving various real-time complex issues [18]. In this paper, the author discusses GWO which is a hybrid algorithm using different methods. It is also a comprehensive standard GWO algorithm used to solve E-commerce prediction real-time problems with different characteristics & applications of the GWO algorithm.

Machine Learning

Machine learning & Artificial intelligence are the whole & soul of information technology services in coming future challenges. It creates algorithms or programs that can access and learn from data, without any human interruption. A machine learning algorithm is very beneficial to reduce inaccuracy & prediction errors. The challenge faced by machine-learning methodology is to obtain connections between different chunks of information, clean noisy data from dirty datasets, & eliminate gaps of incomplete data.

"Machine learning is defined as the science of a computer's ability to learn without being explicitly programmed. We are exposed to machine learning daily in our lives through applications such as Google search, speech recognition & self-parking cars even without being aware of it" (Fig. 1).

The three categories of Machine Learning (ML) algorithm are depicted below:

[A] Supervised Learning
[B] Unsupervised Learning
[C] Reinforcement Learning

Supervised Learning creates a program that can be useful for estimating by referring to training data also for prediction & estimation, a machine learning algorithm finds different patterns & resolve the relations between different parameters.

Unsupervised learning is nothing but whenever some known parameters are available for the output, not necessary the output would be the correct answer. The process of detecting pattern and identifying the outliers completely depends on the algorithm. Formal statistical framework is used for developing Machine learning techniques for

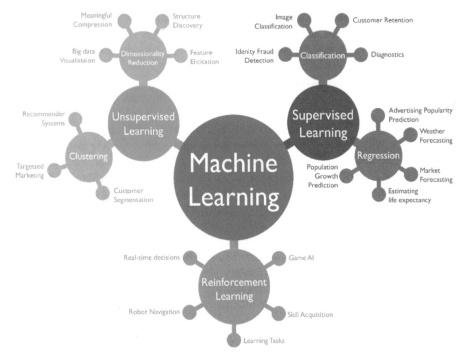

Fig. 1. Building blocks of ML

building statistical modelling. It has capability to utilize the collections of insights, learn from data, develop and test algorithms for developing models that are accurate in prediction analysis.

2 Literature Review

The paper focuses on a detailed analysis of different techniques applied in E-commerce market prediction with different applications. Niccolò G., Valerio V. (2016) [27] introduced the development of a model called churn prediction methodology used for customizing the B2B E-commerce which has been used for testing the predicting capacity of newmodel, the SVM setup on the AUC parameter-selection technique (SVMauc).

Daniela S., Patricia M., Oscar C. (2017) [30] presented a human recognition system for Modular Granular Neural Networks using GWO, it worked for the output that are compared with genetic algorithm along with firefly algorithm for analysing the techniques which provide good outcome when implemented to recognize human.

Seyedali M., Faris., Hossam., Ibrahim A., Mohammed Azmi Al-Betar., (2018) [13] presented the survey of the GWO algorithm which thoroughly described inspiration, its mathematical model, and analysis of an algorithm & investigation of its performance have explained in terms of exploration and exploitation.

Leonardo V., David H., Mauro C., Aleÿs P. (2018) [24], proposed a system for predicting default customer in E-Commerce using artificial intelligence, also developed expert systems along with its applications, genetic programming (GP) for Credit scoring applications.

María B., Pilar G., Jorge S. (2018) [25], this paper is on digital marketing techniques & based on recommendation, by allowing industries to implement a predictive model which is expanded using an amalgamation of big data, data science techniques, and machine learning methods, so as to customize financial incentives of users depending upon the quality of new customers available on the website for cashback.

Sneha V., Shrinidhi R., Sunitha S, Mydhili N. (2019) [28] came up with a methodology for Sparse Data by merging Collaborative filtering which is based on the recommender system that executes regression and GWO for prediction analysis of the ratings unrated items depending upon the history of the ratings given by user and its similarity with other users. which can merge with GWO for the application of loss function optimization.

3   GWO Algorithm

GWO algorithm has its hierarchical behavior between the wolves which are main and recognizable also, its an efficient hunting mechanism. Alpha is a dominant wolf work as a leader and beta, delta, and omega the other three wolves in the pack have an adaptable mechanism that is used to search, approach, and finally hunt the prey. This method gives global optimum solutions in various fields through complex research Spaces. GWO is a metaheuristic developed by Mirjalili [13]. The GWO algorithm represents the intelligent techniques of grey wolves, & the unique features of hunting. The figure gives a brief idea about a hierarchical pattern, & a pack of wolves follows it, in which various types of wolves uses different mechanisms for hunting (Fig. 2).

Fig. 2.   Hierarchy level

Searching and hunting are the two major activities that can be entitled as exploration and exploitation which has to be accomplished in a manner that the local stagnation of the solution doesn't occur. numerous enlargements in the standard GWO has presented by various researchers using multiple alternatives and operators applied for applications of GWO in various fields.

Features & Functionality of GWO
GWO Also called Nature Inspired Algorithm (NIA has received much popularity in the dynamic world, consisting of multiple applications for optimizing issues. To simulate the leadership hierarchy Alpha, Beta, Delta, and Omega are assured. For obtaining the prey, core steps include hunting, encircling the prey, and finally grab the prey. a meta-heuristic is a higher-level functioning for finding, generating, or selecting a heuristic that might give a decently good and optimized result to a particular problem (Fig. 3).

Fig. 3. General structure of GWO algorithm

GWO Algorithm is having leadership & hunting mechanism of grey wolves, the model represents:

Chasing/Pursuing: Once the target is entered the territory, the pack will chase the prey to make it kill. the wolves pack focuses on selecting and targeting the one which is delicate i.e. (weak prey) so it can be easy to kill the grabbed prey. The first best solution is α, accordingly, β ranks second, and δ ranks third respectively. In the optimization process α, β and δ is used for calculating the location of prey, & the rest wolves simultaneously keep on updating their locations depending upon the locations of α, β, δ respectively.

Encircling: Grey wolves reaches the prey & encircle it, the hunting is accomplished by attacking the prey till the time it stops moving. Here α, β, and δ agents must have complete information regarding prey's position to create a group that symbolizes the similar behavior of grey wolves for hunting. further calculations of the prey position in the search area or specified space and according to ω wolf locate itself around the prey.

1. The distance between a grey wolf and its prey is given by Eq. (1).

$$D = \left| \vec{C} \cdot \vec{X}_{p(t)} - \vec{X}_{(t)} \right| \tag{1}$$

t - current iteration, $\vec{X}_p$, and $\vec{X}$ have position vectors of prey & grey wolf respectively, and $\vec{C}$ - coefficient vector, it shows by Eq. (2).

$$\vec{C} = 2 \cdot \vec{r_1} \tag{2}$$

In Eq. (2) $\vec{r_1}$ - random vector having interval [0, 1].

2. Equation (3) gives recognition of prey location.

$$\vec{X}_{(t+1)} = \vec{X}_{p(t)} - \vec{A} \cdot \vec{D} \tag{3}$$

$$\vec{A} = 2 \cdot \vec{d} \cdot \vec{r_2} - \vec{d} \tag{4}$$

Where $\vec{A}$ is a coefficient vector, $\vec{r_2}$ is a random vector in interval of [0, 1], and components of $\vec{d}$ are linearly decreasing from 2 to 0 throughout iterations (Fig. 4).

Fig. 4. Encircling mechanism of wolves

3. Location of a grey wolf is given in Eq. (5).

$$\vec{D_\alpha} = \left| \vec{C_1} \cdot \vec{X_\alpha} - \vec{X} \right| \tag{5}$$

$$\vec{D_\beta} = \left| \vec{C_2} \cdot \vec{X_\beta} - \vec{X} \right| \tag{6}$$

$$\vec{D_\delta} = \left| \vec{C_3} \cdot \vec{X_\delta} - \vec{X} \right| \tag{7}$$

α, β, δ together with the distance between each search agent has been measured by Eqs. (5), (6) & (7)

$$\vec{X1} = \vec{X\alpha} - \vec{A_1} \cdot \left(\vec{D_\alpha} \right) \tag{8}$$

$$\vec{X2} = \vec{X\beta} - \vec{A_1} \cdot \left(\vec{D_\beta} \right) \tag{9}$$

$$\vec{X3} = \vec{X\delta} - \vec{A_1} \cdot \left(\vec{D_\delta} \right) \tag{10}$$

$$\overrightarrow{X_{(t-1)}} = \frac{\vec{X_1} + \vec{X_2} + \vec{X_3}}{3} \tag{11}$$

Attacking: The grey wolves conclude hunting by striking the prey until its movement is stopped. The Framework is determined using different values of $\vec{A}$. The range of $\vec{A}$ is in the interval $[-2a, 2a]$ and declines by $\vec{a}$ from 2 to 0 all over the interval $[-2a, 2a]$ and decline by $\vec{a}$ from 2 to 0 all over repetitions. This parameter has focuses on exploration and exploitation. If the difference between current values of any location and the location of the prey occurs, It is clear that if $|A| < 1$ indicates to attack the prey, incidental pressure for prey to non-linearly emphasize $(C > 1)$ or deemphasize $(C < 1)$ in the impact of prey of deciding the distance.

4   GWO Algorithm & Classification Analysis

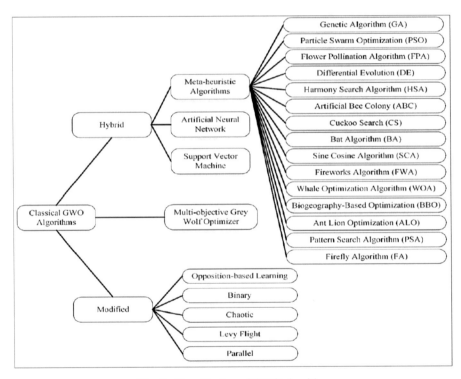

Fig. 5. Classification of GWO algorithm

The study includes, the analysis of the GWO algorithm based on Fig. 5, where the classical GWO algorithms have been classified into hybrid, multi-objective, and modified.

The hybridization of meta-heuristic algorithms - has represented the amalgamation of GWO & the GA to reduce a molecule's perfectly clear model of the energy function. Total three procedures are implemented,

1) application of Grey Wolf Optimizer for balancing exploration & the exploitation process.

2) To increase the variety of the search in the algorithm, utilization of the dimensionality reduction & the population segregation methods by splitting up the population into sub-populations and use the arithmetical crossover operator in every sub-population has been carried out.

3) By applying the genetic modification operator in the overall community to avoid early concurrences, to cut off local minima. And its efficiency can be verified for solving molecular potential energy functions [33].

ANN hybridization is beneficial in utilizing pattern identification or data categorization, through a learning process. Neurons consist of conjugating connections which can be incorporated due to learning in biological structures. For example, In breast cancer classification, a feed foreword neural network is implemented & it is trained by a grey wolf optimization algorithm. it gave a very vigorous, productive, and accurate categorization as compared to other classifiers [34].

GWO-SVM approach has used for automatic seizure detection is convenient for the decomposition of EEG bifurcate into five sub-band elements, the discrete wavelet transform (DWT) assist in utilizing extracted features set. Then, the features are available to upskill the SVM with radial basis function (RBF) & kernel function. here, GWO sorting out the remarkable analysis subset also the foremost parameters of SVM, to acquire a successful EEG classification [35].

MOGWO algorithm uses a leader selection strategy,& selects from three leaders: α, β, and δ. In the individual iteration of the optimization, for the decision of three non-dominated solutions. These reference points updates all solutions in the population [36]. Then lastly, needs to insert into the archive.

The last type that is modified versions of GWO has different applications in various fields. for example, a Modified GWO algorithm works for transmission network expansion operation problem, which develops its modified version & undergoes testing on 20 standard benchmark test functions & the result came out in the form of the accuracy as well as proficiency of the proposed algorithm [31].

The responsibility of a hybrid approach for optimization is gaining popularity remarkably and the focus is on performing the improvement of classical algorithms in terms of the idea of hybridizing the categories from other optimization strategies. research confirms the working of the GWO algorithm upgraded through the incorporation of other operators from meta-heuristic techniques [32].

5 Performance Comparison of Different Applications of GWO Algorithm

The distribution of published research articles on GWO concerning hybridizations, Multi-objective GWO, and modifications are represented in Table 1. The grey wolf algorithm (GWO) has multiple functional applications that cover a huge area. It has implemented as a power conserver in "an application of GWO for optimistic power flow of wind integrated power systems" utilized to conserve the resources for "energy-efficient

Table 1. Comparative applications of GWO

Sr No.	Type	Concept	Application	Outcome
1	PSO-GWO (Hybridization with Meta-Heuristic Algorithm)	Grid-Connected optimal design [9]	Reverse osmosis desalination merged with a plant grid-connected hybrid renewable energy supply natural water to a residential group	The performance of the hybrid version of PSO-GWO is demonstrated as the optimization parameters are proved to perform better than various optimization methods when implemented in isolation
2	Levy Flight-based Grey Wolf Optimizer (LFGWO)	FS for Image Steganalysis [13]	For selecting the prominent characteristics for the step analysis algorithm from a set of original characteristics, a novel Levy Flight-based Grey Wolf Optimization method is used	Better converging precision and effectively reducing irrelevant and redundant features at the same time by maintaining the high classification accuracy rate compared to other characteristics selection methods
3	EOG signal, GWO; HCI; NN classifier (Hybridization with Artificial Neural Networks)	Eye Movement Recognition [17]	For minimizing the error function of the outcome of classifier GWO is used	The performance of the proposed methodology is better validated using performance examination in terms of precision, accuracy, sensitivity, specificity, FPR, FNR, FDR, NPV and F1 score

(continued)

Table 1. (*continued*)

Sr No.	Type	Concept	Application	Outcome
4	Kernel extreme learning, Grey wolf optimization (Hybridization with Support Vector Machine)	Bankruptcy Prediction [18]	Adjusting the SVM specifications utilizing the GWO algorithm	Superiority of model generated is validated using the results in terms of Type I & Type II error, classification accuracy (validation, training, test), an area that comprises under operating attribute curve (AUC) at receiver
5	Oppositional Learning-based Grey Wolf Optimizer (OGWO)	Cloud Environment using machine Schedule methods [20]	Oppositional learning is used along with standard GWO for enhancing the convergence and the speed of computation	The proposed method provides quality schedules with less memory utilization and computation time
6	Robot Path Planning Optimization (Multiobjective GWO) (MOGWO)	Optimal Robot Path Planning (ORPP) [21]	The robot path planning issue considering the characteristics of distance and smooth path have been transformed for the fitness function into minimization	Proposed method guides the robot to reach its aim by avoiding collision with obstacles and also provides an alternative method for the robot planning using optimization

train operations using nature-inspired algorithm". Its application includes scheduling of robots and thresholding. The main methodologies used are search, encircle, and hunt till the optimized target of hunting is achieved. For evaluating the performance, calculation of fitness factor along with the functional optimizers are considered as medium function. In the matter of conserving power, GWO searches the optimal path in which it is essential that the windmills operate in a manner so that the energy is consumed efficiently. The train scheduling algorithm assures that the path followed by the train is optimistic and the accidents are avoided and hence the money is conserved GWO handles all obstacles that occur while obtaining an optimal path of an issue, obtain a proper allocation of existing resources for conserving the quantity which is essential for that program. It also searches the path which can be utilized by robots for properly utilizing its characteristics.

5.1 In Binary Problems

Punithavathani & Sujatha [2] proposed a 'method that removes the noisy characteristics by combining multi-focused images with decision data of optimized individual characteristics.

Manikandan et al. presented an approach which can be used for selecting gene from microarray data [4]. Binary GWO and Mutated GWO are the names of the proposed approaches.

For solving a nonconvex problem like a profit-making power transmission, which is a non-continuous and non-linear issue, Jayabarathi et al. [16] implemented crossover & mutation with grey wolf optimizer. Total economic dispatched issues solved were four and outcomes outclass the other meta-heuristics.

5.2 Dynamic & Restricted Problems Having Constraints

Jangir and Jangir proposed the non-dominated sorting GWO [15]. In this, the gathering of entire pareto optimal solutions is done till the last iteration evolved gathers all the pareto optimal solutions in the database. Execution time & high coverage are the parameters used for determining the efficacy rate of the algorithm.

Korayem et al. [17] depicted a methodology in which the combination of original GWO with the K is carried out. It represents that derived capacitated version of K-GWO algorithm exhibits the efficient performance.

For solving the dynamic security-constrained optimal power flow (DSCOPE) problem Teeparthi & Kumar [20] presented the GWO algorithm. To obtain the target, the power system actives the power production and is rescheduled against the constraints which can be static and dynamic or both.

5.3 In Multi-criterion Problem

For successfully automating the power generation control of two areas' interconnected power generation, Gupta & Saxena [19] demonstrated the extend of GWO for obtaining the synchronizing frameworks by performing the comparison analysis of the application of the two objective functions – [a] Integral square error and [b] Integral time absolute error.

To solve real-world welding scheduling problems Lu et al. [13] proposed an algorithm. An integer programming model is combined with it and it produces better outcomes.

Mustaffa et al. [9] proposed a model called hybrid forecasting, with the help of Least Squares Support Vector Machines by optimizing LSSVM's hyperparameters.

To reduce attributes by implementing MOGWO was proposed by Emery et al. [8] It works by finding an optimal characteristic subset that can achieve the description of data with minimal redundancy along with the performance classification.

6 Performance Comparison of Different Techniques of E-commerce Market Prediction

Thus, depending upon the available models and the systematic implementation of the modules I conclude that one of the current issues of E-commerce Prediction processes can be solved by these many different hybrid combinations of algorithms. Specifically, the GWO algorithm is best suited for all kinds of hybridizations and by observing the survey of these comparisons it is distinctly defined that "GWO is a better performing algorithm in this area". This research conducted a precise, broad (not comprehensive) survey to get the important writing on the hybridizations, alterations, and applications

Table 2. Comparative techniques of GWO algorithm

Sr No.	Type	Concept	Technique	Outcome
1	GWO & Modular granular Neural Networks (MGNN) [30]	Modular Granular Neural Networks (MGNN) with GWO for Human Recognition	Optimized granulation of data along with the representation of modular neural network (MNN) architectures is produced for human recognition	The effectiveness of MGNN (modular granular neural network) is proved with its benchmark databases of the biometric measures of ear, iris, and face, is used to carry out tests and comparing it with other related works
2	GA & Fuzzy logic [29]	Predicting the ratings of review using Fuzzy-based method in e-commerce	Review rating prediction; sentiment analysis; dimensionality reduction; data mining; genetic algorithm; fuzzy c-means	It is found that the performance and prediction accuracy in rating the reviews by the system produces better outcomes when compared with the related works
3	GWO & Collaborative filtering [28]	Collaborative Filtering-based Recommender System that uses regression & GWO algorithm for Sparse Data	Grey wolf optimization; sparse matrix; Collaborative Filtering; recommender system; regression; evaluation metrics; hit rate	It is found that the accuracy rate, user coverage, and different variants of hit rate of the proposed techniques works well, in highly sparse data scenario
4	GWO & Fuzzy Logic [26]	MOGWO and quantum-inspired possibilistic fuzzy C-means for robust product recommendation system	Grey wolf optimizer, Possibilistic fuzzy C-means, Latent Dirichlet allocation, recommendation system Lemmatization	The performance evaluation metrics like accuracy rate, precision and recall of the system proposed performed better in recommending product when compared to the available systems

of the GWO algorithm when utilized to illuminate issues of high dimensionality in completely different spaces (Table 2).

7 Conclusion & Future Scope

The natural phenomenon which inspiring the algorithm is presented. It was accompanied by a mathematical model and the investigation of its performance using evaluation metrics like exploration & exploitation is done. Varied versions of GWO along with its usability along with parameter tuning, varied approaches for selecting features, and performing classification is elaborated and then its hybridized versions are also discussed.

The areas that will be investigated in the future are as follows:

- Updating the locations of the non-dominated solutions in dynamic multi-purpose optimization issues.
- feeding variations engaged in constraints, objective tasks, and outcomes.
- verifying practicable outcomes of a diverse quantity of wolves assigned to all iterations.
- Simplification and removal of the parameters engaged will be done altogether.

References

1. de Souza, J.G.C., et al.: Generating e-commerce product titles and predicting their quality. In: Proceedings of the 11th International Conference on Natural Language Generation, pp. 233–243 (2018)
2. Kumar, P., Kumar, V., Thakur, R.S.: A new approach for rating prediction system using collaborative filtering. Iran J. Comput. Sci. 2(2), 81–87 (2019)
3. Sujatha, K., Shalini Punithavathani, D.: Optimized ensemble decision-based multi-focus image fusion using binary genetic Grey-Wolf optimizer in camera sensor networks. Multimedia Tools Appl. 77(2), 1735–1759 (2018)
4. Davis, L.: Handbook of Genetic Algorithms Van Nostrand Reinhold New York Google Scholar (1991)
5. Manikandan, S.P., Manimegalai, R., Hariharan, M.: Gene selection from microarray data using binary grey wolf algorithm for classifying acute leukemia. Curr. Signal Transduct. Therapy 11(2), 76–83 (2016)
6. Dorigo, M., Gambardella, L.M.: Ant colony system: a cooperative learning approach to the traveling salesman problem. IEEE Trans. Evol. Comput. 1(1), 53–66 (1997)
7. Pal, N.R., Corchado, E.S., Kóczy, L.T., Kreinovich, V.: Advances in Intelligent Systems and Computing (2015)
8. Emary, E., Yamany, W., Hassanien, A., Snasel, V.: Multi-objective gray-wolf optimization for attribute reduction. Procedia Computer Science 65, 623–632 (2015). https://doi.org/10.1016/j.procs.2015.09.006
9. Abdelshafy, A.M., Hassan, H., Jurasz, J.: Optimal design of a grid-connected desalination plant powered by renewable energy resources using a hybrid PSO–GWO approach. Energy Convers. Manage. 173, 331–347 (2018)
10. Kumar, A., Pant, S., Ram, M.: System reliability optimization using gray wolf optimizer algorithm. Qual. Reliab. Eng. Int. 33(7), 1327–1335 (2017)

11. Fouad, M.M., Hafez, A.I., Hassanien, A.E., Snasel, V.: Grey wolves optimizer-based local-ization approach in wins. In: 2015 11th International Computer Engineering Conference (ICENCO), pp. 256–260. IEEE (2015)
12. Mirjalili, S.: SCA: a sine cosine algorithm for solving optimization problems. Knowl. Based Syst. **96**, 120–133 (2016)
13. Faris, H., Aljarah, I., Al-Betar, M.A., Mirjalili, S.: Grey wolf optimizer: a review of recent variants and applications. Neural Comput. Appl. **30**(2), 413–435 (2018)
14. Jain, U., Tiwari, R., Godfrey, W.W.: Odor source localization by concatenating particle swarm optimization and Grey Wolf optimizer. In: Bhattacharyya, S., Chaki, N., Konar, D., Chakraborty, U.K., Singh, C.T. (eds.) Advanced Computational and Communication Paradigms. AISC, vol. 706, pp. 145–153. Springer, Singapore (2018). https://doi.org/10.1007/978-981-10-8237-5_14
15. Jangir, P., Jangir, N.: A new non-dominated sorting grey wolf optimizer (NS-GWO) algorithm: development and application to solve engineering designs and economic constrained emission dispatch problem with integration of wind power. Eng. Appl. Artif. Intell. **72**, 449–467 (2018)
16. Jayabarathi, T., Raghunathan, T., Adarsh, B.R., Suganthan, P.N.: Economic dispatch using hybrid grey wolf optimizer. Energy **111**, 630–641 (2016)
17. Mulam, H., Mudigonda, M.: Eye movement recognition by grey wolf optimization based neural network. In: 2017 8th International Conference on Computing, Communication and Networking Technologies (ICCCNT), pp. 1–7. IEEE (2017)
18. Wang, M., et al.: Grey wolf optimization evolving kernel extreme learning machine: application to bankruptcy prediction. Eng. Appl. Artif. Intell. **63**, 54–68 (2017)
19. Gupta, E., Saxena, A.: Grey wolf optimizer based regulator design for automatic generation control of interconnected power system. Cogent Eng. **3**(1), 1151612 (2016)
20. Natesan, G., Chokkalingam, A.: Opposition learning-based grey wolf optimizer algorithm for parallel machine scheduling in a cloud environment. Int. J. Intell. Eng. Syst. **10**(1), 186–195 (2017)
21. Tsai, P.-W., Nguyen, T.-T., Dao, T.-K.: Robot path planning optimization based on multi-objective grey wolf optimizer. In: Pan, J.-S., Lin, J.C.-W., Wang, C.-H., Jiang, X. H. (eds.) ICGEC 2016. AISC, vol. 536, pp. 166–173. Springer, Cham (2017). https://doi.org/10.1007/978-3-319-48490-7_20
22. Mirjalili, S., Lewis, A.: The whale optimization algorithm. Adv. Eng. Softw. **95**, 51–67 (2016)
23. Imran, A.M., Kowsalya, M.: A new power system reconfiguration scheme for power loss minimization and voltage profile enhancement using fireworks algorithm. Int. J. Electr. Power Energy Syst. **62**, 312–322 (2014)
24. Vanneschi, L., Horn, D.M., Castelli, M., Popovič, A.: An artificial intelligence system for predicting customer default in e-commerce. Exp. Syst. Appl. **104**, 1–21 (2018)
25. Ballestar, M.T., Grau-Carles, P., Sainz, J.: Predicting customer quality in e-commerce social networks: a machine learning approach. Rev. Manag. Sci. **13**(3), 589–603 (2018)
26. Kolhe, L., Jetawat, A.K., Khairnar, V.: Robust product recommendation system using modified grey wolf optimizer and quantum inspired possibilistic fuzzy C-means. Clust. Comput. **24**(2), 953–968 (2020)
27. Gordini, N., Veglio, V.: Customers churn prediction and marketing retention strategies. An application of support vector machines based on the AUC parameter-selection technique in B2B e-commerce industry. Ind. Mark. Manage. **62**, 100–107 (2017)
28. Sneha, V., Shrinidhi, K.R., Sunitha, R.S., Nair, M.K.: Collaborative filtering based recom-mender system using regression and grey wolf optimization algorithm for sparse data. In: 2019 International Conference on Communication and Electronics Systems (ICCES), pp. 436–441. IEEE (2019)
29. Velvizhy, P., Pravi, A., Selvi, M., Ganapathy, S., Kannan, A.: Fuzzy-based review rating prediction in e-commerce. Int. J. Bus. Intell. Data Mining **17**(1), 101–116 (2020)

30. Sánchez, D., Melin, P., Castillo, O.: A grey wolf optimizer for modular granular neural networks for human recognition. Comput. Intell. Neurosci. **2017**, 1–26 (2017)
31. Khandelwal, A., Bhargava, A., Sharma, A., Sharma, H.: Modified grey wolf optimization algorithm for transmission network expansion planning problem. Arab. J. Sci. Eng. **43**, 2899–2908 (2017)
32. Ouhame, S., Hadi, Y., Arifullah, A.: A hybrid grey wolf optimizer and artificial bee colony algorithm used for improvement in resource allocation system for cloud technology. Int. J. Onl. Biomed. Eng. **16**, 4–17 (2020)
33. Tawhid, M.A., Ali, A.F.: A Hybrid grey wolf optimizer and genetic algorithm for minimizing potential energy function. Memet. Comput. **9**(4), 347–359 (2017)
34. Pal, S.S.: Grey wolf optimization trained feed foreword neural network for breast cancer classification. Int. J. Appl. Indust. Eng. **5**(2), 21–29 (2018)
35. Hamad, A., Houssein, E.H., Hassanien, A.E., Fahmy, A.A.: A hybrid EEG signals classification approach based on grey wolf optimizer enhanced SVMs for epileptic detection. In: Hassanien, A.E., Shaalan, K., Gaber, T., Tolba, M.F. (eds.) AISI 2017. AISC, vol. 639, pp. 108–117. Springer, Cham (2018). https://doi.org/10.1007/978-3-319-64861-3_10
36. Ibrahim, R.A., Elaziz, M.A., Lu, S.: Chaotic opposition-based grey-wolf optimization algorithm based on differential evolution and disruption operator for global optimization. Exp. Syst. Appl. **108**, 1–27 (2018)

Applying Extreme Gradient Boosting for Surface EMG Based Sign Language Recognition

Shashank Kumar Singh[1]([✉]), Amrita Chaturvedi[1], and Alok Prakash[2]

[1] CSE Department, IIT (BHU), Varanasi 221005, India
{shashankkrs.rs.cse17,amrita.cse}@itbhu.ac.in
[2] School of Bio Medical Engineering, IIT (BHU), Varanasi 221005, India
alokp.rs.bme15@itbhu.ac.in

Abstract. Sign languages are used as a non-verbal form of communication that helps people to exchange information. Various sign language recognition systems have been developed to assist healthy and differently-abled people in understanding and conveying messages using sign languages. These recognition systems are mostly developed using computer vision-based algorithms, which have a limitation that their efficiency depends heavily upon surrounding lighting conditions. On the other hand, surface Electromyography is least affected by surrounding lightning conditions, so can be used as an alternative to computer vision based methods. In this paper, we have focused on surface Electromyography signal's ability to build a reliable sign language recognition system. We have applied boosting based algorithm to build an accurate and reliable classifier to recognize American Sign Language (ASL) using surface Electromyography signals. For, this a new data-set was made by collecting surface Electromyography signals from ten adult-subjects. The classification model was trained using Extreme gradient boosting (XGBoost) algorithm and we obtain an accuracy of 99.09%.

Keywords: Surface Electromyography · Sign language recognition system · Machine learning · Boosting algorithm

1 Introduction

Sign language as a field in linguistic was established in early 1970. They are a non-verbal form of communication that helps healthy as well as differently-able people to exchange information. Particularly, they are beneficial for mute and deaf, and hard of hearing people. However, in many scenarios, due to certain disabilities, it's not easy for differently able people to generate and understand the sign language gesture. In such conditions, computer-assisted systems can play a vital role. They can automate sign language recognition and act as an interpreter for communication involving differently able people. Generally, such systems are known as sign language recognition system (SLR). These SLR are

© The Author(s), under exclusive license to Springer Nature Switzerland AG 2022
R. Misra et al. (Eds.): ICMLBDA 2021, LNNS 256, pp. 175–185, 2022.
https://doi.org/10.1007/978-3-030-82469-3_16

broadly classified based on data acquisition techniques [6,16,22]. Mostly, for these systems, the data acquisition is performed using cyber gloves, Electromyography sensors, cameras, or depth-sensing technology such as Microsoft kinetic sensors [3]. However, the area of sign language recognition is mainly dominated by the computer-vision based paradigm [22]. In the same paper the authors listed that among total number of research articles related to SLR, about 68% of Indian Sign Language, 33% of Mexican Sign Language, 60% of Arabic Sign Language, 67% of Thai Sign Language, 67% of British Sign Language, 44% of American Sign Language (ASL), 34% of Greek Sign Language, 67% of Malaysian Sign Language and have used vision-based techniques for the respective languages.

Despite the dominance of computer vision based methods, these methods have a drawback that their system efficiency depends on environmental lighting conditions. In fact, these systems are easily affected by visibility, camera orientation, background, camera angle, intensity, and other lighting conditions. However, sensor-based signals such as surface Electromyogram (sEMG) are invariant to lighting conditions [11]. They can be used as an alternative to a computer vision based method to build sign language recognition systems. For building an efficient sEMG based sign language recognition system, an accurate classifier is required to identify the signals generated for different sign language gestures. Generally, Machine learning techniques can be deployed to achieve such functionality. The details of these approaches are described in Sect. 2.

Boosting-based algorithms are very efficient in classification task [19]. However, their capability to classify sEMG signals is still much unexplored. In this paper, we have tried to investigate the XGBoost algorithm as a possibility to accurately classify sEMG signals generated for American sign language (ASL). The advantage of having a greater classification accuracy is that we could build a more accurate real-time SLR. Already work has been done to apply machine learning for sEMG based Sign Language recognition; however, to the best of our knowledge, no paper has evaluated the XGBoost in classifying the sEMG based signals. The XGBoost shows promising results and achieves a classification accuracy of 99.02%. This paper aims to evaluate the capability of the XGBoost algorithm in classifying sEMG signals generated for American sign language. The paper further describes related work, Evaluation metrics used, experimental setup, followed by a conclusion

2   Related Work

Various papers have applied traditional as well as machine learning techniques for classifying the ASL signs.

In [21], the authors extracted different features based on the time-domain, average power, frequency domain for different ASL sEMG signals. They further used principal component analysis for finding uncorrelated features among the various extracted features. The classification was done using the ensemble method and Support Vector Machine. Their method achieved classification accuracy 54.64% to 79.35% for various datasets. Another paper [7] applied Naive Bayes, Support Vector Machine(SVM) and random forest on sEMG data and achieved classification accuracy up to 92.25%. While the paper [23] used an inertial measurement unit along with sEMG signals to classify 80 ASL signs collected from four subjects. The author achieved 96.6% and 85.24% average accuracies. Another paper [24] used sEMG signals along with an inertial sensor for classifying the ASL signs. They used customized hardware with a 9-axis motion sensor along with a sensor to measure the acceleration, angular velocity, and magnetic axis. Their framework obtained a 95.94% recognition rate. In [20], the authors perform the multi-class classification using a support vector machine. They were able to obtain an offline recognition rate of 91% for different signs of ASL.

In the paper [24], the authors use Morkov hidden model for classifying the ASL signs. They translate these signs to text using Microsoft visual studio 2010. The authors achieved d a recognition rate of 94% for the single character. In [25] authors extracted 16 different features from the sEMG signals using Mahalanobis distance. Their classification method achieves an accuracy of 97.7%. Another paper [9] uses the support vector machine and deep neural network to classify sEMG signals and achieves the classification accuracy of 80.30% and 93.81% for the methods, respectively. The paper [15] used force-sensitive resistors and sEMG signals for classifying the signs of ASL. The dataset was collected for 3 sets and achieved 96.7% accuracy by applying 16 force-sensitive resistors. The authors [12] used single as well as a double-handed gesture. They applied a support vector machine and achieved an accuracy of 59.96% and 33.66% for various sets of data.

3 Methodology

3.1 Experimental Setup

Low-cost wearable sensors were used for collecting surface sEMG signals. The skin's surface was cleaned, and eight sEMG sensors were placed on the subject's forearm. To make subjects aware of ASL signs, proper illustration and audio visual means were used. The data from these sensors were collected on a computer system over a Bluetooth connection. The sEMG signals were recorded and stored in CSV format. To ensure a better quality of signals, real-time monitoring was performed by an online tool [18]. Figure 1 shows the complete experimental setup with a subject wearing sensors and performing poses for an ASL signs.

3.2 Dataset

A new dataset was made by collecting sEMG signals from 10 subjects consisting of eight male and two female participants. The dataset was collected for all the digits of ASL. At a time, eight different signals were recorded from eight different wearable sEMG sensors. These signals were collected 200 Hz with a sampling time of 2 s for each gesture. We took the sampling time as used in the paper [10]. Multiple samples of each sign were taken. All these sEMG signals were stored and processed in digital format (CSV format). Our labeled data-set consisting of z samples with n tuples can be described as:

$$H = \{(y_i, l_i)\} \, (|H| = z, y_i \in R^n, l_i \in R) \tag{1}$$

For classification, we framed 10 classes for each digits of ASL. We defined the classes C_i as

$$C_i = \{s_{1,k}, s_{2,k+1}, s_{3,k+2}, s_{4,k+5}, s_{5,k+6}, ...s_{n,k+n-1}, l_i\} \tag{2}$$

Where, C_i is the number of classes corresponding to ten digits of ASL which varies from $0 \leq i \leq 9$ and $s_{i,k}$ are the data-points derived using semg signals corresponding to each sign of ASL. The l_i, $0 \leq i \leq 9$, refers to the class label specific for each class C_i.

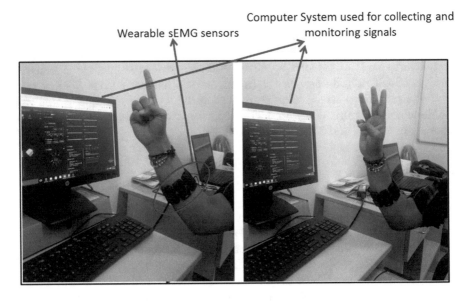

Fig. 1. Experimental set up used for collecting and monitoring the sEMG signals for ASL

3.3 Evaluation Metrics

For evaluating the effectiveness of the classifier we have used the standard metrics which are as follows:

Accuracy: It is defined as a ratio of total no of correct prediction to the total number of predictions done for that class.

F1-Score: It is formulated as a harmonic mean between precision and recall.

Recall: It is calculated as the ratio of true positive samples to the summation of true positive and false negative samples.

Sensitivity: It is also known as a true positive rate. Is the ratio of true positive samples to the summation of false-negative and true positive samples.

Specificity: Also known as a false-positive rate. It is the ratio of false-positive to the summation of false positive and true negative.

Precision: Precision is the ratio of true positive to the summation of true positive and false-positive samples.

The value of these evaluation metrics varies between the range [0–1]. For an ideal classifier value of accuracy, the F1 score and precision should be near to 1 whereas the value of the False-negative rate and False positive rate near to 0 is considered efficient.

3.4   Algorithm

We framed the sign language recognition as a supervised learning problem where based on a training dataset (multiple feature x_n) tried to predict a target class label y_i. For this, we used a recent boosting tree based method known as Extreme Gradient boosting (XGBoost). Basically, the algorithm is an efficient modification over the traditional gradient boosting method with which it can achieve multiple times computational efficiency compared to the previous technique. The objective function for the algorithms is defined as:

$$\chi(\phi) = \sum_j \nu(y_j, \hat{y}_j) + \sum_k \Theta(f_k) \tag{3}$$

whereas,

$$\Theta(f) = \alpha T + \frac{1}{2}\beta \|w\|^2 \tag{4}$$

and,

$$\hat{y} = \sum_{k=1}^{K} (f_k(x_i), f_k \in \varphi) \tag{5}$$

The above Eq. 3, Eq. 4 and Eq. 5 are derived from the paper [8]. Here, the loss function ν measures the difference between the target and the predicted value. And, φ denotes tree space (CART). Each f_k represent individual structure of the tree. T is the number of leaves, w represent the leaf weights, K is the total number of additive function used for prediction. $\hat{y}$ is the predicted value of a sample by the algorithm.

Efficiency of the XGBoost depends upon the algorithm optimization and modification done over traditional gradient boosting. Theses modification includes a regularized learning objective, weighted quartile stretch, sparsity aware algorithm, and out of core computation. The regularisation term helps to deal with overfitting and model complexity. Apart from this, XGBoost also uses the column subsampling and shrinkage method too to prevent model over-fitting. However, if the regularization parameter is made to zero, the objective function tends to traditional gradient boosting.

Weighted quartile stretch is a novel technique used in XGBoost, which allows the algorithm to accurately approximate the objective function for a weighted dataset. The XGBoost also proposed a sparsity aware algorithm, which helps the algorithm to have linear computational complexity to the number of the non-missing entries in the dataset. The algorithm assigns a default direction for the missing data, which is further used for classifying them. The out-of-core computation features allow XGBoost to efficiently execute the program with a larger size than the main memory. The algorithm uses two approaches, mainly block compression and Block Sharding. The complete description of optimizations modification can be found in [13,14].

4   Result and Discussion

We applied the Extreme gradient boosting on our new sEMG signal dataset. The dataset consists of sEMG signals collected by placing the eight wearable sensors at the subjects' right hand. The sEMG signals corresponding to each of the eight sensors were collected in CSV format. Several data processing and data transformations were applied before applying the sEMG data into the machine learning pipeline. Most of the model's implementation was done in python using sci-kit learn library [1]. The experiment was performed with the motive of getting a higher classification accuracy. We performed a very comprehensive parameter optimization for the XGBoost algorithm by applying the grid search method of sci-kit learn [4]. Further, for the generalization of the model, we applied k-fold cross-validation. The value of k was taken as 10. Rest of the parameters after optimization for this problem were obtained as follows: min-child-weight = 11, max-depth = 6, learning rate = 0.5, n-estimator = 1000, colsample-bytree = 0.7, sub-sample = 0.8. The model performs very well and shows an overall classification accuracy of 99.09%.

To ensure the classification's reliability, we calculated evaluation metrics such as Precision, Recall, F1 score, and sensitivity for each of the classes C_i. We found that for the different classes, C_i values of Precision vary in the range of 0.97 to 1.00, while Recall values range from 0.97 to 1.00, the F1 score was in the range of 0.98 to 1.00, and Sensitivity being 0.99 to 1.00. Considering these values, we can say our trained classifier was reliable and accurate for this classification task. The values of the evaluation matrices can be found in Table 1. Taking all classes C_i into account, we also computed a box-plot Fig. 4 for each evaluation matrices to visualize better and understand the range of these evaluation matrices. The plot describes the various quartiles and min-max values for each evaluation matrices. Considering the values of Table 1, we can say the classifier efficiently predicted the label for most of the classes. Among all the classes C_i, the classifier was highly efficient for class 2 with Precision, Recall, and Sensitivity calculated as 1, while it was least effective for class 6. Based on the above mentioned statistics, we can conclude XGBoost is an efficient algorithm for sEMG based classification problems. Also, to ensure the trained model is least affected by training biases, a considerable amount of data was collected. We also used ten k-fold validation to ensure the reliability of classification.

Table 1. Precision, Recall, F1 Score and Sensitivity for different classes C_i of digits of American Sign Language

Classes	Precision	Recall	F1 Score	Sensitivity
Class 0	1.00	1.00	1.00	0.99
Class 1	1.00	1.00	1.00	0.99
Class 2	1.00	1.00	1.00	1.00
Class 3	1.00	1.00	1.00	0.99
Class 4	1.00	1.00	1.00	0.99
Class 5	0.99	0.97	0.98	0.99
Class 6	0.97	0.99	0.98	0.99
Class 7	1.00	1.00	1.00	0.99
Class 8	0.98	0.97	0.98	0.99
Class 9	0.98	0.99	0.98	0.99

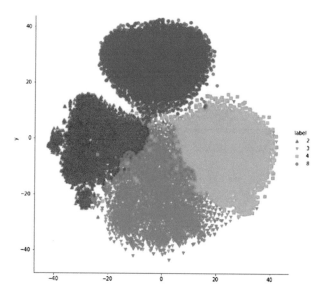

Fig. 2. Data visualization using t-SNE

The high accuracy on our dataset can be supported by the fact that similar sEMG based ASL data-set have achieved comparable classification accuracy [7,25]. Our work is just an incremental improvement in the classification accuracy on these type of data set. Some of the possible reasons for higher accuracy were quality of the data-set and the different distinctive patterns of sEMG signals which were generated for each class C_i respective to signs of ASL. These patterns were very efficiently recognized by the Extreme Gradient boosting algorithm. Further, to visualize the quality we plot the subset of our of data set in two dimensional space using t-sne tools [17] Fig. 2.

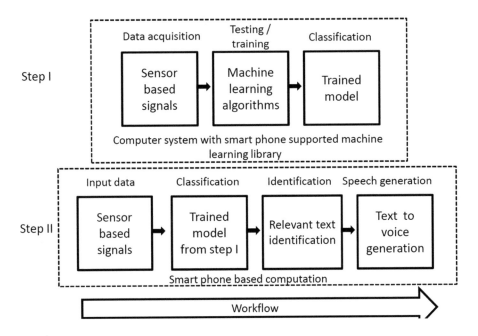

Fig. 3. Possible smartphone based cost effective sign language recognition system

Further, the use of smartphone supported machine learning libraries can help us to build a cost-effective sign recognition system. The model can be trained and eventually be deployed on mobile phones using these frameworks [5]. Also, few commercial sEMG sensors can interact with smartphones using their specific interface and libraries [2]. So combining all the above mentioned technologies, there is a possibility of building an effective sign language recognition system that could be reliable, accurate and cost effective. A road-map to such a system can be under stood from Fig. 3.

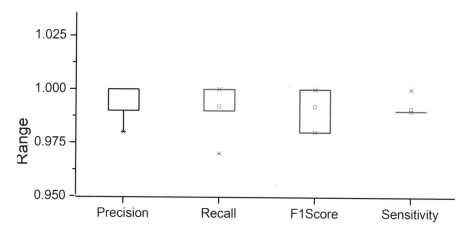

Fig. 4. Boxplot depicting the range of Precision, Recall, F1 score and Sensitivity for different classes C_i of sEMG signals

5   Conclusion and Future Work

Boosting-based algorithms have been very successful while doing the classification task. However, the efficiency of these algorithms to classify sEMG signals is not much explored. In this paper, we tried to study the efficiency of XGBoost, a boosting-based technique to classify sEMG signals for American Sign Language. The algorithm shows good classification accuracy and can be further used to build more accurate and reliable classifiers for sign language recognition systems. Another advantage of using these algorithms is that they are computationally efficient and can be used with a more extensive data set. Despite the advancement in sign language recognition, a cost-effective sign language recognition system is not available commercially. A smartphone-based sign recognition system can be a better alternative. Recent mobile-based machine learning libraries can help build these systems, maintaining the tradeoff between cost and reliability. One of the limitations of our work is that we have used independent static signs of ASL. However, we must expand the study further to classify ASL signs corresponding to continuous words and dynamic signs of ASL. We have left this for our future work.

References

1. https://scikit-learn.org/stable/. Accessed 22 July 2020
2. Getting started with your Myo armband. https://support.getmyo.com/hc/en-us/articles/203398347-Getting-started-with-your-Myo-armband. Accessed 22 July 2020
3. Kinetic. https://en.wikipedia.org/wiki/Kinect. Accessed 1 Jan 2021

4. scikit-learn 0.24.0: https://scikit-learn.org/stable/modules/generated/sklearn. model_selection.GridSearchCV.html. Accessed 1 Jan 2021
5. Abadi, M., et al.: TensorFlow: large-scale machine learning on heterogeneous systems (2015). https://www.tensorflow.org/. Software available from tensorflow.org
6. Al-Ahdal, M.E., Nooritawati, M.T.: Review in sign language recognition systems. In: 2012 IEEE Symposium on Computers & Informatics (ISCI), pp. 52–57. IEEE (2012)
7. Bian, F., Li, R., Liang, P.: SVM based simultaneous hand movements classification using sEMG signals. In: 2017 IEEE International Conference on Mechatronics and Automation (ICMA), pp. 427–432. IEEE (2017)
8. Chen, T., Guestrin, C.: XGBoost: a scalable tree boosting system. In: Proceedings of the 22nd ACM SIGKDD International Conference on Knowledge Discovery and Data Mining, pp. 785–794 (2016)
9. Chong, T.W., Lee, B.G.: American sign language recognition using leap motion controller with machine learning approach. Sensors 18(10), 3554 (2018)
10. Chuan, C.H., Regina, E., Guardino, C.: American sign language recognition using leap motion sensor. In: 2014 13th International Conference on Machine Learning and Applications, pp. 541–544. IEEE (2014)
11. De Luca, C.J.: The use of surface electromyography in biomechanics. J. Appl. Biomech. 13(2), 135–163 (1997)
12. Derr, C., Sahin, F.: Signer-independent classification of American sign language word signs using surface EMG. In: 2017 IEEE International Conference on Systems, Man, and Cybernetics (SMC), pp. 665–670. IEEE (2017)
13. XGBoost developers: XGBoost Documentation (2020). https://xgboost. readthedocs.io/en/latest/index.html. Accessed 01 Jan 2021
14. dmlc/xgboost: xgboost (2020). https://github.com/dmlc/xgboost. Accessed 01 Jan 2021
15. Jiang, X., Merhi, L.K., Xiao, Z.G., Menon, C.: Exploration of Force Myography and surface electromyography in hand gesture classification. Med. Eng. Phys. 41, 63–73 (2017)
16. Kudrinko, K., Flavin, E., Zhu, X., Li, Q.: Wearable sensor-based sign language recognition: a comprehensive review. IEEE Rev. Biomed. Eng. 14, 82–97 (2020)
17. Van der Maaten, L., Hinton, G.: Visualizing data using t-SNE. J. Mach. Learn. Res. 9(11), 2579–2605 (2008)
18. Myo: Myodiagnostics. http://diagnostics.myo.com/. Accessed 1 Jan 2021
19. Prokhorenkova, L., Gusev, G., Vorobev, A., Dorogush, A.V., Gulin, A.: CatBoost: unbiased boosting with categorical features. In: Advances in Neural Information Processing Systems, pp. 6638–6648 (2018)
20. Savur, C., Sahin, F.: Real-time American sign language recognition system using surface EMG signal. In: 2015 IEEE 14th International Conference on Machine Learning and Applications (ICMLA), pp. 497–502. IEEE (2015)
21. Savur, C., Sahin, F.: American sign language recognition system by using surface EMG signal. In: 2016 IEEE International Conference on Systems, Man, and Cybernetics (SMC), pp. 002872–002877. IEEE (2016)
22. Wadhawan, A., Kumar, P.: Sign language recognition systems: a decade systematic literature review. Arch. Comput. Methods Eng. 28(3), 785–813 (2021)
23. Wu, J., Sun, L., Jafari, R.: A wearable system for recognizing American sign language in real-time using IMU and surface EMG sensors. IEEE J. Biomed. Health Inform. 20(5), 1281–1290 (2016)

24. Wu, J., Tian, Z., Sun, L., Estevez, L., Jafari, R.: Real-time American sign language recognition using wrist-worn motion and surface EMG sensors. In: 2015 IEEE 12th International Conference on Wearable and Implantable Body Sensor Networks (BSN), pp. 1–6. IEEE (2015)
25. Yun, L.K., Swee, T.T., Anuar, R., Yahya, Z., Yahya, A., Kadir, M.R.A.: Sign language recognition system using SEMG and hidden Markov models. Ph.D. thesis, Universiti Teknologi Malaysia (2012)

Automated Sleep Staging System Based on Ensemble Learning Model Using Single-Channel EEG Signal

Santosh Kumar Satapathy[1]([✉]), Hari Kishan Kondaveeti[2], and Ravisankar Malladi[3]

[1] Pondicherry Engineering College, Pillaichavadi, Puducherry, India
santosh.satapathy@pec.edu
[2] Computer Science and Engineering, VIT University, Amaravati, Andhra Pradesh, India
[3] Department of CSE, Gudlavalleru Engineering College (A), Vijayawada, Andhra Pradesh, India

Abstract. Sleep is essential for people health and well-being. However, numerous individuals face sleep problems. These problems can lead to several neurological and physical disorder diseases, and therefore, decrease their overall life quality. Artificial intelligence methods for automated sleep stage classification (ASSC) are a fundamental approach to evaluate and treat this public health challenge. The main contribution of this paper is to present the design and development of an automated sleep staging system based on the ensemble techniques using single-channel of EEG signal. In this study, a novel method is applied for signal preprocessing, feature screening and classification models. In signal preprocessing we obtain the Online Streaming Feature Selection (OSFS). In feature extraction, we obtain a total of 28 features based on both time and frequency domain features and non-linear features. The important contribution of this research work is establishes two-layers an ensembling learning model. The base learning model consists of Random forest (RF), Gradient Boosting Decision Tree (GDBT), and Extreme Gradient Boosting (XGBoost) and the second layer is Logistic Regression. We obtained the ISRUC-Sleep subgroup-III subjects sleep recordings for our proposed experimental work. Comparing with the recent contributions on sleep staging performances, it has seen that our proposed ensemble learning model was reported best sleep staging classification accuracy performance for five sleep stages classification task (CT-5). The overall classification accuracy reported as 96.86% for OSFS selected features with SG-III dataset.

Keywords: Sleep scoring · Electroencephalography · Stacking model · Machine learning

1 Introduction

Sleep is one of the important physical activities for maintaining good human health. In general, each human-occupied one-third of his/her time as sleeping. It has been seen that longer periods of unhealthy sleep may lead to so many neurological disorders [1, 2]. The

© The Author(s), under exclusive license to Springer Nature Switzerland AG 2022
R. Misra et al. (Eds.): ICMLBDA 2021, LNNS 256, pp. 186–202, 2022.
https://doi.org/10.1007/978-3-030-82469-3_17

sleep experts considered five metrics for good sleep such as total sleep time, continuity of sleep, total sleep duration, awakening, and quality. Most of the sleep metrics can be obtained from polysomnography (PSG) analysis. Sometimes sleep diseases affect physical and mental disabilities and their impacts directly affect professional activities and quality of life. Some sleep diseases like insomnia, hypersomnia, sleep apnea, parasomnia, sleep-related breathing, cardiovascular diseases, sleep movement disorders threaten human health. The main important diagnosis and treatment step of sleep-related diseases is sleep stages classification [3] and the entire process is called sleep scoring and sleep staging.

Sleep staging is scored by using physiological signals such as electroencephalogram (EEG), electromyogram (EMG), electrooculogram (EOG). Secondly, the recorded signals are segmented into 30 s epochs, and each epoch assigned into one particular sleep stage: wake (W), rapid eye movement (REM), non-REM stage1 (N1), non-REM stage2 (N2), non-REM stage3 (N3), and non-REM stage4 (N4) as edited by the Rechtschaffen and kales sleep manual [4] (or) followed the AASM sleep manual [5], in which the stage N3 and N4 merged into one stage called as N3 (or) slow-wave sleep (SWS) stage. Thirdly, the experts observed the changes sleep behavior occurred during different stage of sleep such as spontaneous arousals, sleep spindles, k-complexes and repository events. The most important step during the sleep stage is step-2, where the sleep experts need more concentrates upon each segmented epochs and observed the sleep characteristics, and annotated one sleep stage. The entire manual visual-inspection process takes a lot of time, labor-intensive, and requires more human interpretation, which limits the efficiency of the PSG analysis [6]. The other main drawback concerning manual sleep staging is variations on the sleep staging result because of the expert's experience. These all drawbacks demand to development of an automated sleep staging system. Many of the authors have proposed several contributions on sleep staging based on the ML techniques [7, 8].

In the recent research developments, it has been seen that two types of ML approach [9, 10] used: one is traditional machine learning techniques and deep-learning-based techniques. The traditional ML methods required hand-crafted features for recognizing the sleep characteristics and classifying the sleep stages. On the other hand, the deep learning models do not require any types of explicit features and is automatic extracts the features from the input signal by obtaining different deep neural models [11]. It has been observed that many of the automated sleep staging systems are based on EEG signals instead of PSG signals. The main reason behind this was for the convenient and reliable during the sleep data recordings with less connectivity.

1.1 Related Work

In this section, the authors analyze the related research studies available in the literature. Most of the proposed studies were implemented using EEG signals. These studies recommend the extraction of features from the representative input signals. Moreover, these studies also suggest different feature selection algorithms. Finally, different classification techniques have been used to analyze EEG signals considering two to six sleep stages.

Oboyya et al. [12] proposed wavelet transform techniques to extract the features from EEG signals, and a fuzzy c-means algorithm used to classify the sleep stages. The overall classification accuracy reported was 85%.

In [13], obtained feature weighting method using K-means clustering is proposed. The welch spectral transform was considered for feature extraction, and the selected features have been used with K-means and decision trees techniques. The proposed study was reported that the overall accuracy was reported as 83%.

Aboalayon [14] used a Butterworth bandpass filter to segment the EEG signals. The extracted features have been used with an SVM classifier. The research work reported 90% classification accuracy.

The authors of [15] have proposed architecture using bootstrap aggregating for classification. The methods have been applied on single-channel EEG and achieved an accuracy of 92.43%.

In [16], the author proposed sleep stage classification based on time-domain features and structural graph similarity. The experimental work uses a single channel of EEG signals. The proposed SVM classifier presents an average classification accuracy of 95.93%.

Kristin M. Gunnarsdottir et al. [17] have designed an automated sleep stage scoring system using PSG data. In this study, the authors only have considered healthy individual subjects with no prior sleep diseases, and the extracted properties were classified through decision table classifiers. The overall accuracy reported was 80.70%.

Sriraam, N. et al. [18] used a multichannel EEG signal from ten healthy subjects. The proposed automatic sleep stage scoring model considers sleep stage 1. In this study, spectral entropy features have been extracted from input channels to identify irregularities in different sleep stages. The multilayer perceptron feedforward neural network has been used, and the overall accuracy with 20 hidden units was reported as 92.9%. Moreover, using 40, 60, 80, and 100 hidden units, the proposed method reported 94.6, 97.2, 98.8, and 99.2, respectively.

Memar, P. et al. [19] have proposed a system to classify sleep and wake stages. The authors have selected 25 suspected sleep subjects and 20 healthy subjects for experimental tests. In total, 13 features are extracted from each eight (alpha, theta, sigma, beta1, beta2, gamma1, and gamma2) sub-band epochs. The extracted features have been validated through the Kruskal-Wallis test. The overall accuracy reported using random forest with 5-fold cross-validation and subject-wise cross-validation is 95.31% and 86.64%, respectively.

Da Silveira et al. [20] used a discrete wavelet transform for decomposing the signal into different sub-bands. The Skewness, Kurtosis, and variance features have been extracted from respective input channels. The Random Forest classifiers have been tested for discriminating the sleep stages with an overall accuracy of 90%.

Xiaojin Li et al. [21] have proposed a hybrid automatic sleep stage classification based on a single channel of EEG signals. In this study, different extracted features from multiple domains such as time, frequency, and non-linear features are forwarded into Random Forest classifiers and the resulted accuracy is 85.95%.

Zhu, G et al. [22] obtained the EEG signals are mapped into visibility graphs and a horizontal graph to detect gait-related movement recordings. Finally, the extracted nine features from the input signals are forwarded to SVM classifiers considering multiple stages of sleep stages. The proposed method presents an accuracy of 87.50%.

Eduardo T. Braun et al. [23] proposed a portable sleep staging classification system using a different combination of features from EEG signals. The proposed method presents the best classification accuracy of 97.1%.

In this proposed sleep staging study, we have considered the ensemble learning techniques for classifying the sleep stages for five sleep states. The main objective of this study is to improvement on sleep staging performance. In this paper, we have used ensemble techniques instead of applying individual traditional ML models. For analyzing the changes sleep characteristics with respect to time and frequency domain, we extracted the both time and frequency domain features. For feature selection, we have considered the Online Streaming Feature Selection technique, which is more suitable to identify the strong relevant features with high-dimensional data. Though the brain EEG signals are more complicated and to choose suitable features OSFS is more effective. It has been found that the OSFS algorithm is more effective in comparison to other selection techniques due to determining the statistical significance of the differences between the selected features among different sleep stages. Finally the sleep staging results proved that the proposed model is more efficient incomparable to the other existing contributions.

Further, the rest of the work is organized as follows: Sect. 2 describes detailed the experimental data preparation. Section 3, we describe the proposed methodology used in this paper for sleep staging evaluation. Section 4 discusses the obtained experimental results from the proposed methodology from two subgroups subject's sleep recordings. Section 5, we briefly discuss our proposed methodology results, advantages, and limitations and make a result analysis with the state-of-the-art methods. Section 6 ends with concluding remarks with future work descriptions.

2   Experimental Data

In this study, the authors use healthy controlled category of the subjects with different medical conditions. The required sleep data retrieved from the ISRUC-Sleep database, which was prepared by the set of sleep experts in the sleep laboratory at the Hospital of Coimbra University [25]. The three different sections data contained in this database. The first subsection includes 100 subjects, with one recording session per subject. The second subsection consists of 8 subjects with two recording sessions performed per subject. Finally, the third subsection includes information from 10 healthy subjects with one recording session per subject. In this study, the authors used ISRUC-Sleep subgroup-III sleep recordings and the sleep behavior monitored through the C3-A2 channel for computing sleep stage classification. Table 1 provides detailed information on the sleep records and respective subjects in this study.

Table 1. Distribution of the sleep epochs

Subject number/ subgroups	W	N1	N2	N3	R
Subject-1 (SG-III)	149	91	267	158	85
Subject-2 (SG-III)	89	120	274	149	118
Subject-5 (SG-III)	67	65	287	251	80
Subject-6 (SG-III)	54	111	261	247	77

3   Methodology

This study presents a novel ensemble-based automated sleep staging system using single-channel EEG signals. The entire proposed methodology was executed with help of four phases such as preprocessing the raw signal, extracting the features, screening the features, and classification. The raw signals have been preprocessing, the features extracted, and the best features have been selected. The classification methods have been applied considering time and frequency domain features information. Figure 1 represents the framework architecture of the proposed method using a single-channel of EEG signal. The authors have used the single channel of EEG signal and each length of epochs is 30 s. The proposed automatic sleep stage detection method study follows the AASM standards with concern to annotations of sleep stages. The raw channels are filtered using a Butterworth bandpass filter to remove the irrelevant signal compositions, eye, and muscle artifacts. The C3-A2 channel was selected as input. The features have been extracted from each subject to distinguish the sleep irregularities occurrences during sleep. The frequency and time domain features are considered since the physiological state of human sleep changes with frequency fluctuations reflected in different stages of sleep. Consequently, both domains of features have been considered, and a total of 28 features have been extracted. The proposed automatic sleep staging study follows the AASM standards with concern to annotations of sleep stage.

3.1   Features Extraction

It is one of the important steps during signal analysis, though the recorded sleep recordings are highly unstable throughout the sleep cycle. It's highly necessary to study the changes in sleep characteristics in both the time and frequency ranges. Time-domain feature analysis is one of the direct traditional methods for analyzing the signals to analyze the changes amplitude, duration, mean, and so on. It helps to discriminate the signal behaviour concerning changes in time during sleep hours. But time-domain analysis ignores the changes in frequency characteristics in the recorded EEG signals. For that reason in this work, we also extracted the frequency-domain properties from the EEG signals, which supports to helps us further discriminating the changes in characteristics in the different ranges of frequency levels. In this work, as a whole, we have obtained 28 features. A brief description of the extracted feature is described in Table 2.

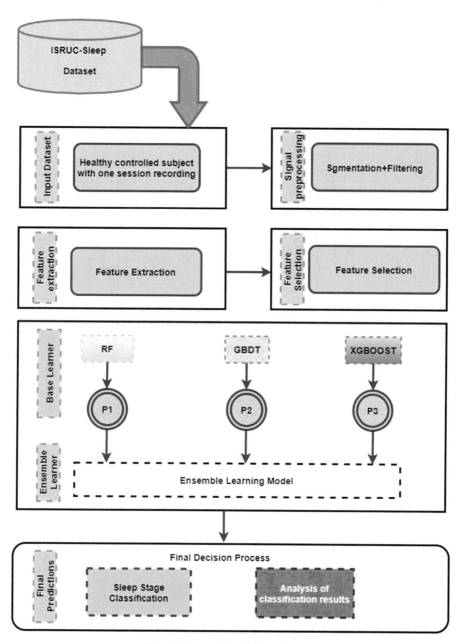

Fig. 1. Proposed model architecture

Table 2. Extracted features for this proposed study

Feature domain	Extracted feature	Feature number
Time domain	Mean	1
	Maximum	2
	Minimum	3
	Standard deviation	4
	Median	5
	Variance	6
	75 percentile	7
	Signal skewness	8
	Signal kurtosis	9
	Signal activity	10
	Signal mobility	11
	Signal complexity	12
Frequency domain	Relative spectral power δ band	13
	Relative spectral power θ band	14
	Relative spectral power α band	15
	Relative spectral power β band	16
	Power ratio δ/β	17
	Power ratio δ/θ	18
	Power ratio θ/α	19
	Power ratio θ/β	20
	Power ratio α/β	21
	Power ratio α/δ	22
	Power ratio $(\theta + \alpha)/(\alpha + \beta)$	23
	Band power in δ	24
	Band power in θ	25
	Band power in α	26
	Band power in β	27
	Zero-crossing rate	28

3.2 Feature Selection

Next, to extract the features, there should be another challenging task is to identify the most suitable features, which helps during the classification of the sleep stages. It has been seen earlier that, sometimes the performance of the classification model affects due to improper selection of the feature set. Though our input dataset is highly in randomness concerning changes in time and frequency domain. So it's essential to screen the features before fed into the classification model. In this work, we considered the feature screening algorithms as online streaming feature selection (OSFS) for identifying the most suitable features from the extracted feature vector. Mainly this algorithm identifies an optimal subset through in two ways that is relevance and redundancy analysis. During relevance analysis, majorly discriminate in between the strong and weak relevant features. The selected features are included into a separate list called as Best Candidate Feature (BCF). While a new feature enter into BCF, that time redundancy analysis phase work starts and check whether the new features relevance with respect to its class label, if it is suitable then included into BCF and if it has found irrelevant then it rejected [28].

3.3 Classification

In this proposed research work,we obtained three base layer classifiers as random forest (RF) [29], Gradient Boosting Decision Tree (GBDT) [30], and Extreme Gradient Boosting (XGBoost) [31]. In the meta-classification layer, we used logistic regression classifier. The proposed classification model is an ensemble-based learning technique; the main principle of this technique is to obtain all the decisions made by the base classification models and those decisions again made as input to a meta-classifier and finally, the meta-classifier predicts the final decision with regards to multiple classifications [32].

4 Experimental Results

In this research work, we have only considered single-channel of EEG signal as C3-A2 for extracting the sleep behaviour of the subjects, according to recent sleep staging, it has found that C3-A2 channel is most useful to analyze the brain behaviour for sleep studies in terms classification accuracy performances because it provides central part of brain information with related to research the brain behaviour during sleep [39–53]. The full information received from the central part of the brain so that here we have considered C3-A2 channel for our proposed work. After pre-processing the signals using Butterworth bandpass filter. Further, we have obtained feature extraction step, where we extracted both time and frequency domain features for characterizing the sleep behaviour in terms of frequency and time-oriented properties. The dimension size of the feature vectors represented in the form of no of epochs × total number of sample points. Each feature vector size is 750 × 6000. The detailed description of extracted features list is presented in Table 4. The extracted feature vectors are forwarded into OSFS feature selection techniques for further processing to select the suitable features for the classification models. In this research work; we have conducted three individual experiments for sleep staging. The proposed ensemble learning stacking model used for classifying the sleep stages. In the proposed sleep study, we have conducted two-five sleep states (SSC-2 TO SSC-5) classification problems. Finally, we have computed the performance of the proposed methodology, we have considered the performance metrics such as accuracy [33], sensitivity [34], specificity [35], precision [36], F1Score [37], and Cohen's kappa score [38].

4.1 Feature Selection Results

The extracted features were screening through OSFS. The selected features are presented in the Table 3 for SG-III datasets.

Table 3. Selected features results with SG-III dataset

Algorithm	Selected features
Online stream feature selection (OSFS)	1,3,4,5,8,9,13,14,15,16,17,18,22,2,10,6,12,7,11,19,20,25,23,24,21,28,27,26

4.2 Performance of Sleep Staging Using (SG-III) Dataset

In this experiment, we have obtained completely four healthy controlled subjects sleep recordings, who were not affected any types of sleep problems during sleep hours. The same methodology was implemented for sleep staging with this dataset. The same properties were extracted for discriminating the sleep classes. The selected properties using OSFS are mentioned in Table 3. The performance achieved from base learning classifiers using SG-III dataset is presented in Tables 4, 6, and 8. The reported confusion matrix with respect to RF, GBDT, and XGBoost are presented in Tables 5, 7, and 9.

Table 4. Accuracy results for RF classifiers with SG-III dataset

Feature selection algorithm	CT-2	CT-3	CT-4	CT-5
OSFS	98.97%	97.89%	95.41%	96.56%

4.3 Performance of Sleep Staging Using Proposed Ensemble Stacking Algorithm Using Subgroup-III Datasets

Finally we have deployed our proposed ensemble techniques for sleep scoring, using integration of base layers classification model. In this study we have used three classification techniques such as RF, GBDT, and XGBoost for taking the first predictions on sleep scoring. This computed first layer predictions output considered as to the second layer, called as meta classification layer. We obtained logistic regression classification

Table 5. Confusion matrix results with SG-III dataset using OSFS selected features with RF classifier

True/predicated	W (0)	N1 (1)	N2 (2)	N3 (3)	R (5)
W (0)	601	9	5	4	0
N1 (1)	4	350	24	11	0
N2 (2)	3	3	1030	7	3
N3 (3)	3	3	8	580	1
R (5)	1	2	8	4	336

Table 6. Summary of the accuracy results for GBDT classifiers with SG-III dataset

Feature selection algorithm	CT-2	CT-3	CT-4	CT-5
OSFS	98.17%	97.49%	96.41%	96.3%

Table 7. Confusion matrix results with SG-III dataset using OSFS selected features with GBDT classifier

True/predicated	W (0)	N1 (1)	N2 (2)	N3 (3)	R (5)
W (0)	603	7	7	1	1
N1 (1)	16	316	14	17	22
N2 (2)	3	3	1030	7	3
N3 (3)	3	3	8	572	9
R (5)	37	2	8	4	300

Table 8. Summary of the accuracy results for XGBoost classifiers with SG-III dataset

Feature selection algorithm	CT-2	CT-3	CT-4	CT-5
OSFS	98.77%	97.69%	96.61%	93.96%

Table 9. Confusion matrix results with SG-III dataset using OSFS selected features with XGBoost classifier

True/predicated	W (0)	N1 (1)	N2 (2)	N3 (3)	R (5)
W (0)	601	13	5	8	8
N1 (1)	27	295	16	17	22
N2 (2)	14	20	980	12	13
N3 (3)	3	8	13	565	6
R (5)	1	2	8	10	330

algorithm in the second layer, which integrates the previous layers predictions and generate the final decisions on sleep staging. In this proposed study, the stacking model consists two layers of classifiers. We integrated RF, GBDT and XGBoost for designing stacking models. Hence we considered first three classifiers RF, GBDT and XGBoost are base learning classifiers and the second layer of this proposed stacking model contains logistic classification model.

In this model, we have applied the same feature selection techniques combinations, which we have finally obtained by performing the feature selection analysis. The classification results of this proposed ensemble learning stacking model using SG-III datasets for five sleep states classification problem are presented in Table 10 and the result of the confusion matrix is presented in Table 11.

Table 10. Summary of the accuracy results for stacking model classifier with SG-III dataset

Feature selection algorithm	CT-2	CT-3	CT-4	CT-5
OSFS	99.17%	98.79%	97.61%	95.26%

Table 11. Confusion matrix result with SG-III dataset using OSFS selected features with ensemble stacking model

True/predicated	W (0)	N1 (1)	N2 (2)	N3 (3)	R (5)
W (0)	593	7	11	3	5
N1 (1)	4	371	3	6	5
N2 (2)	10	8	1000	11	17
N3 (3)	3	5	3	563	8
R (5)	11	7	9	6	331

4.4 Summary of Results

In this paper, we were used SG-III dataset of ISRUC-Sleep for multi-class sleep staging. In all the experimental part conducted for this study using single-channel EEG signal. The first three experiment conducted based on base layers learning classification model and the final experiment executed with the proposed stacking model using three feature selection techniques such as FS, RF and OSFS. Table 12 presents the summary results that were obtained through different classification models of base layers and ensemble learning of stacking model with the health categories of the subjects sleep recordings using three feature selection techniques.

The highest sleep staging was obtained using ensemble learning stacking model from all the three subgroups dataset. The highest classification accuracy was reported as 96.86% for five-sleep states classification problem with SG-III dataset.

The authors have compared the proposed system performance with other studies available in state of the art. Therefore, the authors have selected studies with similar

Table 12. Summary results of classification accuracy for various classification models and subgroups datasets

Data set	Classification model	OSFS
SG-III	RF	96.56%
	GBDT	96.3%
	XGBoost	93.96%
	Ensemble stacking model	96.86%

datasets according to our proposed study and based on a single channel. In Table 13, the features used in the proposed research work are compared to others used by related works using single-channel EEG signals of ISRUC-Sleep dataset. The comparisons with other similar research proposals available in the literature must take into consideration the use of single-channel EEG, different features and classification models is presented in Table 14.

Table 13. Comparisons of CAs (in %) of our proposed model with other state-art-of the techniques used same features and datasets

Author	Classifier	Classification accuracy
Khalighi et al. [58] 2011	SVM	95%
Hugo Simoes et al. [59] 2010	Bayesian classifier	83%
Khalighi et al. [60] 2016	SVM	93.97%
Sousa et al. [61] 2015	SVM	86.75%
Khalighi et al. [62] 2013	SVM	81.74%
KD Tzimourta et al. [63] 2018	Random forest	75.29%
Najdi et al. [64] 2017	Stacked sparse auto-encoders	82.3%
Kalbkhani, H. et al. [65] 2018	SVM	83.33%
Proposed	**Ensemble model (SG-III)**	**96.86%**

Tables 13 and 14 presents a comparisons of the overall accuracies presented by the proposed methods and the available studies in the literature. The results reported, but the authors using the ensemble learning stacking model for SG-III present the highest results when compared with the other methods available in the literature.

The presented study shows that it is possible to achieve high classification accuracies through three different sleep data subgroups and considering different patients with distinct medical conditions. The authors validate the effectiveness and usefulness of the proposed method by presenting a detailed comparative analysis with similar research proposals available in the literature. The authors have analyzed the classification results of our proposed work with other 8 similar studies that include the same dataset.

Table 14. Comparisons of sleep staging accuracy (%) between the proposed methodology and the recent methods

Author	Classifier	Classification accuracy
Huang, W. et al. [66] 2019	SVM	92.04%
Dhok, S. et al. [67] 2020	GSVM	87.45%
Wang, Q. et al. [68] 2019	Stacking model	96.6
Sharma, M. et al. [69] 2019	SVM	91.5%
Hassan, A.R. et al. [70] 2016	Bagging	90.69%
Da Silveira et al. [20] 2016	RF	91.5%
Memar, P. et al. [19] 2018	RF	95.31%
Santaji, S. et al. [71] 2020	RF	97.8%
Proposed	**Ensemble model (SG-III)**	**96.86%**

5 Conclusion

This paper has presents an ensembling learning stacking classification model according AASM sleep standards for multiple sleep stage classification using a single channel of EEG signals. The proposed methodology has been analyzed 6000 epochs composed from three different medical conditioned subject's dataset.

There are basically three main important contributions in this research study. The first one is obtained Butterworth bandpass filter techniques for eliminating the various artefacts exists in the input signals and followed by we have extracted numerous features from time domain, frequency domain and non-linear features. These set of features supports to analysis sleep EEG parameters and its characteristics. It has observed that multi feature extraction improves the sleep staging accuracy.

Secondly, the proposed research work obtained feature screening techniques, which directly useful for identifying the most relevant features from extracted feature vectors. Thirdly, this proposed research work establishes an ensemble learning model, which integrates multiple classification models which implements in two layers. In the first base layers, we obtain RF, GBDT, and XGBoosting classifiers, and the second layer contained logistic regression techniques. The proposed stacking model reported high recognition rates for multiple sleep staging classification tasks.

References

1. Panossian, L.A., Avidan, A.Y.: Review of sleep disorders. Med. Clin. N. Am. **93**, 407–425 (2009). https://doi.org/10.1016/j.mcna.2008.09.001
2. Smaldone, A., Honig, J.C., Byrne, M.W.: Sleepless in America: inadequate sleep and relationships to health and well-being of our nation's children. Pediatrics **119**, 29–37 (2007)
3. Hassan, A.R., Hassan Bhuiyan, M.I.: Automatic sleep scoring using statistical features in the EMD domain and ensemble methods. Biocybern. Biomed. Eng. **36**(1), 248–255 (2016). https://doi.org/10.1016/j.bbe.2015.11.001

4. Aboalayon, K., Ocbagabir, H.T., Faezipour, M.: Efficient sleep stage classification based on EEG signals. In: Systems, Applications and Technology Conference (LISAT), pp. 1–6 (2014)
5. Iber, C., Ancoli-Israel, S., Chesson, A.L., Quan, S.F.: The AASM Manual for the Scoring of Sleep and Associated Events: Rules, Terminology and Technical Specification. American Academy of Sleep Medicine, Westchester (2007)
6. Fiorillo, L., et al.: Automated sleep scoring: a review of the latest approaches. Sleep Med. Rev. **48**, 101204 (2019)
7. Acharya, U.R., et al.: A deep convolutional neural network model to classify heartbeats. Comput. Biol. Med. **89**, 389–396 (2017)
8. Cheng, J.-Z., et al.: Computer-aided diagnosis with deep learning architecture: applications to breast lesions in US images and pulmonary nodules in CT scans. Sci. Rep. **6**, 24454 (2016)
9. Talo, M., Baloglu, U.B., Yıldırım, Ö., Rajendra Acharya, U.: Application of deep transfer learning for automated brain abnormality classification using MR images. Cogn. Syst. Res. **54**, 176–188 (2019)
10. Acharya, U.R., Oh, S.L., Hagiwara, Y., Tan, J.H., Adeli, H.: Deep convolutional neural network for the automated detection and diagnosis of seizure using EEG signals. Comput. Biol. Med. **100**, 270–278 (2018)
11. Macaš, M., Grimová, N., Gerla, Václav., Lhotská, L.: Semi-automated sleep EEG scoring with active learning and HMM-based deletion of ambiguous instances. Proceedings **31**(1), 46 (2019)
12. Obayya, M., Abou-Chadi, F.: Automatic classification of sleep stages using EEG records based on Fuzzy c-means (FCM) algorithm. In: Radio Science Conference (NRSC), pp. 265–272 (2014)
13. Güneş, K.P., Yosunkaya, Ş: Efficient sleep stage recognition system based on EEG signal using k-means clustering based feature weighting. Exp. Syst. Appl. **37**, 7922–7928 (2010)
14. Aboalayon, K., Ocbagabir, H.T., Faezipour, M.: Efficient sleep stage classification based on EEG signals. In: Systems, Applications and Technology Conference (LISAT), pp. 1–6 (2014)
15. Hassan, A.R., Subasi, A.: A decision support system for automated identification of sleep stages from single-channel EEG signals. Knowl.-Based Syst. **128**, 115–124 (2017)
16. Diykh, M., Li, Y., Wen, P.: EEG sleep stages classification based on time do-main features and structural graph similarity. IEEE Trans. Neural Syst. Rehabili. Eng. **24**(11), 1159–1168 (2016)
17. Gunnarsdottir, K.M., Gamaldo, C.E., Salas, R.M.E., Ewen, J.B., Allen, R.P., Sarma, S.V.: A novel sleep stage scoring system: combining expert-based rules with a decision tree classifier. In: 40th Annual International Conference of the IEEE Engineering in Medicine and Biology Society (EMBC) (2018)
18. Sriraam, N., Padma Shri, T.K., Maheshwari, U.: Recognition of wake-sleep stage 1 multi-channel EEG patterns using spectral entropy features for drowsiness detection. Austr. Phys. Eng. Sci. Med. **39**(3), 797–806 (2018)
19. Memar, P., Faradji, F.: A novel multi-class EEG-based sleep stage classification system. IEEE Trans. Neural Syst. Rehabil. Eng. **26**(1), 84–95 (2018)
20. da Silveira, T.L.T., Kozakevicius, A.J., Rodrigues, C.R.: Single-channel EEG sleep stage classification based on a streamlined set of statistical features in wavelet domain. Med. Biol. Eng. Comput. **55**(2), 343–352 (2016)
21. Wutzl, B., Leibnitz, K., Rattay, F., Kronbichler, M., Murata, M., Golaszewski, S.M.: Genetic algorithms for feature selection when classifying severe chronic disorders of consciousness. PLOS ONE **14**(7), e0219683 (2019)
22. Zhu, G., Li, Y., Wen, P.: Analysis and classification of sleep stages based on difference visibility graphs from a single-channel EEG signal. IEEE J. Biomed. Health Inf. **18**(6), 1813–1821 (2014)

23. Braun, E.T., De Jesus, A., Kozakevicius, T.L., Da Silveira, T., Rodrigues, C.R., Baratto, G.: Sleep stages classification using spectral based statistical moments as features. Rev. Inf. Teór. Appl. **25**(1), 11 (2018)
24. Scholkopf, B., Smola, A.: Learning with Kernels. MIT Press, Cambridge (2002)
25. Khalighi, S., Sousa, T., Santos, J.M., Nunes, U.: ISRUC-sleep: a comprehensive public dataset for sleep researchers. Comput. Methods Prog. Biomed. **124**, 180–192 (2016)
26. Robnik-Šikonja, M., Kononenko, I.: Theoretical and empirical analysis of ReliefF and RReliefF. Mach. Learn. **53**, 23–69 (2003). https://doi.org/10.1023/A:1025667309714
27. Jin, X., Bo, T., He, H., Hong, M.: Semisupervised feature selection based on relevance and redundancy criteria. IEEE Trans. Neural Netw. Learn. Syst. **28**, 1974–1984 (2016)
28. Eskandari, S., Javidi, M.M.: Online streaming feature selection using rough sets. Int. J. Approx. Reason. **69**, 35–57 (2016)
29. Shabani, F., Kumar, L., Solhjouy-fard, S.: Variances in the projections, resulting from CLIMEX, boosted regression trees and random forests techniques. Theor. Appl. Climatol. **129**(3–4), 801–814 (2016)
30. Xie, J., Coggeshall, S.: Prediction of transfers to tertiary care and hospital mortality: a gradient boosting decision tree approach. Stat. Anal. Data Min. **3**(4), 253–258 (2010)
31. Chen, T., He, T., Benesty, M., Khotilovich, V., Tang, Y.: xgboost: extreme gradient boosting (2016)
32. Tang, B., Chen, Q., Wang, X., Wang, X.: Reranking for stacking ensemble learning. In: Wong, K.W., Sumudu, B., Mendis, U., Bouzerdoum, A. (eds.) ICONIP 2010. LNCS, vol. 6443, pp. 575–584. Springer, Heidelberg (2010). https://doi.org/10.1007/978-3-642-17537-4_70s
33. Sanders, T.H., McCurry, M., Clements, M.A.: Sleep stage classification with cross frequency coupling. In: 36th Annual International Conference of the IEEE Engineering in Medicine and Biology Society (EMBC), pp. 4579–4582 (2014)
34. Bajaj, V., Pachori, R.: Automatic classification of sleep stages based on the time-frequency image of EEG signals. Comput. Methods Prog. Biomed. **112**(3), 320–328 (2013)
35. Hsu, Y.-L., Yang, Y.-T., Wang, J.-S., Hsu, C.-Y.: Automatic sleep stage recurrent neural classifier using energy features of EEG signals. Neurocomputing **104**, 105–114 (2013)
36. Zibrandtsen, I., Kidmose, P., Otto, M., Ibsen, J., Kjaer, T.W.: Case comparison of sleep features from ear-EEG and scalp-EEG. Sleep Sci. **9**(2), 69–72 (2016)
37. Berry, R.B., et al.: The AASM Manual for the Scoring of Sleep and Associated Events: Rules, Terminology and Technical Specifications. American Academy of Sleep Medicine (2014)
38. Sim, J., Wright, C.C.: The kappa statistic in reliability studies: use, interpretation, and sample size requirements. Phys. Therap. **85**(3), 257–268 (2005)
39. Liang, S.-F., Kuo, C.-E., Yu-Han, H., Pan, Y.-H., Wang, Y.-H.: Automatic stage scoring of single-channel sleep EEG by using multiscale entropy and autoregressive models. IEEE Trans. Instrument. Measur. **61**(6), 1649–1657 (2012)
40. Kim, J.: A comparative study on classification methods of sleep stages by using EEG. J. Korea Multimed. Soc. **17**(2), 113–123 (2014)
41. Peker, M.: A new approach for automatic sleep scoring: combining Taguchi based complex-valued neural network and complex wavelet transform. Comput. Methods Prog. Biomed. **129**, 203–216 (2016)
42. Subasi, A., Kiymik, M.K., Akin, M., Erogul, O.: Automatic recognition of vigilance state by using a wavelet-based artificial neural network. Neural Comput. Appl. **14**(1), 45–55 (2005)
43. Tagluk, M.E., Sezgin, N., Akin, M.: Estimation of sleep stages by an artificial neural network employing EEG, EMG and EOG. J. Med. Syst. **34**(4), 717–725 (2010)
44. Hassan, A.R., Bhuiyan, M.I.H.: An automated method for sleep staging from EEG signals using normal inverse Gaussian parameters and adaptive boosting. Neurocomputing **219**, 76–87 (2017)

45. Hassan, A.R., Bhuiyan, M.I.H.: Automated identification of sleep states from EEG signals by means of ensemble empirical mode decomposition and random under sampling boosting. Comput. Methods Prog. Biomed. **140**, 201–210 (2017)
46. Diykh, M., Li, Y.: Complex networks approach for EEG signal sleep stages classification. Exp. Syst. Appl. **63**, 241–248 (2016)
47. Diykh, M., Li, Y., Wen, P.: EEG sleep stages classification based on time domain features and structural graph simi-larity. IEEE Trans. Neural Syst. Rehabil. Eng. **24**(11), 1159–1168 (2016)
48. Mohammadi, S.M., Kouchaki, S., Ghavami, M., Sanei, S.: Improving time–frequency domain sleep EEG classification via singular spectrum analysis. J. Neurosci. Methods **273**, 96–106 (2016)
49. Şen, B., Peker, M., Çavuşoğlu, A., Çelebi, F.V.: A Comparative study on classification of sleep stage based on EEG signals using feature selection and classification algorithms. J. Med. Syst. **38**(3), 1–21 (2014)
50. Burioka, N., et al.: Approximate entropy in the electroencephalogram during wake and sleep. Clin. EEG Neurosci. **36**(1), 21–24 (2005)
51. Obayya, M., Abou-Chadi, F. E. Z.: Automatic classification of sleep stages using EEG records based on Fuzzy c-means (FCM) algorithm. In: 2014 31st National Radio Science Conference (NRSC), pp. 265–272 (2014)
52. Fraiwan, L., Lweesy, K., Khasawneh, N., Fraiwan, M., Wenz, H., Dickhaus, H.: Classification of sleep stages using multi-wavelet time frequency entropy and LDA. Methods Inf. Med. **49**(03), 230–237 (2018)
53. Herrera, L.J., et al.: Combination of heterogeneous EEG feature extraction methods and stacked sequential learning for sleep stage classification. Int. J. Neural Syst. **23**(3), 1350012 (2013)
54. Radha, M., Garcia-Molina, G., Poel, M., Tononi, G.: Comparison of feature and classifier algorithms for online automatic sleep staging based on a single EEG signal. In: 36th Annual International Conference of the IEEE Engineering in Medicine and Biology Society, pp. 1876–1880 (2014)
55. Jo, H.G., Park, J.Y., Lee, C.K., An, S.K., Yoo, S.K.: Genetic fuzzy classifier for sleep stage identification. Comput. Biol. Med. **40**(7), 629–634 (2010)
56. Herrera, L.J., Mora, A.M., Fernandes, C.M.: Symbolic representation of the EEG for sleep stage classification. In: 11th International Conference on Intelligent Systems Design and Applications, pp. 253–258 (2011)
57. Vanbelle, S.A.: New interpretation of the weighted kappa coefficients. Psychometrika **81**, 399–410 (2016)
58. Khalighi, S., Sousa, T., Oliveira, D., Pires, G., Nunes, U.: Efficient feature selection for sleep staging based on maximal overlap discrete wavelet transform and SVM. In: Annual International Conference of the IEEE Engineering in Medicine and Biology Society (2011)
59. Simões, H., Pires, G., Nunes, U., Silva, V.: Feature extraction and selection for automatic sleep staging using EEG. In: Proceedings of the 7th International Conference on Informatics in Control, Automation and Robotics, vol. 3, pp. 128–133 (2010)
60. Khalighi, S., Sousa, T., Santos, J.M., Nunes, U.: ISRUC-Sleep: a comprehensive public dataset for sleep researchers. Comput. Methods Prog. Biomed. **124**, 180–192 (2016)
61. Sousa, T., Cruz, A., Khalighi, S., Pires, G., Nunes, U.: A two-step automatic sleep stage classification method with dubious range detection. Comput. Biol. Med. **59**, 42–53 (2015)
62. Khalighi, S., Sousa, T., Pires, G., Nunes, U.: Automatic sleep staging: a computer assisted approach for optimal combina-tion of features and polysomnographic channels. Exp. Syst. Appl. **40**(17), 7046–7059 (2013)
63. Tzimourta, K.D., Tsilimbaris, A.K., Tzioukalia, A.T., Tzallas, M.G., Tsipouras, L.G.: EEG-based automatic sleep stage classification. Biomed. J. Sci. Tech. Res. **7**(4) (2018)

64. Najdi, S., Gharbali, A.A., Fonseca, J.M.: Feature transformation based on stacked sparse autoencoders for sleep stage classification. In: Camarinha-Matos, L.M., Parreira-Rocha, M., Ramezani, J. (eds.) DoCEIS 2017. IAICT, vol. 499, pp. 191–200. Springer, Cham (2017). https://doi.org/10.1007/978-3-319-56077-9_18

65. Kalbkhani, H., Ghasemzadeh, P., Shayesteh, M.G.: Sleep stages classification from EEG signal based on stockwell transform. IET Signal Process. **13**(2), 242–252 (2019)

66. Huang, W., et al.: Sleep staging algorithm based on multichannel data adding and multifeature screening. Comput. Methods Prog. Biomed. **187**, 105253 (2019). https://doi.org/10.1016/j.cmpb.2019.105253

67. Dhok, S., Pimpalkhute, V., Chandurkar, A., Bhurane, A.A., Sharma, M., Acharya, U.R.: Automated phase classification in cyclic alternating patterns in sleep stages using Wigner-Ville Distribution based features. Comput Biol Med. **119**, 103691 (2020). https://doi.org/10.1016/j.compbiomed.2020.103691

68. Wang, Q., Zhao, D., Wang, Y., Hou, X.: Ensemble learning algorithm based on multi-parameters for sleep staging. Med. Biol. Eng. Comput. **57**(8), 1693–1707 (2019). https://doi.org/10.1007/s11517-019-01978-z

69. Sharma, M., Patel, S., Choudhary, S., Acharya, U.R.: Automated detection of sleep stages using energy-localized orthogonal wavelet filter banks. Arab. J. Sci. Eng. **45**(4), 2531–2544 (2019). https://doi.org/10.1007/s13369-019-04197-8

70. Hassan, A.R., Bhuiyan, M.I.H.: Computer-aided sleep staging using complete ensemble empirical mode decomposition with adaptive noise and bootstrap aggregating. Biomed. Signal Process. Control **24**, 1–10 (2016). https://doi.org/10.1016/j.bspc.2015.09.002

71. Santaji, S., Desai, V.: Analysis of EEG signal to classify sleep stages using machine learning. Sleep Vigilance **4**(2), 145–152 (2020). https://doi.org/10.1007/s41782-020-00101-9

Histopathological Image Classification Using Ensemble Transfer Learning

Binet Rose Devassy[1,2(✉)] and Jobin K. Antony[3]

[1] APJ Abdul Kalam Technological University, Thiruvananthapuram, Kerala, India
[2] Department of Electronics and Communication Engineering, Sahrdaya College of Engineering and Technology, Thrissur, Kodakara, Kerala, India
[3] Department of Electronics and Communications, Rajagiri School of Engineering and Technology, Rajagirivalley P.O., Kochi 682039, Kerala, India

Abstract. Histopathological image analysis of biopsy sample provides an accurate diagnosis method for cancer. Usually Pathologists examine the microscopic images of biopsy sample manually for the detection and grading of cancer. Automation in this field helps the pathologists to take a second opinion before confirming with their findings. We propose an effective method of automated cancer detection by combining the effect of transfer learning and ensemble learning. Six pre-trained models such as Densenet121, Resnet 50, Xception, EfficientNet B7, MobileNetV2, and VGG19 are used for preparing an ensemble model. A dataset contains 5547 H&E stained histopathological images of malignant and benign tissues are used to train and validate each models individually and obtained an accuracy of 77.9%, 79%, 79.8%, 78.3%, 77%, and 76% respectively. Based on the accuracy, best performing three models Resnet50, Xception, and EfficientnetB7 are selected to form an ensemble model. Then the layers and weights of these models are freezed and the output layers are concatenated to make an ensemble model. New dense layers are added to the ensembled model to provide a single output for binary classification. The model is compiled by an Adam optimizer with a learning rate of 0.001. The images are again applied to this ensemble model to classify the malignant and benign tissues and obtained an accuracy of 96% and precision of 96%.

Keywords: Deep learning · Transfer learning · Ensemble learning · Histopathological images

1 Introduction

Nowadays, cancer is one of the major threatening diseases to the public health and is the second leading cause of death worldwide. Since cancer patients are having very less immunity power, they are one of the most critical communities in the current scenario of pandemic COVID-19. As per the key statistics of WHO, it is reported that the number of new cancer cases is on the rise globally. Deaths from cancer worldwide are projected to reach over 13 million in *2030* [1]. Between 2008 and 2030 the number of new cancer cases is expected to increase more than 80% in low-income countries [2], which is double the

© The Author(s), under exclusive license to Springer Nature Switzerland AG 2022
R. Misra et al. (Eds.): ICMLBDA 2021, LNNS 256, pp. 203–212, 2022.
https://doi.org/10.1007/978-3-030-82469-3_18

rate expected in high-income countries (40%). Early detection and diagnosis of Cancer helps to provide the right treatment for these patients. Biopsy test is the final word of cancer diagnosis. Even though other tests such as blood test, Ultrasound, MRI and CT scan are used for cancer detection, biopsy test can give an accurate decision for cancer diagnosis. During a biopsy test, the pathologists remove a small amount of tissue from the affected part of the body. Then it is fixed and dehydrated. These samples are stained by Haematoxyllin and Eosin dye. It stains nuclei into blue colour and cytoplasm into pink colour. Histopathology refers to microscopic examination of these biopsy samples. Histopathological images are the digital images of these samples scanned by whole slide digital scanner. These digitized images can be used for automated detection of cancer based on deep learning methods.

Deep learning methods [3] are used to extract features and information from the input data and learn abstract representations of data by itself. They can solve the problems of traditional feature extraction and have been successfully applied in bio-image classification problems such as cancer detection. Only low-level features of images are extracted by traditional feature extraction methods, but in Deep learning methods high-level abstract features can be extracted automatically [4]. To speed up the training and to improve the performance of deep learning model, transfer learning can be introduced [5]. Transfer learning is a machine learning technique where a trained model on one task can be reassigned on another related task. This optimization technique helps to speed up and improve performance while training on the second task. Transfer learning allows us to start with the learned features on the ImageNet dataset and adjust the weights of hidden layers to suit the new dataset instead of starting the learning process on the data from scratch with random weight initialization.

Ensemble learning [6] is a machine learning method that combines multiple base models in order to produce more accurate predictive model. In ensemble modeling, a set of models are concatenated parallelly and whose outputs are combined with a decision fusion strategy. In this paper we introduce an efficient method of ensemble learning by combining three pre-trained models and averaging their result to come to a better predictive result on cancer detection.

2   Methodology

2.1   Transfer Learning Method

Transfer learning is the method of applying a pre-trained deep learning model and reusing the same model for a new but similar task. In this approach, the pre-trained model which is trained on a large annotated image database (such as ImageNet) are used to start a training process. The weights of layers are getting fine-tuned with respect to new problem. This accelerates the training process and improves the performance of pre-trained deep learning model while applying on a new problem. In this paper, transfer learning method is used to reuse the layers and weights of a pre trained model.

In the proposed method, five pre trained deep learning models namely MobileNet V2, Xception, Resnet 50, DenseNet 121, EfficientNet B7, VGG-16.have been used for training. The parameters of all these models are pre-trained on the ImageNet dataset.The internal layers and weights of these networks are freezed and additional dense layers are added according to the classification required for our model. The hyper parameters for all the above models are assigned same as shown in the Table 1.

MobileNets [7] use depth wise separable convolutions which include depth wise convolution followed by point wise convolutions. It introduces two new global hyper parameter width multiplier and resolution multiplier which helps to reduce latency and increase accuracy. MobileNet has 28 layers and 4.2 million parameters.

Xception [8] use the modified depth wise separable convolution which include the point wise convolution followed by a depth wise convolution. It means 1×1 convolution is done first before any $n \times n$ spatial convolutions. The Xception architecture has 36 convolutional layers and 23 million parameters.

The ResNet-50 model [9] consists of 5 stages each with a convolution and Identity block. Convolution block and identity block contain convolution layers but identity block will not change the dimension of the input. The concept of skip connection is included in Resnet. This helps to reduce vanishing gradient problem and higher layers can perform as good as lower layers. The ResNet-50 has over 23 million trainable parameters.

VGG-16 is a network with 16 layers and uses 3×3 filters with stride of 1 in convolution layer and uses SAME padding in pooling layers 2×2 with stride of 2. This network is a pretty large network and it has about 138 million (approx.) parameters.

Dense Convolutional Network (DenseNet) [10], which connects each layer to every other layer in a feed-forward fashion. In convolutional networks with L layers have L connections but DenseNet has L (L + 1)/2 direct connections. For each layer, the feature-maps of all preceding layers are used as inputs, and its own feature-maps are used as inputs into all subsequent layers. DenseNets have several good features such as reduce the vanishing-gradient problem, strengthen feature propagation, encourage feature reuse, and substantially reduce the number of parameters.

EfficientNet [11] uses a new scaling method that uniformly scales all dimensions of depth, width and resolution of the network. They used the neural architecture search to design a new baseline network and scaled it up to obtain a family of deep learning models, called EfficientNets, which achieve much better accuracy and efficiency as compared to the previous Convolutional Neural Networks.

2.2 Ensemble Learning Method

There are different methods for combining the result of individual model to obtain the predictive result of ensemble model. The different methods are bagging, boosting and stacking [12, 13]. Bagging combines the independent models parallelly and combines their result based on some kind of deterministic averaging process. In boosting method, the models are fitted iteratively, learn them sequentially and combine them following a deterministic strategy. Each model in the sequence is giving more effort on the observations in the dataset which was badly handled by the previous models.Stacking fits the models in parallel and combines them by training a meta-model to output a prediction based on the different weak models predictions. Split the dataset to train the base models and combine the result of each model for training a meta-model. To build a stacking model, the two specifics are need: the base models that need to fit and the meta-model that combine them.

We have a model $f_t(y/x)$ an estimate of the probability of class y given input x. For a set of these, $\{t = 1 \ldots .T\}$ the ensemble probability estimate is,

$$f(y/x) = \sum_{l=1}^{L} w_t \, f_t(y/x) : \qquad (1)$$

If the weights $w_t = \frac{1}{T}, \bigvee t$.

The uniform averaging of the probability estimates is applied in this paper.

3 Evaluation of Proposed Ensemble Transfer Learning Model

3.1 Proposed Ensemble Transfer Learning Method

Deep feature extraction and classification of histopathological images is done using the ensemble transfer learning model [14]. The ensemble model is adopted on the concept that generalization ability of an ensemble model is usually higher than that of a single learner [15]. Fine-tuning of pre-trained CNN models carried out independently [16]. Finally, their individual classifier outputs are averaged to provide final result (Fig. 1)

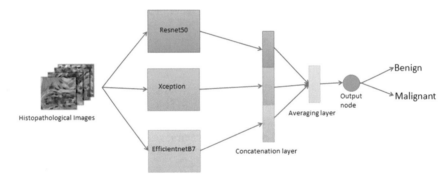

Fig. 1. Proposed Architecture of Ensemble transfer learning model

3.2 Evaluation of Proposed Ensemble Transfer Learning Method

To test the effectiveness of the proposed Ensemble transfer learning method, Histopathological image dataset of lymphoma is applied from kaggle database. There are 5547 images from malignant and benign samples. In this, 2759 images without cancer and 2788 images with cancer with 50.26% positive images. Among them, some of the images are plotted below (Fig. 2)

The images with label 0 indicate benign and label 1 indicate malignant samples.

All these images are reshaped to a size of $75 \times 75 \times 3$. The different pre-trained models are trained and tested to the image dataset using transfer learning. Initially pre-trained models such as Densenet121, Resnet 50, Xception, EfficientNet B7, MobileNetV2, and

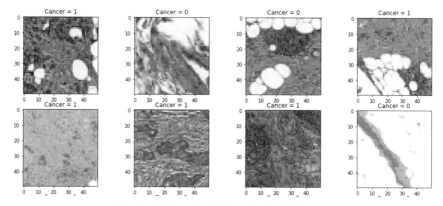

Fig. 2. H&E stained histopathological images

VGG19 are trained and tested over the image dataset. These models are pre-trained over ImageNet dataset. While applying these models into new image dataset of Lymphoma, the weights of initial layers are frozen and final layers are trained more for abstract and specific features of image dataset.

The hyper parameters are mentioned below in Table 1.

Table 1. Hyper parameters

Hyper parameters	Value
Input size	$75 \times 75 \times 3$
Batch size	32
Epoch	10
Learning rate	0.001
Optimizer	Adam

The training and validation accuracy versus number of epochs are obtained as shown in Figs. 3–8.

From these individual models, best performing 3 models such as Resnet50, Xception, and EfficientnetB7 are selected based on their validation accuracy. Layers of all these models are frozen with their weights. Outputs of all these models are concatenated and added a dense layer to merge the output. The final dense layer of ensemble model is given with single output of sigmoid activation for binary classification. The ensemble model is optimized by Adam optimizer with learning rate of 0.001. The accuracy of ensemble model is increased to 96%.

Fig. 3. Densenet 201

Fig. 4. Resnet 50

Fig. 5. Xception

Fig. 6. EfficientNetB7

For evaluating the model performance, Precision, Recall and F1 Score are calculated from the confusion matrix. Precision is the ratio of correctly predicted True positive (TP) predictions to the total positive observations (TP + FP). Recall is the ratio of predicted True positive (TP) cases to the actual positive cases (TP + FN). F1 Score is a better measure to obtain a harmonic mean between Precision and Recall.

$$Precision = \frac{TP}{TP + FP} \quad Recall = \frac{TP}{TP + FN}$$

$$F1\ Score = 2 * \frac{Precision * Recall}{Precision + Recall} \tag{2}$$

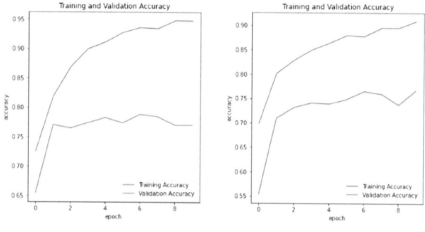

Fig. 7. MobileNetV2   **Fig. 8.** VGG19

4   Result

The proposed model which is obtained by ensembling Resnet50, Xception and Efficient-NetB7 has obtained a validation accuracy of 96%. The training and validation accuracies are plotted in Fig. 9.

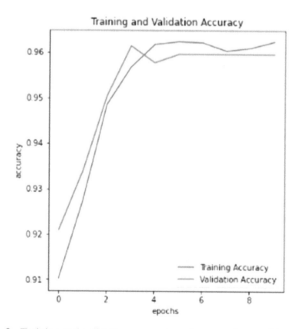

Fig. 9. Training and validation accuracies of proposed ensemble model

For evaluating the performance of the model, confusion matrix is plotted and performance measures are calculated from that. For the ensemble transfer learning model, the following performance measures are obtained as shown in Table 2.

Table 2. Performance measures

Performance measures	Result
Accuracy	96%
Precision	96%
Recall	92%
F1 Score	95%

Comparison of validation accuracies for different models with ensemble transfer learning model is plotted in Fig. 10.

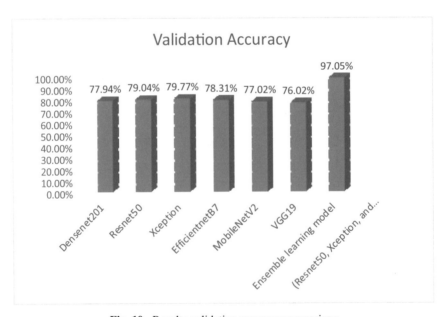

Fig. 10. Results-validation accuracy comparison

5   Conclusion

In this paper, six different CNN models are trained and tested for lymphoma dataset using transfer learning. Performance of each model is evaluated based on their classification accuracy for the equal number of epochs and iterations. From the tested six pre-trained models, three best performed CNN models of Resnet50, Xception, and EfficientnetB7 are ensembled. The ensemble model is optimized by an Adam optimizer for a learning rate of 0.001 with binary cross entropy loss function and obtained an accuracy of 96.05%. The ensemble model performs better than individual CNN models with the same batch size, epochs and learning rate. The ensemble model resulted precision with 96%, Recall with 92% and F1score with 95%. The proposed method of ensemble transfer learning model of Resnet50, Xception, and EfficientnetB7 exhibited the stronger classification capability for the cancer detection using histopathological images.

References

1. Boyle, P., Levin, B.: World Cancer report 2008: IARC Press. International Agency for Research on Cancer (2008)
2. https://www.who.int/nmh/publications/ncd_report_chapter1.pdf
3. Xie, J., Liu, R., Luttrell, J., Zhang, C.: Deep learning based analysis of histopathological images of breast cancer. Front. Genet. **10**(80), 1–19 (2019)
4. Bello, M., Nápoles, G., Sánchez, R., Bello, R., Vanhoof, R.: Deep neural network to extract high-level features and labels in multi-label classification problems, Neurocomputing, **413**, 259–270 (2020). ISSN 0925–2312
5. Hussain, M., Bird, J.J., Faria, D.R.: A study on CNN transfer learning for image classification. In: Lotfi, A., Bouchachia, H., Gegov, A., Langensiepen, C., McGinnity, M. (eds.) UKCI 2018. AISC, vol. 840, pp. 191–202. Springer, Cham (2019). https://doi.org/10.1007/978-3-319-97982-3_16
6. Nanni, L., Ghidoni, S., Brahnam, S.: Ensemble of convolutional neural networks for bioimage classification, Appl. Comput. Inf. (2020). ISSN 22108327
7. Howard, A.G., et al.: MobileNets: efficient convolutional neural networks for mobile vision applications (2017). https://arxiv.org/abs/1704.04861
8. Chollet, F.: Xception: deep learning with depthwise separable convolutions. In: 2017 IEEE Conference on Computer Vision and Pattern Recognition (CVPR), pp. 1800–1807. Honolulu, HI, USA (2017)
9. https://towardsdatascience.com/understanding-and-coding-a-resnet-in-keras-446d7ff84d33
10. Huang, G., Liu, Z., Weinberger, K.Q.: Densely connected convolutional networks. https://arxiv.org/abs/1608.06993
11. Tan, M., Le, Q.V.: EfficientNet: rethinking model scaling for convolutional neural networks. In: Proceedings of International Conference on Machine Learning (ICML) (2019)
12. Huang, F., Xie, G., Xiao, R.: Research on ensemble learning. In: Proceedings of the 2009 International Conference on Artificial Intelligence and Computational Intelligence. vol. 3. IEEE Computer Society (2009)
13. Opitz, D., Maclin, R.: Popular ensemble methods: an empirical study. J. Artif. Intell. Res. **11**, 169–198 (1999)
14. Zhu, Y., Brettin, T., Evrard, Y.A., et al.: Ensemble transfer learning for the prediction of anti-cancer drug response. Sci. Rep. **10**, 18040 (2020). https://doi.org/10.1038/s41598-020-74921-0

15. Xue, D., et al.: An application of transfer learning and ensemble learning techniques for cervical histopathology image classification. IEEE Access **8** (2020)
16. Kandaswamy, C., Silva, L.M., Alexandre, L.A., Santos, J.M.: Deep transfer learning ensemble for classification. In: Rojas, I., Joya, G., Catala, A. (eds.) IWANN 2015. LNCS, vol. 9094, pp. 335–348. Springer, Cham (2015). https://doi.org/10.1007/978-3-319-19258-1_29

A Deep Feature Concatenation Approach for Lung Nodule Classification

Amrita Naik[1], Damodar Reddy Edla[1($\boxtimes$)], and Ramesh Dharavath[2]

[1] National Institute of Technology, Goa, India
dr.reddy@nitgoa.ac.in
[2] Indian Institute of Technology, Dhanbad, India

Abstract. Lung cancer is the most common cancer around the world, with the highest mortality rate. If the malignant tumors are diagnosed at an early stage, the patient's survival rate can be improved. Early diagnosis is possible with the help of lung cancer screening using low-dose CT scans. Identifying the malignant nodules in CT scans is quite challenging at an early stage, and hence there is a need of machine learning architecture that can effectively identify malignant and benign lung nodule in lung CT scans. This study combines the deep features extracted from Alexnet and Resnet deep learning models to classify the malignant and non-malignant nodule in CT scan images. The proposed deep learning architecture was experimented on LUNA 16 dataset and achieved an accuracy, sensitivity, specificity, positive predictive value, and Area under Curve (AUC) score of 94.3%, 95.52%, 91.11%, 89.52%, and .96%, respectively.

Keywords: Lung nodule classification · Deep learning · Alexnet · Resnet · Convolution neural network

1 Introduction

Lung cancer is the leading cause of death in males and females, which is 18.4% of the world's total cancer deaths [1]. Delay in the diagnosis of Lung cancer can increase the risk of death [2]. Hence, patients can be screened at regular intervals using low dose CT scans to reduce death risk. The radiologist assesses each slice of the CT scan images to identify the malignancy in nodules. However, few malignant tumors can be missed out due to observer's error. Hence we can automate the task of identifying the cancerous nodules using various machine learning architectures to assist the radiologist in scoring the malignancy.

Multiple researchers have used several machine learning approaches for identifying lung nodule malignancy. Among them, the deep learning (DL) approach has shown better results concerning accuracy. A deep learning model automatically extracts the features from the images and learns them to classify the images into cancerous or benign. The researcher proposed several deep learning architectures [3–18] to classify the lung nodules into benign or cancerous. Some of the architecture used in the identification of malignant nodules is deep belief network [3], convolutional neural network (CNN) [6, 7,

© The Author(s), under exclusive license to Springer Nature Switzerland AG 2022
R. Misra et al. (Eds.): ICMLBDA 2021, LNNS 256, pp. 213–226, 2022.
https://doi.org/10.1007/978-3-030-82469-3_19

9, 10, 12, 13, 15], autoencoders [4], and deep reinforcement learning [8]. CNN is widely used among all deep learning architectures as it provides better accuracy with lesser false positives, which helps in accurate diagnosis of disease at an early stage. CNN does not require an explicit feature extraction and selection phase, as CNNs includes both stages sequentially in their architecture. The performance of CNN features is promising when compared to hand-crafted features. The performance was further improved by combining deep features with handcrafted features [19]. CNN features can be either two dimensional or three dimensional. Three dimensional CNN can extract the spatial information but requires higher computational cost. Hence we have a 2D CNN network to extract deep features from the medical images. Several CNN architectures were implemented to categorize the nodule into benign and malignant. Some of the CNN architecture used in Lung Nodule classification are Alexnet [13, 26, 27], Resnet [20, 28, 29], Densenet [18], Googlenet [21], VGG-net [22], U-net [23] and Le-net [19]. Among them, Alexnet and Resnet are most commonly used to extract deep features from medical images.

In the proposed work, we combine the high-level features extracted from the Alexnet and Resnet model to gather significant features from both models. Since feature extraction is a very important phase in the classification system, we plan to combine both the model features to provide better accuracy in lung nodule classification. We also plan to train the model from scratch rather than using the transfer learning approach. Transfer learning works well if the source and the target problems belong to the same domain. If the target's training data differs from the source, then the performance of trained models might decrease. Most of the pre-trained models are trained on generic datasets rather than medical datasets. Hence we train the Alexnet and Resnet model from scratch with the LUNA CT scan images for lung nodule classification.

2 Preliminaries

2.1 AlexNet Architecture

Alexnet [24] consist of a sequence of five convolution layer followed by three fully connected layers. The network was deeper with many filters per layer. ReLu activation was added after every convolution and fully connected layer. The model was also trained using overlapping pooling layers. The overfitting problem was addressed using dropout instead of regularization. Dropout was applied before the first and the second fully connected layers. The deep features thus extracted from the previous layers are sent to a fully connected layer for classification. Finally, the images are classified into benign or malignant using the softmax activation function. A brief description of the Alexnet model is shown in Fig. 1.

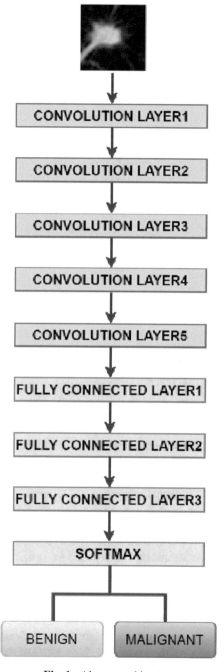

Fig. 1. Alexnet architecture

2.2 ResNet Architecture

Resnet [25] is a CNN architecture that supports deeper networks without compromising the model's performance. Usually, the network's performance is reduced with the inclusion of very deep network due to the vanishing gradient problem. ResNet solved this problem by skipping few training layers and thus creating a residual block. Each residual block consists of a convolution layer, batch normalization, and activation layer. Our study used ResNet-20, ResNet-56, and ResNet-164, which uses 3, 9, and 27 residual blocks, respectively. The segmented lung nodule is fed to a series of the residual block. At the start of every stage, the number of feature maps is halved but the number of filters are doubled. Finally, the dimensionality of the feature set is reduced using average pooling. Resnet gains a significant increase in accuracy with an increase in depth and is also easy to optimize. A brief description of the ResNet model is shown in Fig. 2.

Each Resnet layer mentioned in Fig. 2 consist of two convolution unit of size 3×3. A batch normalization and ReLU activation function follow every convolution unit, as shown in Fig. 3.

3 Proposed Work

A combination of deep features from the Alexnet and Resnet model was implemented in this paper to classify the lung nodule based on malignancy. Figure 4 gives a brief description of the architecture.

The proposed model takes a cropped lung nodule as input. The image is trained in parallel with the resnet and alexnet model to generate a concatenated feature vector and finally fed to the softmax layer. The input to Alexnet and Resnet model is an image of size 50×50. The Alexnet model uses five convolution and three fully connected layers. A nodule of size 50×50 is fed as an input to the convolution layer of Alexnet architecture. The kernel size in the first layer is 11×11 with a stride of 4. The output is then sent to batch normalization and ReLU layer followed by max-pooling layer of size 2×2. The second convolution layer has a kernel size 5×5 with a stride of 1. The output is then fed to the batch normalization and ReLU layer. The kernel size of the third, fourth and fifth convolution layer is 3×3 with a stride of 1. Max pooling is applied to the first, second, and fifth convolution layer. The convolution layers are followed by three fully connected layers. ReLU is the activation layer used in each of the layers, and the complexity of a fully connected layer is reduced using a dropout of 0.4 in each fully connected layer. We finally extract a feature map of size ten from the Alexnet model. The input image of size 50×50 is also passed to the Resnet model. The lung nodule is fed to the convolutional layer utilizing a convolution of size 1×1, followed by a series of residual blocks. Each residual block consists of convolution, batch normalization, and activation. A residual block skips one or two layers and concatenates the previous layer's feature with the output of the next layer. This concept of skipping a connection and allowing the gradient to flow through the shortcut path solves the problem of vanishing gradient and also ensures that the higher layers perform better. In this study, we used three residual blocks. A kernel of size 3×3 was used in every convolution layer with a

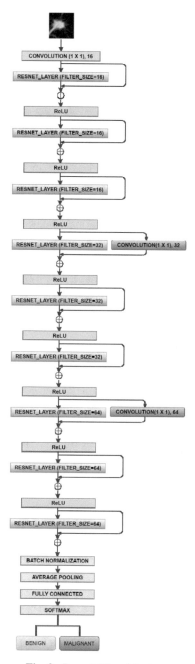

Fig. 2. Resnet-20 architecture

stride of 1. The number of filters in each block is doubled. The number of filters is 16, 32, and 64 in the first, second and third blocks. Each convolution layer is followed by

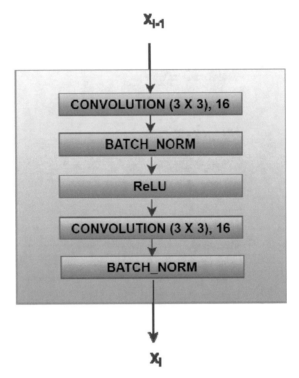

Fig. 3. Resnet layer (filter size = 16)

batch normalization and ReLU activation. An average pooling layer follows a sequence of 3 residual blocks to extract a feature map of size 64. The feature map extracted from Alexnet and Resnet is finally concatenated and are used to categorize the lung nodules into malignant or benign using a fully connected layer and softmax activation function. We use adam optimizer with cross-entropy as a loss function.

A brief explanation of the training and testing phase is shown in Fig. 5. The dataset was divided into 3 sets: Training, Testing, and Validation sets. In the training phase, training images are fed to the alexnet and resnet model. The features extracted from both the model are concatenated, and the concatenated features are used to train the model using a softmax classifier. The validation sets are used to tune the hyperparameters. The ground truth labels of training and testing images is obtained from the LUNA [30] data set. In testing phase, a set of the test images are passed through the trained model, to classify the lung nodule into malignant/benign. The results obtained from the testing phase were used to evaluate the performance of the proposed model using metrics like accuracy, specificity, sensitivity and positive predictive value.

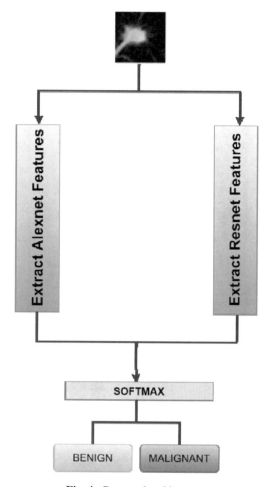

Fig. 4. Proposed architecture

Since the medical imaging datasets are very limited, the problem of overfitting could exist. The overfitting problem in the architecture can also be solved using data augmentation and downsampling of the negative sets. As there is a huge imbalance in positive and negative sets, we reduce the number of negative samples. Also, we increase the number of positive samples by rotating the positive images by an angle of 90° and 180° and thus upsampling the positive sets. Batch normalization and drop out used in both the models also address the overfitting problem.

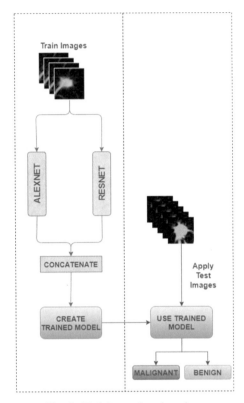

Fig. 5. Training and testing phase

4   Experiment Analysis

4.1   Data Description

We extract the nodules from the publicly available LUNA16 [30] dataset and have divided it into ten equal-sized subsets. LUNA16 has 888 low-dose lung CT scans. The dataset maintains around 200 slices of two dimensional CT images for each patient. Each slice has a dimension of 512 * 512 pixels. LUNA extracts annotations from LIDC (Lung image database consortium), which obtains the ground truth of nodule by averaging four doctors' diagnosis. We study only those nodules whose size is greater than 3 mm. Each of the nodules is scored by 4 radiologists as either malignant (1) or benign (0). If the final score obtained by averaging the score assigned by each radiologist is less than 3 (uncertain about malignant or benign), the nodules were considered irrelevant and were not used in the study. If the score of the nodule is greater than or equal to 3, i.e., 3 or more radiologists have labeled the nodule as malignant, they are labeled as malignant.

4.2 Data Pre-processing

Lung CT images in the LUNA [30] dataset are stored in a meta-image (mhd/raw) format. Each.mhd file has an associated.raw binary file. An annotation file named candidates.csv file is also maintained by the dataset, which contains information about the UID of the scan, the x, y, and z position of each nodule, and the class labels. We cropped the images based on the coordinates mentioned in the annotation file to reduce the training time. The nodule was cropped to a size of 50 × 50 pixels.

4.3 Data Augmentation

The dataset has a total of 551065 annotations. Among them, 1351 were labeled as malignant, and 549714 nodules were labeled as benign. As there is a huge class imbalance between benign and malignant samples, malignant nodules were augmented by rotating at a particular angle, and the benign samples were downsampled. Positive samples were augmented by rotating the image by 90° and 180°. Data augmentation reduces the need for regularization.

4.4 Training

The dataset consists of training, testing, and validation sets. The number of training, testing, and validation sets were 6131, 1903, and 1534 respectively. The model was implemented using Keras with TensorFlow as a backend. The training iteration is set to 70 epochs; the batch size was 100. Since our model deals with unbalanced datasets, we select cross-entropy as the loss function.

4.5 Simulation

Our proposed network is trained using Adam (Adaptive Moment Estimation) optimizer. While training the model, we recorded the training and the validation accuracy of the model for 70 epochs. Training accuracy has to increase as it learns new data. The training accuracy was 99%, and the validation accuracy was 90.10%.

There are two types of error while training the model: 1) Training error/loss 2) Validation error/loss. Training error is the error that is noted during the training process, and Validation error is an error that occurs while testing the model with validation data not used in training. Both Training and Validation loss has to be as minimum as possible. Figure 6 shows that with the increase in epoch, the accuracy of the model also increases, whereas Fig. 7 shows that with an increase in epoch, the training and validation error decreases.

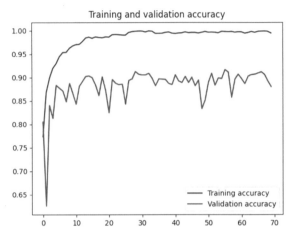

Fig. 6. Validation accuracy and training accuracy using a combination of deep feature from Alexnet and Resnet model

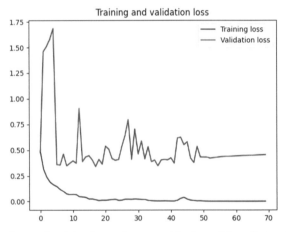

Fig. 7. Validation loss and training loss using a combination of deep feature from Alexnet and Resnet model

We also compared our results with other convolutional networks like alexnet, CNN, resnet-164, resnet-56, and resnet-20. A comparison between the proposed work and the existing work in literature for nodule classification is shown in Table 1. To evaluate our model, we use four metrics: accuracy, sensitivity, specificity, positive predictive value. Failure to detect a life-threatening nodule can increase the risk of the patient, and hence the sensitivity of our classifier has to be high. The classifier also has to provide a low false-positive rate and hence sensitivity and positive predictive value has to be evaluated.

Table 1. Comparison of lung nodule classification with existing work in literature

Model	Accuracy (%)	Sensitivity (%)	Specificity (%)	Positive predictive value (%)
Basic CNN [24]	87.17	91.01	78.13	90.74
Alexnet [24]	93.4	94.57	90.07	**96.26**
Resnet20 [25]	86.33	91.66	74.95	88.65
Resnet56 [25]	84.44	83.80	85.96	93.42
Resnet164 [25]	90.80	93.35	84.72	93.56
Proposed work	**94.3**	**95.52**	**91.11**	89.52

The formula for each of the metrics is mentioned below:

$$\text{Accuracy} = \frac{TP + TN}{TP + TN + FP + FN} \tag{1}$$

$$\text{Sensitivity} = \frac{TP}{TP + FN} \tag{2}$$

$$\text{Specificity} = \frac{TN}{TN + FP} \tag{3}$$

$$\text{Positive Predictive Value} = \frac{TP}{TP + FP} \tag{4}$$

FP, FN, TP, TN is a false positive, false negative, true positive, and true negative respectively obtained from the confusion matrix.

We achieved a testing accuracy of 94.3%, sensitivity of 95.52%, and specificity of 91.11% from the proposed work, which is comparatively better than the rest of the network. The proposed work's sensitivity is quite high, indicating that the malignant nodules were correctly classified and will not generate false negatives. The number of false positives was also less. Table 1 shows that the accuracy of the deep learning model increases with the concatenation of CNN models.

A ROC (Receiver Operating Characteristic Curve) curve is a plot of true positive rate with false-positive rate and acts as a diagnostic tool to determine the model's correctness. AUC (Area under Curve) measures the total area under this curve, and the value ranges from 0 to 1. If the ROC curve lies well above the diagonal, it shows that model can correctly classify the images into benign and malignant. The ROC curve in Fig. 8 shows that the proposed model has shown better test results as the ROC curve has moved away from the diagonal. And the AUC score of the model is 0.96, indicating that the model has a better ability to classify benign and malignant images.

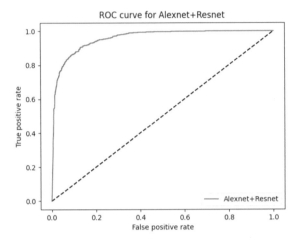

Fig. 8. ROC curve for the proposed work

5 Conclusion

This paper has classified the lung nodules into malignant and benign using a combination of Alexnet and Resnet architecture. The accuracy of lung nodule classification using the concatenation of Alexnet and Resnet architecture was 94.3% using Adam optimizer. The model showed a sensitivity, specificity, and positive predictive value of 95.52%, 91.11%, and 89.52%, respectively. It is also evident from the results that concatenation of deep features from deep models has shown better results than results from deep models alone. The proposed architecture also showed less false positive rate and a better AUC score of .96. Our work suggests that the deep learning model's concatenation shows improvement in the accuracy and sensitivity of the models.

References

1. Bray, F., Ferlay, J., Soerjomataram, I., Siegel, R.L., Torre, L.A., Jemal, A.: Global cancer statistics 2018: GLOBOCAN estimates of incidence and mortality worldwide for 36 cancers in 185 countries. CA: Cancer J. Clin. **68**(6), 394–424 (2018)
2. Byrne, S.C., Barrett, B., Bhatia, R.: The impact of diagnostic imaging wait times on the prognosis of lung cancer. Can. Assoc. Radiol. J. **66**(1), 53–57 (2015)
3. Lakshmanaprabu, S.K., Mohanty, S.N., Shankar, K., Arunkumar, N., Ramirez, G.: Optimal deep learning model for classification of lung cancer on CT images. Future Gen. Comput. Syst. **92**, 374–382 (2019)
4. Sun, B., Ma, C., Jin, X., Luo, Y.: Deep sparse auto-encoder for computer aided pulmonary nodules CT diagnosis. In: 2016 13th International Computer Conference on Wavelet Active Media Technology and Information Processing (ICCWAMTIP), Chengdu, pp. 235–238 (2016)
5. Xie, Y., Zhang, J., Xia, Y., Fulham, M., Zhang, Y.: Fusing texture, shape and deep model-learned information at decision level for automated classification of lung nodules on chest CT. Inf. Fusion **42**, 102–110 (2018)
6. Wei Shen, M., et al.: Multi-crop convolutional neural networks for lung nodule malignancy suspiciousness classification. Pattern Recogn. **61**, 663–673 (2017)

7. Cao, P., et al.: A_2, 1 norm regularized multi-kernel learning for false positive reduction in Lung nodule CAD. Comput. Methods Prog. Biomed. **140**, 211–231 (2017)
8. Ali, I., et al.: Lung nodule detection via deep reinforcement learning. Front. Oncol. **8**, 108 (2018)
9. Dou, Q., Chen, H., Lequan, Y., Qin, J., Heng, P.-A.: Multilevel contextual 3-D CNNs for false positive reduction in pulmonary nodule detection. IEEE Trans. Biomed. Eng. **64**(7), 1558–1567 (2017)
10. Jiang, H., Ma, H., Qian, W., Gao, M., Li, Y.: An automatic detection system of lung nodule based on multigroup patch-based deep learning network. IEEE J. Biomed. Health Inform. **22**(4), 1227–1237 (2018)
11. Singh, G.A.P., Gupta, P.K.: Performance analysis of various machine learning-based approaches for detection and classification of lung cancer in humans. Neural Comput. Appl. **31**(10), 6863–6877 (2018)
12. Causey, J.L., et al.: Highly accurate model for prediction of lung nodule malignancy with CT scans. Sci. Rep. **8**, 9286 (2018)
13. Sun, W., Zheng, B., Qian, W.: Automatic feature learning using multichannel ROI based on deep structured algorithms for computerized lung cancer diagnosis. Comput. Biol. Med. **89**, 530–539 (2017)
14. Yuan, J., Liu, X., Hou, F., Qin, H., Hao, A.: Hybrid-feature-guided lung nodule type classification on CT images. Comput. Graph. **70**, 288–299 (2018)
15. Xie, H., Yang, D., Sun, N., Chen, Z., Zhang, Y.: Automated pulmonary nodule detection in CT images using deep convolutional neural networks. Pattern Recogn. **85**, 109–119 (2019)
16. Gu, Y., et al.: Automatic lung nodule detection using a 3D deep convolutional neural network combined with a multi-scale prediction strategy in chest CTs. Comput. Biol. Med. **103**, 220–231 (2018)
17. Silva, G.L.F., Valente, T.L.A., Silva, A.C., Paiva, A.C., Gattassa, M.: Convolutional neural network-based PSO for lung nodule false positive reduction on CT images. Comput. Methods Prog. Biomed. **162**, 109–118 (2018)
18. Zhan, J., Xia, Y., Zeng, H., Zhang, Y.: NODULe: combining constrained multi-scale LoG filters with densely dilated 3D deep convolutional neural network for pulmonary nodule detection. Neurocomputing **317**, 159–167 (2018)
19. Xie, Y., Zhang, J., Liu, S., Cai, W., Xia, Y.: Lung nodule classification by jointly using visual descriptors and deep features. In: Henning Müller, B., et al. (eds.) MCV/BAMBI -2016. LNCS, vol. 10081, pp. 116–125. Springer, Cham (2017). https://doi.org/10.1007/978-3-319-61188-4_11
20. Nóbrega, R.V.M.D., Peixoto, S.A., da Silva, S.P.P., Filho, P.P.R.: Lung nodule classification via deep transfer learning in CT lung images. In: 2018 IEEE 31st International Symposium on Computer-Based Medical Systems (CBMS), Karlstad, pp. 244–249 (2018)
21. Chen, J., Shen, Y.: The effect of kernel size of CNNs for lung nodule classification. In: 2017 9th International Conference on Advanced Infocomm Technology (ICAIT), Chengdu, 2017, pp. 340–344 (2017)
22. Paul, R., et al.: Deep feature transfer learning in combination with traditional features predicts survival among patients with lung adenocarcinoma. Tomography **2**(4), 388–395 (2016)
23. Zhao, C., Han, J., Jia, Y., Gou, F.: Lung nodule detection via 3D U-net and contextual convolutional neural network. Int. Conf. Netw. Netw. Appl. **2018**, 356–361 (2018)
24. Krizhevsky, A., Sutskever, I., Hinton, G.E.: ImageNet classification with deep convolutional neural networks. In: Proceedings of the 25th International Conference on Neural Information Processing Systems - Volume 1 (NIPS 2012). Curran Associates Inc., Red Hook, pp. 1097–1105 (2012)
25. He, K., Zhang, X., Ren, S., Sun, J.: Deep residual learning for image recognition. In: CVPR (2016)

26. Jin, T., Cui, H., Zeng, S., Wang, X.: Learning deep spatial lung features by 3D convolutional neural network for early cancer detection. In: 2017 International Conference on Digital Image Computing: Techniques and Applications (DICTA), Sydney, NSW, pp. 1–6 (2017)

27. Wang, Z., Xu, H., Sun, M.: Deep learning based nodule detection from pulmonary CT images. In: 2017 10th International Symposium on Computational Intelligence and Design (ISCID), vol. 1, pp. 370–373 (2017)

28. Kuan, K., Ravaut, M., Manek, G., Chen, H.: Deep Learning for Lung Cancer Detection: Tackling the Kaggle Data Science Bowl 2017 Challenge. https://arxiv.org/pdf/1705.09435.pdf

29. Xie, Y., et al.: Knowledge-based collaborative deep learning for benign-malignant lung nodule classification on chest CT. IEEE Trans. Med. Imaging **38**(4), 991–1004 (2018)

30. Setio, A.A.A., et al.: Validation, comparison, and combination of algorithms for automatic detection of pulmonary nodules in computed tomography images: the LUNA16 challenge. Med. Image Anal. **42**, 1–13 (2017)

A Deep Learning Approach for Anomaly-Based Network Intrusion Detection Systems: A Survey and an Objective Comparison

Shailender Kumar, Namrata Jha[✉], and Nikhil Sachdeva

Delhi Technological University, Delhi, India
shailenderkumar@dce.ac.in

Abstract. In this information age, data is of utmost importance. With the ongoing rapid digitization of multitude of services, a large variety of databases, often containing highly sensitive data, have come into existence. This magnitude of crucial data is accompanied by concerns, which have grown into an undeniable and immediate necessity for an Intrusion Detection System for Database Security. Detection of both privilege abuse (insider attack) as well as outsider intrusions, has become the foremost concern for maintaining scalability, dynamicity and reinforcement of these databases. Our approach using Outlier Analysis (Anomaly Detection) by the means of Data Mining is intended to efficiently perform detection and prevention of intrusive transactions within the database environment and therefore reinforces the security of critically sensitive information. The database, to be mined for patterns and rule extraction, contains user provided queries, previously generated user roles and transaction profiles. These sequential patterns and extracted rules shall be used to devise a tool for classification of transactions as malicious and non-malicious.

Keywords: Intrusion Detection Systems (IDS) · Artificial Neural Networks (ANN) · Cyberattacks · Autoencoders · LSTM

1 Introduction

Since malware attacks keep changing, given their dynamic nature, datasets should regularly be updated and benchmarked with the latest IDS developed using machine learning methods. Thus, the system for detection must be resilient and efficacious to detect and categorize novel and unforeseeable cyberattacks. This type of research study enables the identification of the most effective algorithm to detect cyberattacks in future.

Intrusions are most often instigated by users without authorization whom we call attackers. The primary aim of an attacker is to get remote access of a computer through the Internet or to render the service inoperable. This is called a malware attack. Detection of intrusion and its classification accurately requires understanding the approach to attack a system successfully.

© The Author(s), under exclusive license to Springer Nature Switzerland AG 2022
R. Misra et al. (Eds.): ICMLBDA 2021, LNNS 256, pp. 227–235, 2022.
https://doi.org/10.1007/978-3-030-82469-3_20

Research has shown that potential for better representation with exact feature extraction lies with deep learning resulting in higher accuracy from the data to create models with higher efficiency. Furthermore, we studied the performance of this neural network in two kinds of classification - binary and multiclass classification and have analyzed the metrics for precision calculations such as Accuracy, Precision, True Positive Rates, False Positive Rates and F-1 Score.

The goal is to create a comparative study of the previous approaches to the IDS problem by researchers. We have studied performance of previous work in this field which includes - the multi-layer perceptron, convolutional neural networks, recurrent neural networks, LSTMs among various other machine learning methods have been taken into consideration to carry out multi-class classification on the benchmark dataset in the field of IDS i.e., the NSL-KDD dataset.

The birth of computer architectures gave rise to research on security issues related to various domains of IDS. In recent studies, a myriad of machine learning solutions to NIDS have been devised by security researchers and specialists.

2 Theory

2.1 Network-Based Intrusion Detection Systems (NIDS)

The fundamental job of an IDS is the monitoring of network traffic. This can help in detecting suspicious activities and known threats. Once these are detected, the system can raise an alarm. It can be visualized as a packet detector which detects malicious data packets which are anomalous in nature travelling along various channels. The primary aim of IDS is to.

- Monitor all components of a system such as firewalls, key management servers, routers and files
- View the logs of the system. These can include - audit trails of operating systems to calibrate systems which will make sure to better protect against harmful attacks
- Recognize the attack patterns.

The system then matches attack signature databases with previously stored information on systems. An IDS can be categorized into two classes- Host Intrusion Detection System (HIDS) and Network Intrusion Detection System (NIDS). This classification is done on the basis of the placement of IDS sensors - network or host. NIDS analyses and monitors network traffic for suspicious behaviour by scrutinizing the content and network information of data packets which move through the network. These are helped by NIDS sensors which are placed at important points in the network. These sensors examine data packets from all devices on the network. On the other hand, HIDS aims to monitor and analyze configuration of systems and activity of applications for enterprise networks.

NIDS can further be categorized in two kinds based of the method of detection used - (1) signature-based (misuse-based) NIDS (SNIDS): this category of NIDS performs pattern-matching in the network traffic for the pre-existing signatures that are installed. This helps in the detection of an intrusion in the network, and (2) anomaly detection-based NIDS (ADNIDS): ADNIDS detects an intrusion in the network traffic, it occurs when a variance from the usual network pattern is detected in network traffic [1]. ADNIDS is used for the detection of novel and unknown attacks. On the contrary, SNIDS results in a lesser rate of accuracy in the detection of unforeseen attacks. It is caused due to the limitation of signatures based on attack patterns that can be pre-installed in the IDS. Moreover, SNIDS is labour intensive as it is dependent on the manual updating of the signature database for the NIDS, which not only inhibits SNIDS from having an exhaustive database, but also makes it prone to human error. However, SNIDS results in high accuracy in detecting attacks which have already been logged before and records a greater rate of accuracy supported by low rates of false-alarm. Even though ADNIDS produces false-positives of higher magnitude, it is considered to be the potential algorithm for the identification of novel attacks. This encourages its adoption as a field of research. In our study, we majorly focus on building effective Anomaly-based Network Intrusion Detection Systems.

Commercially, NIDS is usually implemented for measures of statistical nature to calculate thresholds on dataset features such as flow size, length of packet, inter-arrival time and other parameters related to network traffic. This enables effective modelling of the dataset within a specific time-window. Studies have shown that NIDS tends to suffer from higher rates of false alerts, both false positives and false negatives. Greater number of false negative alerts usually signal that the rate of failure of NIDS is high. A greater rate of false positive alarms indicate that NIDS can raise an alarm in the system without any reason to give false scares in normal activity. These commercial NIDSs are not effective in modern systems with high rates of network flow. In this study, we focus on analyzing the accuracy and viability of previously tested traditional machine learning algorithms and various kinds of neural networks in IDS. For this purpose, researchers have used network datasets which are publicly available such as KDDCup 99, WSN-DS, UNSW-NB15, NSL-KDD, and Kyoto.

An extensive literature review in this field and its analysis strongly indicates that studies which employ novel deep learning techniques for implementing NIDS on commonly used datasets like NSL-KDD and its predecessor, KDDCup 99 have been proven to result in efficient IDS.

2.2 NSL-KDD Dataset

In this study, we focus on the use of NSL-KDD dataset as the benchmark to study deep learning models and verify the results in terms of efficacy and accuracy of detecting intrusions. The KDD Cup 99 dataset was improved by analysis to obtain the NSL-KDD dataset. The 1998 DARPA IDS network traffic data resulted in the creation of KDD Cup dataset. This dataset includes network data of two kinds - normal and different kinds of attack traffic. The attacks are further classified in four classes - as Denial of Service, User-to-Root, Probing and Root-to-Local. In order to make the task of intrusion detection task realistic, test data consists of some attacks which were not injected during the phase of collection of training data. It should also be noted that novel attacks in this dataset can be learnt from the attacks which are known. Two sets of two million and five million datasets were processed in TCP/IP connection records. They will constitute test and training data respectively.

The NSL-KDD dataset is a development over the KDD99 dataset and is most often used as the basis of research in the network intrusion analysis research field which is inclusive of various tools and techniques. It shares the goal of developing an efficient and effective IDS. An in-depth study [2] of the NSL-KDD dataset using different techniques of machine learning is done in WEKA tool. A comparison between the relevance of NSL-KDD dataset with its predecessor KDD99 cup dataset can be found in [3] by using Artificial Neural Networks.

NSL-KDD dataset is considered to be an improvised form of its predecessor and still maintains the integrity and essential records of the KDD dataset. It is a refinement of the KDD dataset in essentially three ways. Firstly, it eliminates redundancy in records to ensure that the classifiers developed produce an unbiased result. Second, NSL-KDD partitions data in such a way that adequate numbers of network data vectors are available in both the training and the testing datasets. This enables a thorough analysis on the complete set of records. [4] Thirdly, the records have been chosen by randomly sampling the various records belonging to varying levels of difficulty in such a manner that the number of chosen records is inversely related to their respective shares in the discrete records [5]. This exhaustive pre-processing of the KDDCup-99 dataset spanning across multiple steps has increased the favourability of the enumeration of the records for research on the NSL-KDD dataset. In most cases, the study of the NIDS models is carried out on the benchmark NSL-KDD dataset used by researchers in the field of NIDSs and include a comparison between the performance the NIDS models on the following basis [2]:

1. Binary classification problem: Classifying a network traffic vector record is anomalous or normal.
2. Multiclass classification problem: Five-category classification problem, which determines if the network traffic vector is normal or is it an attack belonging to one of the following four classes:

   - Denial of Service
   - Root to Local
   - Probe
   - User to Root

3   Related Work

b This section describes various research studies conducted in the area of detection systems for intrusions using the techniques of machine learning and deep learning. We conducted a preliminary study of research papers spanning across the last couple of years using different techniques to build their IDS model and use different benchmark datasets to evaluate their performance. We shall now discuss different approaches taken by the researchers in this area to tackle our problem statement at hand. Finally, a few deep learning-based approaches in this area will be discussed.

One of the earliest works found in literature, an IDS was designed using Artificial Neural Networks which had enhanced efficiency using back-propagation in a resilient manner [6]. This study made use of only the training component of the dataset to perform training, validation and testing of the model by partitioning the training dataset into three divisions of 70%, 15% and 15% respectively. In accordance with the expectations, using unlabelled data in the testing phase of the model delivered a reduction of performance.

In [2], an IDS approach based on deep learning which used SoftMax regression and sparse auto-encoder towards building an effective and flexible NIDS. The model was evaluated on the standard NSL-KDD dataset to evaluate the accuracy of outlier detection. It was observed that the NIDS delivered better performance as compared to the NIDSs that had been implemented previously. The researchers suggested that the results could be improved by using a Stacked Auto-Encoder, which is an improvement on the sparse auto-encoder in deep belief networks, for UFL for further classification. This study delivered great results when applied directly to the benchmark dataset. Using Self Taught Learning techniques, this approach was able to showcase an accuracy of 88.39% when tested for two-class classification and 79.10% when tested for five-class classification.

An RNN-based IDS model has been proposed in [7], which has a strong ability to model intrusion detections and demonstrates high values of accuracy in both two-class and multi-class classification. The performance displays a higher rate of accuracy and rate of detection with a low rate of false positives as compared to traditional methods of classification like random forest and naive Bayesian. The model was successfully able to not only detect the category of intrusion but also improved the accuracy of this detection. In case of binary classification, the paper achieved 83.28% accuracy and 81.29% accuracy for multi-class classification.

In [8], a comparative study of the performance of a stacked NDAE model was carried out on the NSL-KDD and KDD Cup99 datasets to evaluate how the model performs on refined and unrefined data sets. The model delivered considerably better results on the KDD Cup99 dataset with an accuracy of 97.85% for five-class classification as compared to 85.42% accuracy when applied on the NSL-KDD dataset.

The paper suggests a simple neural network using deep learning architecture [9] which results in an accuracy of 95% on the NSL-KDD dataset. Whereas in [10], an STL model with feature representation using Convolutional Neural Networks and classification using Weight Dropped LSTM gives a high accuracy of 97.1% on the UNSW-NB15 dataset. A comparison of the techniques employed, datasets used and accuracies delivered can be found in the next section.

4   Relative Study of Techniques Applied in Network Intrusion Detection System

Table 1. Past research on Network Intrusion Detection Systems on benchmark datasets.

Author and Year	Dataset	Technique Used	Accuracy
Javaid, Ahmad, et al. [2], 2016	NSL-KDD	STL	2-class classification: 88.39% 5-class classification: 79.10%
Yin, Chuanlong, et al. [7] 2017	NSL-KDD	RNN	Binary classification: 83.28% Multiclass classification: 81.29%
Shone, et al. [8] 2018	NSL-KDD KDD Cup99	Stacked NDAE model	5-Class KDD Cup'99 Classification: 97.85% 5-Class NSL-KDD Classification: 85.42%
Vinayakumar, et al. [9] 2019	NSL-KDD	DNN	95%
Mohammad Mehedi Hassan, et al. [10]	UNSW-NB15	CNN-WDLSTM	97.1%

5   Implementation Strategies

After thorough study of intrusion detection systems and its previous research, we have deduced a set of common stages in the implementation strategies employed by various researchers. This chapter elaborates upon the various components of the approach of developing a network intrusion detection system model. Furthermore, we elaborate on specific strategies which have been deduced to provide higher accuracies on the basis of the comparison in Table 1. Specifically, the use of the Self-Taught Learning Technique (STL) in the training phase has been discussed (Fig. 1).

5.1   Data Preprocessing

Numericalization. The NSL-KDD dataset is attributed with 3 non-numeric fields. It also has 38 numeric fields. The 3 categorical attributes, 'flag', 'service' and 'protocol_type' need to be encoded into some numerical value before being fed input to the NIDS model, as the expected input is a numerical matrix. We apply one-hot encoding on these attributes. Say for encoding the attribute 'protocol_type', which can have three values 'tcp', 'udp' and 'icmp', the resultant binary vectors are (0, 0, 1), (0, 1, 0)

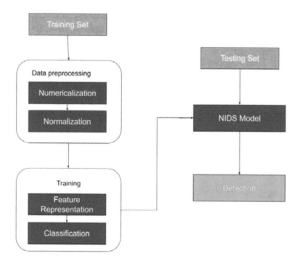

Fig. 1. Block diagram representation of the steps involved in a NIDS implementation

and $(1, 0, 0)$. Similarly,the attributes 'flag' and 'service', have 11 and 70 types of values respectively. Thus, upon encoding all our nominal features, a previously 41-dimensional features map would get transformed into a 122-dimensional features map.

Normalization. There are certain variables in our dataset, NSL-KDD, for which the range is much wider compared to other attributes. For example, we have 'src_bytes' which ranges $[0, 1.3 \times 109]$', 'dst_bytes' which ranges $[0, 1.3 \times 109]$' and 'duration' which ranges $[0, 58329]$'. These features make it a necessity to carry out scaling in order to overcome the huge disparity between their minimum and maximum values. To bring the values of attributes in comparable ranges, as the scaling method we shall apply logarithmic normalization. Furthermore, applying the following equation, the magnitude of each feature is scaled in the range of $[0, 1]$.

5.2 Self-taught Learning (STL)

Self-Taught Learning is an approach to solve classical classification problems, which is executed in two phases of deep learning [2]. The two stages involved in this approach are as below:

- Unsupervised Feature Learning is the process of feature representation. In this step, we extract the features of the dataset while reducing dimensionality of a large set of unlabelled data.
- Subsequently, in the second phase, feature representation learnt in the previous stage is applied to labelled data, xl, and classification purposes.

There must be relevance between the unlabelled and labelled data, in spite of its origins lying in distinct sets of distributions. A variety of approaches are used for the first step i.e., UFL, such as K-Means Clustering, Sparse Auto-Encoder, Gaussian Mixtures and Restricted Boltzmann Machine (RBM) [11]. In our approach, we use feature learning based on non-symmetric deep autoencoder due to its ease of implementation and the performance delivered while using less computing capacity [11]. A sparse autoencoder is an ANN which comprises the following layers - input layer (data input), hidden layer, and output layer (features output). The autoencoder consists of N nodes each in its input and output layers and K nodes in its hidden layer. In the second stage, the learnt representation will be classified using a Weight-Dropped LSTM.

The approach to be undertaken in our project is modelled on STL and has the following two parts -

Feature Representation. In this phase, we compress the feature vector which means the existing features to be used in further stages of the classification are reduced in order to decrease the number of dimensions of large datasets and promote only relevant features to further stages. Unsupervised feature learning can be achieved using autoencoder, gaussian mixtures and K-means clustering.

Classification. In this phase, the learned representation (i.e., the selected features) from the previous phase is fed into a classification model. The classification model can be one of the following types as discussed later in this section - binary classification and multi class classification (5 classes). In recent works in the area, we observed the classifier could be based on various implementations of algorithms in machine learning such as support vector machine (SVM), artificial neural networks and recurrent neural networks.

6 Conclusion

This paper aims to provide a holistic review of the research studies in the field of network intrusion detection systems (NIDS). After thorough review and understanding of the benchmark NSL-KDD dataset, it was observed in the trends of IDS development that models based on ML and DL techniques when classified as binary or multi-class result in higher accuracy. Further, we discuss specific strategies which have demonstrated higher rates of accuracy in classifying network data. The use of self-taught learning (STL) in the training phase, with autoencoders for representing the relevant features of the data and LSTMs for developing the NIDS model shows incredible promise in the field and can be fine-tuned by researchers in future to build more robust models. Moreover, trends also show that there is scope for higher accuracy and efficiency-based models on similar datasets. With the increase of data transfer applications over networks, the scope of such IDS software is never ending. It should also be realized that the datasets in this case also present their limitations and as the transfer evolves, so shall the attacks. Future scope for all NIDS techniques lies in better performance with its evaluation based on its performance on various evaluation metrics like Accuracy, Precisions and F-1 Scores which further interpreted the results delivered by the model in terms of both accuracy and efficiency.

References

1. Javaid, A., Niyaz, Q., Sun, W., Alam, M.: A deep learning approach for network intrusion detection system. In: 9th EAI International Conference Bio-inspired Information Communication Technology (BIONETICS), pp. 21–26. New York, NY, USA (2016)
2. Revathi, S., Malathi, Dr. A.: A detailed analysis on NSL-KDD dataset using various machine learning techniques for intrusion detection, Int. J. Eng. Res. Technol. (IJERT), **2**(12) (2013). ISSN 2278-0181
3. Sanjaya, S.K.S.S.S., Jena, K.: A Detail Analysis on Intrusion Detection Datasets. In: 2014 IEEE International Advance Computing Conference (IACC) (2014)
4. Dhanabal, L., Shantharajah, S.P.: A study on NSL-KDD dataset for intrusion detection system based on classification algorithms. Int. J. Adv. Res. Comput. Commun. Eng. **4**(6), 446–452 (2015)
5. Ferrag, M.A., Maglaras, L., Moschoyiannis, S., Janicke, H.: Deep learning for cyber security intrusion detection: approaches, datasets, and comparative study. J. Inf. Secur. Appl. **50**, 102419 (2020)
6. Naoum, R.S., Abid, N.A., Al-Sultani, Z.N.: An enhanced resilient backpropagation artificial neural network for intrusion detection system. Int. J. Comput. Sci. Netw. Secur. **12**(3), 11–16 (2012)
7. Yin, C., Zhu, Y., Fei, J., He, X.: A deep learning approach for intrusion detection using recurrent neural networks. IEEE Access **5**, 21954–21961 (2017)
8. Shone, N., Tran Nguyen, N., Vu Dinh, P., Shi, Q.: A deep learning approach to network intrusion detection. IEEE Trans. Emerg. Topics Comput. Intell. **2**(1) (2018)
9. Vinayakumar, R., Alazab, M., Soman, K.P., Poornachandran, P., Al-Nemrat, A., Venkatraman, S.: Deep learning approach for intelligent intrusion detection system. IEEE Access **7**, 41525–41550 (2019)
10. Hassan, M.M., Gumaei, A., Alsanad, A., Alrubaian, M.: Fortino, G.: A hybrid deep learning model for efficient intrusion detection in big data environment. Inf. Sci. **513**, 386-396 (2019)
11. Coates, A., Ng, A.Y., Lee, H.: An analysis of single-layer networks in unsupervised feature learning, In: International Conference on Artificial Intelligence and Statistics, pp. 215–223 (2011)

Review of Security Aspects of 51 Percent Attack on Blockchain

Vishali Aggarwal[✉] and Gagandeep

Department of Computer Science, Punjabi University, Patiala, India

Abstract. Blockchain technology has emerged as a secure technology for performing financial transactions. It is a database which stores all the transactions into blocks. It has made the digital data decentralized and thus does not have the problem of single point failure. The newly created transaction is broadcasted to all other nodes in the network. Although blockchain has the property of immutability but it is always prone to the cyber attacks. The 51 percent is the predominant attack encountered over blockchain which is executed forming large mining pools. This is the attack over block addition process where maliciously mined blocks get added to blockchain. This paper encapsulates the critical review of existing solutions in the context of 51 percent attack so as to get the analysis of level of protection and the challenges in the existing methods.

Keywords: Distributed ledger · Proof of work · Proof of stake · Peer to Peer · 51 percent attack

1 Introduction

Blockchain technology is the distributed ledger first explained in 2008 by Satoshi Nakamoto in his research paper about bitcoins (Nakamoto, 2008). In the beginning the technology was limited only to the financial aspect but gradually it has emerged as the layout for secure data storage in many applications like smart contracts, IoT, identity verification etc. It stores the data in the form of transactions into various blocks which are linked to one another forming a chain. The users in the blockchain network have a peer to peer connection. The transactions in the blocks are verified using some consensus mechanism. Blockchain has the feature of immutable ledger as it encrypts the data using hash function which is quite difficult to crack. But since the blockchain technology performs over the internet so the risk of cyber attacks is always there. These attacks can occur over the structure of the blockchain, peer to peer interaction of the various nodes or at the deployment level of blockchain (Saad et al., 2019). Talking about the peer to peer level, the most prominent attack is the 51 percent attack which can severely harm the credibility of blockchain. This attack has been encountered in the real world also many a times. Krypton and Shift are ethereum based blockchains which have suffered from 51 percent attack in 2016. In May 2018, the 26th largest cryptocurrency named Bitcoin Gold suffered from 51 percent attack. Ethereum classic has also suffered from

© The Author(s), under exclusive license to Springer Nature Switzerland AG 2022
R. Misra et al. (Eds.): ICMLBDA 2021, LNNS 256, pp. 236–243, 2022.
https://doi.org/10.1007/978-3-030-82469-3_21

51 percent attack in early 2020. So it becomes the need of the current time to focus over this vulnerability of blockchain so that can become more reliable to use.

The main focus of this paper is to review the blockchain security platforms along with the techniques available for 51 percent attack. The paper is organized as follows: In Sect. 2, background of the 51 percent attack is given. In Sect. 3, the impacts of the said attack over blockchain are presented. In Sect. 4, the types of platforms available for blockchain security are discussed. Then survey of techniques for 51 percent attack is given in the sub section. In the last section, conclusion is followed by references to be referred in throughout the paper.

2   Background

Blockchain is vulnerable because of 51 percent attack because this attack invites many other attacks as well. The miner needs to arrange more than half the computing power to execute this attack successfully. The approach is to take other miners into confidence to mine over the malicious chain of the attacker so that it can grow faster than the actual chain and thus the whole control is shifted towards the attacker's chain (Bae & Lim, 2018). If 51 percent attack is executed over the blockchain, many other attacks like double spending, distributed denial of service and selfish mining can be performed easily over this blockchain by the attacker (Eyal & Sirer, 2018). A malicious node with the intension to execute attack will initially create some normal transaction by spending their coins. At the same time it will start secretly mining a private chain. It will not broadcast the new blocks to the network and keep them private. It will not include its own transaction of spending coins. This is done in order to perform the double spending attack. Now if the malicious chain acquires more than fifty percent of computing power, it will extend faster than the original (honest) chain and it is the time when the malicious chain is broadcasted in the network. The miners automatically switch to the larger chain being unaware of the attack and start mining over this longer chain which is actually the malicious chain. Now the network reconsiders the already spent coins of the attacker as the attacker did not include the transactions into the malicious chain. Thus the attacker can double spend the coins.

3   Impacts of 51 percent Attack

51 percent attack affects the application severely. The attackers get control over the entire network and they can manipulate the transaction processing for their own purposes. As it is not easy to track down the attackers, the attack can remain active for days. There are severe consequences of the said attack especially from financial perspective. The main consequences are as following:

- **Double Spending:** 51 percent attack is the way to execute the further attacks on blockchain like double spending (Karame et al., 2012). If attacker successfully executes 51 percent, he can easily double spend his coins. In double spending the miner can send the same coins to two users and perform the respective transactions. When transaction with one user gets confirmed the malicious miner launches its secret chain with the same coins and can again spend these coins in another transaction.

- **Selfish Mining:** When a single pool controls more than half of the mining power, it will perform selfish mining in order to execute the same attack further (Gemeliarana & Sari, 2018). It can severely affect the mining operation of other miners.
- **Malicious Transactions:** When a miner acquires majority of the mining power, it can control the entire blockchain and can perform transactions for his personal purposes. It can decide which transactions to include and which not. Miner can even block particular transactions and can create empty blocks without any transaction to flood the network.
- **Cryptocurrency loss:** If a miner is performing double spend and blocking transactions after gaining majority hash power in the blockchain network then users of that coin lose the confidence on the coin and prefer not to do any transaction with it and that coin exchange rate might crash down.

4   Review of Literature

In this section, the concise overview of what has been studied, argued, and established about 51 percent attack in blockchain is discussed. Literature survey is organized thematically based upon the various solutions given for 51 percent attack. Second sub-section surveys different techniques for 51 percent attack based upon smart contracts. Similarly the third sub-section surveys various techniques for 51 percent attacks based upon consensus layer.

4.1   Analysis of Blockchain Security Aspects

Various security platforms are proposed with the intension to add on to the security and reliability of blockchain. Smart contracts require more reliability than any other application (Table 1).

Table 1.   Analysis of security issues of blockchain

Reference	Security platform	Security issue	Deployment	Result	Limitations
Kosba et al., 2016	Hawk	Privacy of smart contracts	Zero knowledge proofs	A security framework named Hawk was developed	
Luu et al., 2016	Oyente	Security problem in smart contracts	Ethereum Network	Developed a tool Oyente to analyze smart contracts to detect bugs	
Zhang et al., 2016	Town Crier [19]	Security of smart contracts	Intel's Software Guard Extensions (SGX) combined with ethereum	Developed Town Crier which acts as authenticated data feed for smart contracts	Deployment of Intel Attestation Service is pending

(continued)

Table 1. (*continued*)

Reference	Security platform	Security issue	Deployment	Result	Limitations
Bang & Choi, 2019	Network monitoring system	Illegal transactions	Apache kafka, Apache storm	Efficient storage system for blockchain network monitoring is developed	By monitoring the network the security threats have not been identified
Harikrishnan & Lakshmy, 2019	Blockchain with SHA3	Lack of data security in sensitive blockchains	Zero Knowledge Proofs	Developed Flexible Blockchain having indistinguishable hash functions	The developed system is secure but very slow and not scalable
Liu et al., 2019	Double chain	Information leakage, storage imbalance	Heuristic algorithm followed by customized genetic algorithm	Double chain is developed that improves security if blockchain network	More computation and high memory cost
Marangappanavar & Kiran, 2020	Inter-Planetary File System (IPFS)	Privacy preservation among untrusted Parties	Smart contracts	Fast retrieval of personal health records	Records have to be maintained individually

4.2 Techniques for 51 percent Attack

Various techniques are proposed to avoid 51 percent attack over blockchain. The attack can occur especially on smart contracts to manipulate the terms of the contracts. The existing solutions are implemented over the smartpools, consensus layer or over collaboration of chains i.e. interoperable chains. The following table provides the survey based upon this approach.

Markov chains: A discrete-time Markov chain is a mathematical model that contains of a certain set of states in which the system can tend to exist. The system can only move from one state to another if a transition exists for the same. The system can make transition from one state to multiple others, each with a certain probability. But its behaviour may not be dependent upon the previous behaviour of the model, which is called the Markov property. Continous-time Markov chains allow the system to spend some time in one state. For each transition an expected amount of time for transition is given. A CTMC thus has an exponential distribution. Again, its behaviour does not depend on past behaviour.

SmartPool: SMARTPOOL is a smart contract which implements a decentralized mining pool for Ethereum and runs on the Ethereum network. SMARTPOOL maintains two lists in its contract state — a list for claim is claimList and a verified claim list verClaimList. Whenever a miner submits a set of shares as claim for the current Ethereum block, it is added to the claimList.

ChainLocks: ChainLocks incentivize miners to publish processed blocks immediately, thus minimizing the advantages of secret mining and increasing the difficulty of performing a consensus attack.

Penalty System by Horizon: Penalty is in the form of block acceptance delay in according to the amount of time for which the block has been hidden from the public network.

Delayed Proof of Work (dPoW) by Komodo: Delayed Proof of Work (dPoW) is a security mechanism designed by the Komodo project. It is basically a modified version of the Proof of Work (PoW) consensus algorithm that makes use of Bitcoin's hash power as a way to enhance network security.

Hybrid Consensus: In this technique the authors present a method where all the PoW chains which are generated simultaneously are submitted to a committee. Then the committee decides the best chain and accept it as the main chain. The election of best chain among the committee members is based on a weight calculation. The weight of each committee member is calculated from the PoW power and PoS capability.

Disincentivization Method: Every transaction sent to a trustee is endorsed by observers, i.e. the trustee accepts those transactions only which are endorsed by the observers.

History Weighted Difficulty: This technique takes into account the distribution of miner addresses in the last certain amount of blocks of the blockchain. The assumption is made that in an honest blockchain branch, miners of new blocks will most likely be the miners who mined the previous blocks, and the distribution will reflect the ratio in history (Table 2).

Table 2. Techniques for 51 percent attack

Reference	Technique	Application	Elements used	Result	Limitations
Bastiaan, 2015	Markov chains	Two phase PoW	Bitcoin mining protocol	Large mining pools are divided into smaller pools	Issue of decreasing mining power remains unattended
Luu et al., 2017	SmartPool	Smart Contracts	Data structure called augmented Merkle tree	Solution for distributed pool mining is introduced in the form of smart contracts	A pool may have many users resulting in multiple messages

(continued)

Table 2. (*continued*)

Reference	Technique	Application	Elements used	Result	Limitations
Block, 2018	ChainLocks	Dash	long living masternode quorums (LLMQs)	Locks-in the very first block as a genuine block by discarding any other blocks or chain of blocks	One block confirmation may lead to double-spending with minimal hashing, so the risk level is high
Garoffolo et al., 2018	Penalty system by Horizon	Penalty system for delayed blocks	Fork acceptance delay time	Attacker has to suffer with the penalty for late mining	Violates the longest chain rule of blockchain
Komodo, 2018	Delayed proof of work (dPoW) by Komodo	Delayed proof of work (dPoW)	Unspent transaction output (UTXO)	Adds a security layer to prevent attackers from performing the 51 attack	Does not identify the vulnerability ahead. Coins with few seconds of confirmation time can be at risk
Gupta et al., 2019	Hybrid consensus	PoW and PoS	Strict spaced timestamp	Hybrid PoS-PoW cryptocurrency is created to eliminate 51 percent attack	If a node acquires majority of network processing capability and network stake, it can execute 51 percent attack
Sai & Tipper, 2019	Disincentivization method	Interoperable blockchains		seven analyzed monetary conditions are satisfied during the determination of incentives amount	Rational observer may miss out some double spent transactions
Yang et al., 2019	History weighted difficulty		History weighted information of miners	Cost of 51 percent attack increases by order of two	Small blockchains are still vulnerable

5   Conclusion

Blockchain technology is progressing in every field be it be in finance, healthcare, business, or data management. So it gets extremely important to ensure its security and removing all the possible security breaches. The paper discusses the effects of 51 percent attack over blockchain. Then the existing approaches to put a halt over 51 attack are presented in the tabular form. As per the survey conducted, the hybrid consensus method is adopted by the researchers to avoid 51 percent attack. The governing factor for the transactions is also used in some approaches which is not completely reliable. There are platforms developed in order to enhance the blockchain security but very few are there to keep a check over the cyber attacks like 51 percent attack which is predominant among all the attacks over blockchain. So there is scope in the direction of improving the security of blockchain to avoid the risk of such attacks.

Acknowledgements. This research is being carried out for Ph.D. work. I thank my supervisor Dr. Gagandeep for assisting me in improving this manuscript. I would like to show my gratitude to my supervisor for sharing her pearls of wisdom with me during the entire course of this research work.

Ethical Approval.   This article does not contain any studies with human participants or animals performed by any of the authors.

Informed Consent.   Not applicable.

Conflict of Interest.   Author 1 declares that she has no conflict of interest. Author 2 declares that she has no conflict of interest.

References

Nakamoto, S.: Bitcoin: a peer-to-peer electronic cash system. (2008). https://bitcoin.org/bitcoin.pdf

Karame, G.O., Androulaki, E., Capkun, S.: Double-spending fast payments in bitcoin. In: Proceedings of the 2012 ACM Conference on Computer and Communications Security, pp. 906–917 (2012)

Bastiaan, M.: Preventing the 51-attack: a stochastic analysis of two phase proof of work in bitcoin. In: 22nd Twente Student Conference on IT, pp. 1–10 (2015)

Zhang, F., Cecchetti, E., Croman, K., Juels, A., Shi, E.: Town crier: an authenticated data feed for smart contracts. In: Proceedings of the ACM SIGSAC Conference on Computer and Communications Security, pp. 270–282 (2016)

Luu, L., Chu, D.-H., Olickel, H., Saxena, P., Hobor, A.: Making smart contracts smarter. In: The 2016 ACM SIGSAC Conference on Computer and Communications Security, pp. 254–269 (2016)

Kosba, A., Miller, A., Shi, E., Wen, Z., Hawk, C.P.: The blockchain model of cryptography and privacy-preserving smart contracts. In: IEEE Symposium on Security and Privacy, pp. 839–858 (2016)

Gervais, A., Karame, G.O., Wüst, K., Glykantzis, V., Ritzdorf, H., Capkun, S.: On the security and performance of proof of work blockchains. In: The ACM SIGSAC Conference on Computer and Communications Security, pp. 3–16 (2016)

Luu, L., Velner, Y., Teutsch, J., Saxena, P.: Smart pool: practical decentralized pooled mining. USENIX Security Symposium, pp. 1409–1426 (2017)

Garoffolo, A., Stabilini, P., Viglione, R., Stav, U.: A penalty system for delayed blocksubmission (2018). https://www.horizen.global/assets/files/A-Penalty-System-for-Delayed-Block-Submis sion-by-Horizen.pdf. Accessed 2 Aug 2020

Alexander Block. Mitigating 51 percent attacks with LLMQ-based ChainLocks. 2018. https://blog. dash.org/mitigating-51-attacks-with-llmq-based-chainlocks-7266aa648ec9. Accessed 5 Aug 2020

Gemeliarana, I.G.A.K., Sari, R.F.: Evaluation of proof of work (POW) blockchains security network on selfish Mining. In: International Seminar on Research of Information Technology and Intelligent Systems (ISRITI), pp. 126–130 (2018)

Eyal, I., Sirer, E.G.: Majority is not enough: bitcoin mining is vulnerable. Commun. ACM **61**(7), 95–102 (2018)

Bae, J., Lim, H.: Random mining group selection to prevent 51 percent attacks on bitcoin. In: 48th Annual IEEE/IFIP International Conference on Dependable Systems and Networks Workshops (DSN-W). IEEE, pp. 81–82 (2018)

Komodo: Advanced Blockchain Technology, Focused on Freedom, 2018. https://komodoplatform. com/wp-content/uploads/2018/06/Komodo-Whitepaper-June-3.pdf. Accessed 5 Aug 2020

Yang, X., Chen, Y., Chen, X.: Effective scheme against 51 percent attack on proof-of-work blockchain with history weighted information, IEEE International Conference on Blockchain (Blockchain), pp. 261–265 (2019)

Harikrishnan, M., Lakshmy, K.V.: Secure digital service payments using zero knowledge proof in distributed network. In: 5th International Conference on Advanced Computing & Communication Systems (ICACCS), pp. 307–312 (2019)

Sai, K., Tipper, D.: Disincentivizing double spend attacks across interoperable blockchains. In: First IEEE International Conference on Trust, Privacy and Security in Intelligent Systems and Applications (TPS-ISA), pp. 36–45 (2019)

Bang, J., Choi, M.: Design and implementation of storage system for real-time blockchain network monitoring system. In: 20th Asia-Pacific Network Operations and Management Symposium (APNOMS), pp. 1–4 (2019)

Liu, T., Wu, J., Li, J.: Secure and balanced scheme for non-local data storage in blockchain network. In: IEEE 21st International Conference on High Performance Computing and Communications; IEEE 17th International Conference on Smart City; IEEE 5th International Conference on Data Science and Systems (HPCC/SmartCity/DSS), pp. 2424–2427 (2019)

Saad, M., et al.: Exploring the attack surface of blockchain: a systematic overview, IEEE Communications Surveys & Tutorials, pp. 1–30 (2019)

Gupta, K.D., et al.: A hybrid POW-POS implementation against 51 percent attack in cryptocurrency system. In: IEEE International Conference on Cloud Computing Technology and Science (CloudCom), pp. 396–403 (2019)

Marangappanavar, R.K., Kiran, M.: Inter-planetary file system enabled blockchain solution for securing healthcare records, In: Third ISEA Conference on Security and Privacy (ISEA-ISAP), pp. 171–178 (2020)

Transfer Learning Based Convolutional Neural Network (CNN) for Early Diagnosis of Covid19 Disease Using Chest Radiographs

Siddharth Gupta[1]($\boxtimes$), Avnish Panwar[2], Sonali Gupta[2], Manika Manwal[2], and Manisha Aeri[2]

[1] Computer Science & Engineering Department, Graphic Era Deemed to be University Dehradun, Dehradun, Uttarakhand, India
[2] Computer Science & Engineering Department, Graphic Era Hill University Dehradun, Dehradun, Uttarakhand, India

Abstract. Covid19 is a deadly disease that spreads in the lungs and may result in damaging both the lungs. This infection may be life-threatening if not detected at right time. In this work, chest radiographs were used as the input images. Several pre-trained CNN models such as VGG16 and VGG19 were used for transfer learning. The features were extracted after pre-processing the images. Finally, these images were provided as input to several machine learning classifiers for the classification of images as Covid19 infected or normal chest x-ray images. The classification accuracy of 99.5% was obtained by using the VGG19 model and logistic regression classifier. The results show that the current work will be very useful in the early detection of Covid19 disease to provide in-time treatment to the patients.

Keywords: Coronavirus · Machine learning · Transfer learning · Classification · VGG16 · VGG19

1 Introduction

The immediate escalation of newly discovered coronavirus across the globe has brought aberrant load on the health sector around the world and eradicated the best medical facilities of developed countries [1]. It is a severe infectious disease precipitate due to the SARS-CoV-2 virus [2]. Coronavirus or Covid-19 severely impacted the economy of many countries and has taken away many lives. Due to its deadly causes the World Health Organization (WHO) declared it as a Worldwide pandemic on January 30th, 2020 [3]. According to the reports, till May 11th, 2021 the total patients suffering from Covid19 were 159,699,271 and the total deaths due to Covid19 were 3,319,919 [4]. The common symptoms found in Covid19 positive patients are cold, cough, fever, shaking chills, disturbed throat, dyspnea.

To overcome the dreadful effects of Covid19 and to manage such a large crowd an effective testing technique is an utmost required. The only available technique to

© The Author(s), under exclusive license to Springer Nature Switzerland AG 2022
R. Misra et al. (Eds.): ICMLBDA 2021, LNNS 256, pp. 244–252, 2022.
https://doi.org/10.1007/978-3-030-82469-3_22

detect the presence of coronavirus in the human body is by using a nose swab PCR test called as Reverse Transcription Polymerase Chain Reaction (RTPCR) [5]. However, in developing countries where the population is large, it will not be possible to provide the RTPCR testing kits to everyone. Also, due to the high cost and large diagnosis time (a day or two) of RTPCR makes it uneasy to use [6]. Thus, a financial constraint becomes a big challenge in providing testing for all the citizens especially in under developing countries.

To provide the solution for the aforementioned problems, the admittance of medical imaging techniques such as Computed Tomography (CT), Ultrasounds, Magnetic Resonance Imaging (MRI) were used to examine the infected person. All these image techniques have a unique mode of operations and purposes. Among these the most commonly used image techniques for deep analysis of the study of lungs is by using the chest radiographs [7]. The radiologists apparently diagnose the condition of the lungs such as blockage caused by mucus (greenish/yellow cough), breathlessness, etc. by using chest x-ray scans. Also, major infections like pneumonia, emphysema, etc. can be detected using chest radiographs [8]. The major advantage of using chest radiography is the ease of availability of chest x-ray machines everywhere, low cost, and its high sensitivity [9].

The advancement in medical image techniques these days helps doctors and radiologists to provide early treatment for the speedy recovery of patients. Due to the variation of doctor-patient ratio in the under developing countries require an expert system for early detection of disease is required. Chest radiographs (chest x-ray scans) are used to diagnose viral and bacterial infections accurately. Deep learning models are used for the analysis of chest scans to detect the presence of coronavirus [10]. Deep learning plays a major role by providing help to doctors and radiologists to detect the patterns in medical images. A computer-Aided Diagnosis (CAD) system provides medical assistance to professionals in making clinical decisions [11].

The objective of this study is to employ the different deep learning models for the early analysis of Covid19 patients using chest radiographs. Figure 1 A) shows the chest radiograph of covid19 positive patient and B) shows the normal chest radiograph.

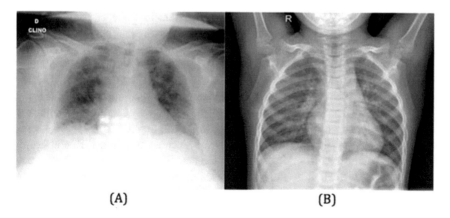

(A)   (B)

Fig. 1. Chest Radiographs A) Covid-19 positive B) Normal chest x-ray

Table 1 consists of the data of ten countries across the world. The total number of covid19 cases, the total deaths, the number of active cases and test per million of population were shown.

Table 1. Covid-19 cases data for ten countries across the world [3]

S. no.	Country name	Total cases	Total death	Active cases	Test/1 M population
1	USA	33,515,308	596,179	6,411,702	1,381,034
2	India	22,992,517	250,025	3,715,188	219,603
3	Brazil	15,214,030	423,436	1,031,469	219,002
4	France	5,780,379	106,684	756,594	1,215,915
5	Turkey	5,044,936	43,311	257,754	582,789
6	Russia	4,896,842	113,976	272,951	903,498
7	UK	4,437,217	127,609	58,909	2,431,711
8	Italy	4,116,287	123,031	373,670	1,013,935
9	Spain	3,581,392	78,895	227,689	1,009,467
10	Germany	3,535,354	85,481	252,973	676,522

2   Related Work

A lot of research was carried out by many researchers to detect the presence of the SARS-CoV-2 virus in the human body. Wang et al. in [12] collected the 1065 CT images of the person who is detected positive for antigen test. Artificial Intelligence methods were opted to diagnose the Covid19 disease. The result shows an accuracy of 73.1%, sensitivity of 74%, and specificity of 67% in the case of external testing. Grewal et al. in [13] used the 77 brain CT images for the detection of a brain hemorrhage. These CT images were passed to various deep learning models such as DenseNet. Also, RAN networks were incorporated. The result shows that using the RADnet 81.82% accuracy in predicting the brain haemorrhages was obtained. Murphy et al. in [14] used a CAD4COVID-Xray to diagnose the Covid19 disease from chest x-ray images. The radiographs were diagnosed by six experts and also with AI system. The results show that the results obtained by the RTPCR test correctly matches with AI system with an AUC of 81%. Xiaowei Xu et al. in [15] used the 618 CT images that were collected from three different hospitals in Zhejiang Province, China. After applying the deep learning models, the accuracy of 86.7% is obtained. Shah et al. in [16] proposed a new deep learning model CTnet-10 for the early diagnosis of Covid19 using the CT scans. The result shows an accuracy of 82.1% is marked by using the CTnet-10 model. Also, different deep learning models like DenseNet-169, VGG16, ResNet-56, InceptionV3, and VGG19 were applied. Out of these models, the highest accuracy of 94.52% is achieved by VGG19 model.

3   Methodology

3.1   Dataset Used and Pre-processing

The dataset images used in this work were extracted from Twitter [17]. These images were collected from my hospital, Spain. Along with these images, some normal chest x-ray images were added. The objective of the author of this dataset is to provide chest x-ray images of Covid-19 positive patients. The given dataset consists of 88 images of Covid19 positive patients and 317 chest x-ray images of normal people. These images are of varying shape and size. Therefore, image pre-processing technique such as image cropping and image resizing is performed to bring all the chest x-ray scans at the same level. Table 2 describes the covid19 chest x-ray images and normal chest x-ray scans used in the work.

3.2   Architecture Used

3.2.1   Convolutional Neural Network Model

In the current work two different model of CNN [18] are compared for the automatic detection of Covid19 infection. Once the dataset images are pre-processed, these images are fed to several deep learning model for the feature extraction.

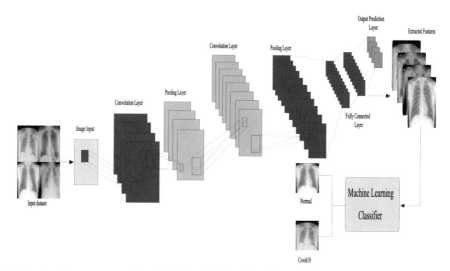

Fig. 2. Architecture containing CNN model for detection of Covid19 disease using chest radiographs

In this work, we have used VGG16 and VGG19 pre-trained model for feature extraction. The VGG16 and VGG19 models are used for their simplicity and ease of use. The VGG16 model is 16 layers deep and comprises a convolution layer that is used for extracting the features (spatial and temporal) from the input image. After this, a Max-pooling layer is present which is used for down sampling the image [19]. This layer results in reducing the dimensionality of the image. Finally, the fully connected layer is used for the classification capabilities. The overall architecture used for the detection of Covid-19 disease using the CNN network can be extracted from Fig. 2. In comparison to the VGG16 model, the VGG19 model performs slightly better as it is 19 layers deep. However, due to the more number of layers, the memory consumption in the VGG19 model is more than that in the VGG16 model.

3.2.2 Transfer Learning

Transfer Learning is defined as a technique where knowledge gained by training one model can be used to implement the other model of the same type. The dataset that constitutes the new problem is smaller to train the CNN from very basic. Transfer learning initiates by training the deep neural network for the distinct tasks by using a large dataset like ImageNet. In our work, the number of images used in the dataset is very few. Therefore, the transfer learning technique is used to increase the number of images to train the model well [20].

3.3 Evaluation Parameters

After the features are extracted from the images using the CNN models, these images were transferred to different machine learning classifiers such as k nearest neighbors (kNN) [21], support vector machine (SVM) [22], random forest (RF), logistic regression (LR) [23], AdaBoost for the classification of images as Covid19 infected or the normal chest x-ray image. Also, to evaluate the performance of the CNN models used in this work different parameters are considered such as accuracy [24], sensitivity [24], specificity, precision, f1 score [24] and AUC curve [25]. Equations (1), (2), (3), (4) gives the values of these parameters.

$$\text{Accuracy} = (TP + TN)/(TP + FN) + (TN + FP) \tag{1}$$

$$\text{Sensitivity} = TP/(TP + FN) \tag{2}$$

$$\text{Specificity} = TN/(FP + TN) \tag{3}$$

$$\text{Precision} = TP/(TN + FP) \tag{4}$$

$$\text{F1 score} = (2 * TP)/(2 * TP + FN + FP) \tag{5}$$

For the aforementioned equations while classifying the covid19 disease and normal images, true positive (TP), true negatives (TN), false positive (FP), false negative (FN) used to denote the total number of images of covid19 classified as covid19, the number of healthy images identified as healthy, the number of healthy images misclassified as covid19 and the number of covid19 images misclassified as healthy images.

4 Results

The primary objective of this work is to use transfer deep learning models for diagnosis of chest x-ray images as Covid19 infected or normal chest x-ray. Initially, the chest x-ray images after pre-processing were fed to VGG16 and VGG19 pre-trained models for feature extraction. Once the features were extracted these images were passed to several machine learning classifier for classifying the images as Covid19 infected or normal chest x-ray image. Table 2 gives the performance of VGG16 model along with the several classifiers used.

Table 2. Covid-19 cases data for ten countries across the world [3]

Model	AUC	Accuracy	F1	Precison	Recall
kNN	0.998	0.993	0.993	0.993	0.993
SVM	0.989	0.938	0.940	0.947	0.938
RF	0.997	0.980	0.980	0.980	0.980
LR	0.998	0.993	0.993	0.993	0.993
AdaBoost	0.978	0.985	0.985	0.985	0.985

Table 3 gives the performance of VGG19 model along with the several classifiers. It can be observed that VGG19 model along with the logistic regression classifiers gives the highest accuracy of 99.5%.

Table 3. Covid-19 cases data for ten countries across the world [3]

Model	AUC	Accuracy	F1	Precison	Recall
kNN	0.997	0.993	0.993	0.993	0.993
SVM	0.991	0.931	0.934	0.944	0.931
RF	0.998	0.992	0.995	0.995	0.995
LR	0.999	**0.995**	0.995	0.995	0.995
AdaBoost	0.951	0.975	0.975	0.975	0.975

The ROC [26] curve demonstrating the performance of the VGG16 and VGG19 models along with different classifiers can be seen from Figs. 3 and 4. The ROC curve is drawn between the sensitivity and the specificity.

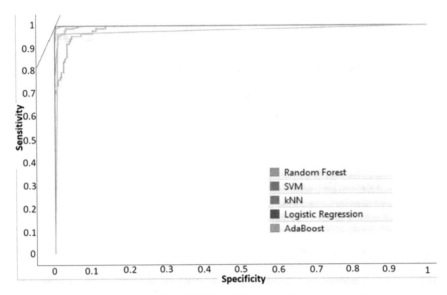

Fig. 3. ROC curve for VGG16 model along with various classifiers

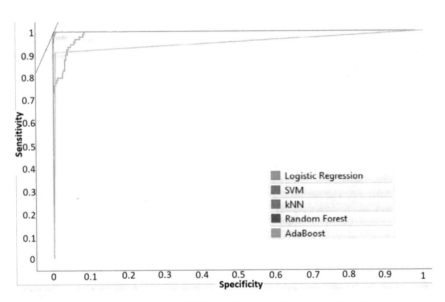

Fig. 4. ROC curve for VGG19 model along with various classifiers

5 Conclusion

The impact of Covid19 has impacted several countries across the world. This deadly disease has taken away many lives, destroyed the economy of many countries, and vanished the best medical facilities. In the current work, a deep CNN-based transfer learning technique is implemented for the automatic diagnosis of Covid19 disease from chest radiographs. It was noted that the VGG19 model used for the feature extraction and the Logistic Regression classifier used for the classification together gives the classification accuracy of 99.5%. The results show that this technique is very much useful in the detection of covid19 disease as the previous techniques such as RTPCR takes a long time in detection of covid19 positive patient. We conclude that using this technique the radiologist may detect the covid19 positive patient early and provide the best in time treatment to save a life. In the future, we will try to enhance the accuracy by either increasing the number of images in the dataset or by implementing the new CNN pre-trained models on the same dataset.

References

1. Chatterjee, P., et al.: The 2019 novel coronavirus disease (COVID-19) pandemic: a review of the current evidence. Indian J. Med. Res. **151**(2–3), 147 (2020)
2. Zhong, N.S., et al.: Epidemiology and cause of severe acute respiratory syndrome (SARS) in Guangdong, People's Republic of China, in February, 2003. Lancet **362**(9393), 1353–1358 (2003)
3. Jebril, N.: World Health Organization declared a pandemic public health menace: a systematic review of the coronavirus disease 2019 "COVID-19." SSRN Electr. J. (2020). https://doi.org/10.2139/ssrn.3566298
4. Covid-19 total cases and total deaths till May 11, 2021. https://www.worldometers.info/coronavirus/
5. Cho, K.-O., Hasoksuz, M., Nielsen, P.R., Chang, K.-O., Lathrop, S., Saif, L.J.: Cross-protection studies between respiratory and calf diarrhea and winter dysentery coronavirus strains in calves and RT-PCR and nested PCR for their detection. Adv. Virol. **146**(12), 2401–2419 (2001). https://doi.org/10.1007/s007050170011
6. Xie, X., Zhong, Z., Zhao, W., Zheng, C., Wang, F., Liu, J.: Chest CT for typical coronavirus disease 2019 (COVID-19) pneumonia: relationship to negative RT-PCR testing. Radiology **296**(2), E41–E45 (2020)
7. Jain, R., Gupta, M., Taneja, S., Hemanth, D.J.: Deep learning based detection and analysis of COVID-19 on chest X-ray images. Appl. Intell. **51**(3), 1690–1700 (2020)
8. Chandra, T.B., Verma, K.: Pneumonia detection on chest X-Ray using machine learning paradigm. In: Chaudhuri, B.B., Nakagawa, M., Khanna, P., Kumar, S. (eds.) Proceedings of 3rd International Conference on Computer Vision and Image Processing. AISC, vol. 1022, pp. 21–33. Springer, Singapore (2020). https://doi.org/10.1007/978-981-32-9088-4_3
9. Mez, J., et al.: Clinicopathological evaluation of chronic traumatic encephalopathy in players of American football. JAMA **318**(4), 360–370 (2017)
10. Bassi, P.R.A.S., Attux, R.: A deep convolutional neural network for COVID-19 detection using chest X-rays. Res. Biomed. Eng. (2021). https://doi.org/10.1007/s42600-021-00132-9
11. Karar, M.E., Hemdan, E.-D., Shouman, M.A.: Cascaded deep learning classifiers for computer-aided diagnosis of COVID-19 and pneumonia diseases in X-ray scans. Complex Intell. Syst. **7**(1), 235–247 (2020)

12. Wang, S., et al.: A deep learning algorithm using CT images to screen for Corona Virus Disease (COVID-19). Eur. Radiol. **31**(8), 6096–6104 (2021)
13. Grewal, M., Srivastava, M.M., Kumar, P., Varadarajan, S.: Radnet: radiologist level accuracy using deep learning for hemorrhage detection in CT scans. In: 2018 IEEE 15th International Symposium on Biomedical Imaging (ISBI 2018), pp. 281–284. IEEE (2018)
14. Murphy, K., et al.: COVID-19 on chest radiographs: a multireader evaluation of an artificial intelligence system. Radiology **296**(3), E166–E172 (2020)
15. Xiaowei, X., et al.: A deep learning system to screen novel coronavirus disease 2019 pneumonia. Engineering **6**(10), 1122–1129 (2020)
16. Shah, V., Keniya, R., Shridharani, A., Punjabi, M., Shah, J., Mehendale, N.: Diagnosis of COVID-19 using CT scan images and deep learning techniques. Emerg. Radiol. **28**(3), 497–505 (2021)
17. Dataset Image. https://twitter.com/ChestImaging/status/1243928581983670272
18. Albawi, S., Mohammed, T.A., Al-Zawi, S.: Understanding of a convolutional neural network. In: 2017 International Conference on Engineering and Technology (ICET), pp. 1–6. IEEE (2017)
19. Bailer, C., Habtegebrial, T., Stricker, D.: Fast feature extraction with CNNs with pooling layers (2018). https://arxiv.org/abs/1805.03096
20. Huh, M., Agrawal, P., Efros, A.A.: What makes ImageNet good for transfer learning? (2016). https://arxiv.org/abs/1608.08614
21. Shaban, W.M., Rabie, A.H., Saleh, A.I., Abo-Elsoud, M.A.: A new COVID-19 Patients Detection Strategy (CPDS) based on hybrid feature selection and enhanced KNN classifier. Knowl. Based Syst. **205**, 106270 (2020)
22. Sethy, P.K., Behera, S.K, Ratha, P.K., Biswas, P.: Detection of coronavirus disease (COVID-19) based on deep features and support vector machine (2020)
23. Bianchetti, A., et al.: Clinical presentation of COVID19 in dementia patients. J. Nutr. Health Aging **24**, 560–562 (2020)
24. Gupta, S., Panwar, A., Goel, S., Mittal, A., Nijhawan, R., Singh, A.K.: Classification of lesions in retinal fundus images for diabetic retinopathy using transfer learning. Int. Conf. Inf. Technol. **2019**, 342–347 (2019)
25. Panwar, A., Semwal, G., Goel, S., Gupta, S.: Stratification of the lesions in color fundus images of diabetic retinopathy patients using deep learning models and machine learning classifiers. In: 26th Annual International Conference on Advanced Computing and Communications (ADCOM 2020), Silchar, Assam, India (2020)
26. Mushtaq, J., et al.: Initial chest radiographs and artificial intelligence (AI) predict clinical outcomes in COVID-19 patients: analysis of 697 Italian patients. Eur. Radiol. **31**(3), 1770–1779 (2020)

Review of Advanced Driver Assistance Systems and Their Applications for Collision Avoidance in Urban Driving Scenario

Manish M. Narkhede[✉] and Nilkanth B. Chopade

Research Center, E&TC Department, Pimpri Chinchwad College of Engineering Pune, Savitribai Phule Pune University, Pune, India

Abstract. Automobile safety systems have become the most important area of research and development in today's world. It is observed that due to increased population and heavy traffic, the number of on-road accidents are also increasing. According to the varied road safety reports, most road accidents occur be-cause of driving error, human behavior, traffic congestion and lane change over, etc. Advanced driver-assistance systems (ADAS) are mainly focusing on automating, and enhancing various vehicle related tasks to provide the better driving experience, which ultimately increases the safety of driver, passenger, and road users as well. The intelligent ADAS system takes appropriate measures to solve problems during transmission by providing automatic controls for varying the speed of the vehicle or stop it during emergency situations. This technique monitors distance between a moving-vehicles, obstacles, etc. With the advancements in technology, today's cars are equipped with many advanced driver assistant systems, which play a significant role in detection of road-side objects including vehicles, cyclist, pedestrians, obstacles, etc. and assist the driver through the navigation. The ADAS system enables safe and comfortable driving, based on intelligent algorithms and sensor technology. ADAS together with a secure human-machine interface, increase both car safety as well as road safety. This paper primarily focusses upon reviewing ADAS and its applications for effective collision avoidance with road objects and users in urban driving scenario.

Keywords: ADAS · Sensor fusion · RADAR · LiDAR · GPS · ACC · ISA · Computer vision · Collision avoidance · VRU · Automotive safety · Machine learning

1   Introduction

ADAS technology is mainly focused on providing the automated solutions to various driving related tasks, to help in avoiding road accidents by minimizing the human error and thereby reduce road fatalities. Many road crashes take place for such reasons as driving error, behavior of road users, traffic congestion, reckless driving and lane change over. Safety systems have become critical for automotive companies because customers have started emphasizing more on safety aspects to be integrated inside a vehicle. With

© The Author(s), under exclusive license to Springer Nature Switzerland AG 2022
R. Misra et al. (Eds.): ICMLBDA 2021, LNNS 256, pp. 253–267, 2022.
https://doi.org/10.1007/978-3-030-82469-3_23

the help of the new technology, we are trying to change the behavioral effects of drivers and implement initiatives relating to driving safety on express roads and highways.

Forward collision alert system tracks the distance between a vehicle, barrier, and the objects around it. Advanced Driver Assistance System (ADAS) is a control system that uses intelligent sensor technology to detect the environment, making the driver relaxed when driving as he would be able to understand both parameters and control during traffic situations. Intelligent ADAS provides automatic lateral and longitudinal control during emergency situations and slows down or stops the car. This ADAS enhances the safety of people during driving as they are able to recognize present scenario and react accordingly. With ADAS systems driving becomes safer and gives more confidence to the drivers.

However, there are many challenges in this driver assistance technology, such as the achievement of data scaling and diversity to ensure safety and reliability in severe conditions, and the testing and execution of algorithms to the best of their ability to provide satisfactory performance in more diverse urban and unstructured semi-urban road conditions.

2 Background and Need of Research

Road traffic accidents have resulted in about 1.35 million fatalities worldwide every year, with non-fatal injuries ranging from 20 to 50 million people. Majority of road accidents and injuries include vulnerable road users, cyclists, pedestrians, etc. Young people and elderly people are especially vulnerable on the high-ways, and large number of deaths have been reported so far in this category. Developing countries have reported comparatively high rates of road accident deaths, with 93% of fatalities. Suffering also entails a major economic strain on victims and their families, both through the cost of care for the wounded or disabled people. More generally, traffic accidents also impact on economies, costing countries around 3% of their annual gross domestic expense [1].

With India's rapid growth and expanding economy, motorization is rising rapidly. However, the country still has severe problems associated with road accidents. According to the Open Government Data Platform (OGD) of India, national highways accounted for 30.4% of total road incidents and 36% of fatalities in 2017. Accidents on State highways and other roads account for 25% and 44.6% respectively. Among the various types of motor vehicles involved in accidents, the highest proportion of two wheelers was 33.9% of the overall accidents in 2017 [2]. According to the 2015 National Highway Safety Administration (NHTSA) statistical report, this causes about 94% of accidents [3]. The fall in road accidents was not so promising. Road accidents in India decreased by just 3.27% in 2017 with 4.65 lacs road accidents compared to 4.81 lacs in 2016 (Fig. 1).

The deaths arising from these incidents saw a decline of just 1.9%. About 1.47 lacs people died in a road crash in 2017 versus 1.51 lacs in 2016. This not so promising data is further undermined by the estimates for road deaths in the first quarter of 2018, which indicate a 1.68% spike over the corresponding previous quarter. Having said that, road accidents are a major concern in India. According to statistics from the NCRB-National Crime Records Bureau, about 35% of incidents registered in the country are road accidents with other unnatural causes, followed by 13% [2, 4].

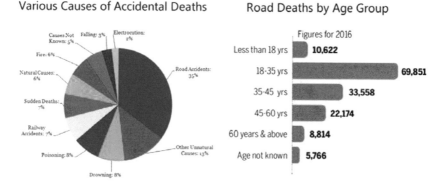

Fig. 1. Accident statistics in India 2016

Currently, there are many ADAS solutions being used in vehicles with varying levels of automation viz, level 0, 1, 2, 3, 4 and 5, as specified by the International Standard of the Society of Automotive Engineers (SAE). The key motivation behind the advancement of Automated Vehicle Technologies (IVT) and Intelligent Transportation Systems (ITS) is the insusceptibility of human related errors that may be caused by distraction, exhaustion, road user actions, misperception due to weather conditions, etc. Intelligent technologies are being developed for autonomous vehicles to effectively identify vulnerable road users in the potentially fatal collisions. However, due to many drawbacks such as applicability and higher costs of the available perception techniques, the majority of vehicles fall under level 1 to level 2 of autonomy [5, 6]. This paper focuses primarily on the review of existing ADAS-based low-cost sensor perception systems to achieve the road safety objective in semi-automated and intelligent vehicles.

3   Advanced Driver Assistance Systems

ADAS are electronic control systems equipped with an interactive interface that support the vehicle driver in a variety of ways. They are designed using built-in ECU modules, signal conditioning devices, onboard MCU, and driver circuits [7]. Automakers regularly update their vehicle models to include innovative technology in their vehicles. Driver assistance system features include easier driving experience with adaptive cruise control, glare-free high beam and pixel illumination, automatic parking, directed navigation system, traffic signal recognition, automotive night vision, blind spot accident warning system, crash avoidance system, etc. All these systems are designed with an objective to avoid road accidents by keeping driver alerted about the potential problems [8, 9].

3.1   Classification of ADAS Systems

The ADAS can be divided into three sections viz, Passive, Active and Cooperative frameworks. The purpose of the Passive system is to mitigate damage after an accident, i.e., to maintain airbags, seatbelts in working condition, while the Active system is intended to avoid crashes or injuries. In the Cooperative framework, V2V communication takes place through custom messages and exchanges of information, so a forward traffic warning is sent to the driver to avoid an accident possibility on busy routes. Cooperative systems are often a mixture of technology, people and organizations that allow communication and collaboration needed to achieve the common purpose of a community that carries out a variety of activities [10, 11]. Various functions of ADAS as per application categories are mentioned in Table 1.

Table 1. Functions of ADAS systems

Driving comfort	Adaptive cruise control, Lane keeping assistance, Automated vehicle control, Automatic parking systems, Glare-free high beam, Rain sensor
Safety	Collision avoidance systems, External vehicle speed control, Crash prevention, Wrong way driving warning
Traffic assistance	Vehicle platooning, Vehicle flow management, Traffic jam dissipation, start-up assist, Vehicle-to-X communication
Lateral motion control	LDWS, Side collision avoidance, LKA, Blind spot monitoring, Lane change assist, Emergency steering intervention, crosswind stabilization
Longitudinal motion control	Adaptive front lighting, Night vision, Safe gap advisory, Deceleration assistance, Active braking, Torque control, Rear collision avoidance

To avoid accidents, we are using ADAS applications based on Information and Communication Technology (ICT), which assists drivers for better driving. A roadmap concept uses:

1. Proprioceptive sensors: They are analyzing the vehicle's behavior and respond accordingly to danger situations;
2. Exteroceptive sensors: They respond on an early stage by predicting possible dangers (e.g., LiDAR, RADAR, ultrasonic sensors, IR and camera sensors);
3. Sensor networks: It further includes application of multisensory platforms and traffic sensor networks [7, 12].

3.2 Sensors Used for Navigation Safety

The ADAS systems use different types of sensors to perceive the environment, and the ECU element then processes the data. Automotive sensors come into two categories: active and passive sensors. Active sensors emit energy in the form of waves and search for artifacts based on the in-formation that reflects back. Passive sensors actually take information from the environment without emitting waves, such as a camera (Table 2).

Figure 2 gives an overview of the sensors used in ADAS-enabled vehicles. The ADAS architecture facilitates the fusion and action of the sensor. Early fusion is helpful in order to evaluate the short-term conditions and a smart centralized processing unit in order to increase the efficiency of detection [13, 14].

Table 2. Sensor technologies and applications

Sensor	Applications
Front Cameras	On-road vehicle detection, Sign recognition
Side and Inside Cameras	Blind spot assistance, Lane change assistance, Driver behaviour monitoring
Infrared Cameras	Pedestrian detection during night time
RADAR Sensor	Audio assistance, Reversing aid
Long Range RADAR	Audio assistance, Obstacle detection, Adaptive cruise control "stop and go"
LIDAR	Distance and speed detection
Ultrasonic Sensors	Parking assistance, Rear crash alert
GPS	Guided navigation, Collision avoidance, Collision mitigation services

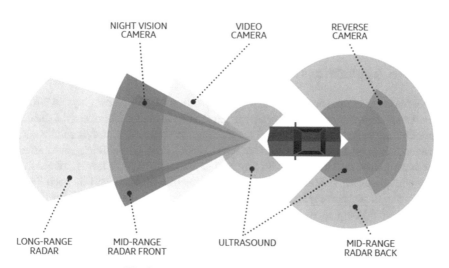

Fig. 2. Sensors used in intelligent vehicles

4   ADAS Architecture

The ADAS architecture contains modules for sensing, processing, learning and decision making. Figure 3 is a general representation of the ADAS method. This framework comprises of sensors; a mixture of CPU-GPUs for information processing.

All ADAS computing takes place in electronic control systems (ECUs). Sensor fusion is the innovation in ADAS by which the inner processing takes feedback from the multiplicity of external sensors and produces a map of potential impediments around the vehicle. This system is taken into account as a close-loop system where the response of the vehicle control actuation is computed according to the sensory data, and the output of the ADAS actuation is fed back into the loop [13].

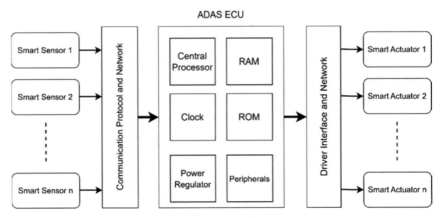

Fig. 3. ADAS architecture

5   ADAS Applications for Collision Avoidance

ADAS plays a significant role in the next generation of AVs. Nowadays, ADAS can make decisions about drivers that override human driving error. For example, the emergency braking system and the LDWS are already incorporated in commercial vehicles. Advanced sensors such as cameras, lidar, RADAR improves the vehicle's environmental perception capability. However, with different autonomous vehicle levels, we are facing a new challenge that requires the ability of artificial intelligence and its socially integrated types to make complex risk mitigation decisions, as they are directly linked to life and death consequences [15].

Extensive research work is being conducted across the globe for developing algorithms for the use of sensor data and integration into the control system [16]. Some low cost ADAS systems rely on a collision prevention method focused on camera vision perception. These systems are used to identify different types of vulnerable road users (VRUs) by algorithms used to detect and track them [17, 18].

Broad estimates of the safety potential of such systems have been claimed to date, but the range of technological and behavioral problems involved with many of the concepts needs complete on-road assessment [19, 20]. To be realistic, most of the proposed systems require a regulated and specified traffic situation where the capacity to minimize accidents is relatively low. Several systems are under development, with a particular emphasis on urban driving [21].

5.1   Adaptive Cruise Control (ACC)

ACC is an enhancement to the current road vehicle cruise control system, a mechanism for maintaining the secure distance of the host vehicle to the adjacent, frontal or preceding vehicle on the road. When the speed of the vehicle ahead is slower than the adjusted speed, the ACC system adjusts the speed of the driven vehicle accordingly to maintain a safe distance [22, 23].

The cruise controller helps to incorporate increasingly complex applications such as vehicle platooning and collision prevention. These systems automatically apply maximum braking and trigger the hazard lights following a collision with the airbag. The goal is to prevent a secondary collision with another vehicle or obstacle. If the driver feels that the braking is likely to increase the risk, it is possible to defeat the mechanism by depressing the accelerator [24, 25].

Lateral control systems keep track of the side areas of vehicle and takes appropriate steps to prevent collision possibility [26]. Lane monitoring and alert systems help the car remain in the lane, stopping drowsy or inattentive drivers from crossing the lane and hitting an obstacle. Longitudinal control systems are used in tracking the front and rear regions of the vehicle and, if necessary, operate on the throttle and brakes in case of potential collisions [27, 28].

5.2 Collision Warning and Avoidance Systems

The forward collision warning system (FCWS) consists of a visual and audible warning that the driver is moving closer to the front vehicle. There are two corresponding "safe" and "critical" alert levels as the distance between vehicles decreases. Reverse assistance system consists of a rear-view camera and a monitor mounted on a stand. With advancements in deep learning technology, object detection techniques using deep learning (DL) have outperformed a variety of conventional strategies in both speed and precision. In the context of DL-based detection, there are mechanisms to boost detection results in an intelligent computational manner [29].

Generally, there are two approaches of image object recognition that use deep learning. One approach is based upon regional proposals. Faster R-CNN is the example. This technique will first run the entire image input through some convoluted layers to get a feature map. From that point on, a different region proposal network, which uses these convolutional features to propose potential regions for detection. Last, the rest of the network provides grouping to these proposed regions. Since there are two parts in the network, one for the bouncing box forecast and the other for the grouping, this form of design can fundamentally restrict the speed of processing [17] (Fig. 4).

Fig. 4. Vehicle detection: short, medium and long range

The second approach uses a single network for both label classification and prediction of future regions. One of the models is that you only look once (YOLO). Given the image data, YOLO will first partition the image to coarse grids. There is a set of base bounding boxes for each grid. YOLO predicts the offset of the true location, the confidence score and the classification score for each base bounding box, if it thinks there is an object in that grid location. Yolo is faster, but now and then it can fail to recognize small objects in the image [30]. As of late, DL-based methodologies have become increasingly popular among researchers, given their powerful learning capabilities and preferences in occlusion management, scale transition, and back-ground switching.

5.3 Vulnerable Road User Detection

VRU includes non-motorized road users like cyclists, motorcyclists, pedestrians, and disabled people or road users with reduced mobility and orientation. Extensive research is being carried all around the world in detection and tracking of obstacles and also development of algorithms to make use of information available from sensor integrated control system [31]. These algorithms are able to predict whether the VRU be vulnerable to the vehicle's path. In case of possibility of potential collision, suitable actions are taken by the collision avoidance system either to deviate from path, slow down speed or to stop the vehicle.

Additionally, deep learning-based solutions are available for detection and recognition of vulnerable road users using SSD-single shot detector and YOLO-you only look once methods [30, 32]. Relatively to the global rapid expansion of the urban area, the proportion of the road accidents occurring at junctions of highly populated urban traffic areas have increased, where sudden obstacle or pedestrian is most likely to appear in the path of the vehicle. At intersections, the collision is mainly happening due to the driver distractions. Vehicle-to-Infrastructure communication can greatly help in reducing such fatal accidents and achieve a better intersection assistance system as denoted in [33].

Intersection collision warning systems continuously communicate with the infrastructure for detecting vehicles crossing an intersection. They allow the vehicle to depart at low speed or to stop, without any driver's intervention by following the front vehicle. They are built on the same premise as the ACC. Intersection Assistance can be useful in the prevention of violation of traffic signal, wrong turnings, collisions along crossing-path as well as with pedestrians.

Regarding vision based vulnerable road user detection, M. Goldhammer, et al. in 2019 proposed movement models supported ML methods like ANN for classifying the present motion state and to predict the next movement trajectory of the vulnerable road users. Both model sorts walkers and cyclists were joined to convey the use of trained motion detectors based upon a consistently refreshed pseudo probabilistic state classifier. This design was utilized for assessing motion specific physical models for 'start' and 'stop' and video-based motion classification of pedestrian [34] (Fig. 5).

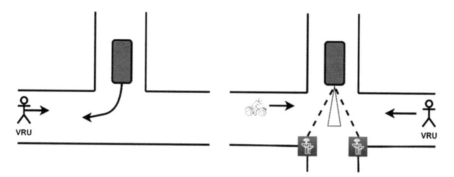

Fig. 5. V2I architecture for VRU safety

5.4 Vision Enhancement and Attention Assistance

These systems are designed to assist drivers in the identification of obstacles or other vehicles during the night. Night vision cameras have the potential to provide supplementary visual information to driver using a separate display. Automotive manufacturers and researchers have developed night vision systems that are generally focused on infrared images. Two methods have been widely used. The first technique is based on near-infrared images which re-enlightens objects present on the road with an infrared light beam. After getting processed, the result image shows illuminated objects. The second technique uses video images from the far-infrared spectrum, which offer thermal chart an atmosphere. No light source is needed. Pedestrians, animals and moving objects in the imagery are hotter than the normal atmosphere [35].

Camera-based night vision support has been continually enhanced through a variety of measures [36]. In 2016, H. Wang et al. proposed a new method for vehicle detection during night time in far infrared images. The vehicle candidate was generated and a DL framework was designed to perform verification of vehicle candidate. This two-step vehicle detection method achieves a maximum vehicle detection rate of 92.3% and a processing time of less than 50 ms per frame compared to existing proven methods, including a maximum distance dependent on local adaptive threshold determination [37, 38] (Fig. 6).

Both FIR-far infrared and NIR-near infrared frameworks are commonly used in most of the systems, as both technologies have their own advantages and disadvantages in processing image, and they are applied to furnish a sharp contrast image. High refresh rates and extremely short latencies are crucial as it helps driver to drive just by viewing the night view image.

5.5 Intelligent Speed Adaptation

According to study reports, it is been observed that drivers appear to reduce speed only for a limited period of time when it comes to law enforcement and speed cameras. In such cases, conventional speed control strategies have proven to be ineffective. The aim of the Intelligent Speed Adaptation system is to maintain the vehicle's speed below the threshold permitted limit. It relies on a GPS system or a communication framework to

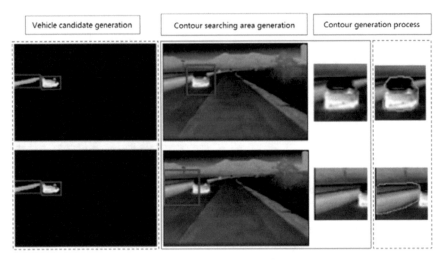

Fig. 6. Vehicle candidate contour generation (night time)

have a local speed limit. Driver shall be alerted about this limit by means of a screen, audio, or a hard to press throttle pedal when it is reached. The Speed limit assistance system senses signals such as 'speed limit' and 'end-of-limit' and shows them to the driver on the instrument cluster screen or on the central monitor. Camera-based recognition is assisted by GPS map data generated by the Automotive Navigation System [39].

In the case of an Intelligent Speed Adaptation, the implementation is either based on static sensors measuring the vehicle speed at the signposts and traffic lights or on the continuous communication of each vehicle speed to the path. In the event of excessive speed, the Center to Car coordination may be used as a guideline for speed reduction or by means of a throttle and/or brake actuator, or where appropriate, traffic lights may be controlled to implicitly stop the speed of the vehicle [40, 41].

6   Discussion and Research Challenges

Alongside the continued and quick growth of vehicle proprietorship, every nation is confronting serious traffic congestion and road safety issues in ever developing urban infrastructure. Based on the review of various ADAS solutions, following research challenges are identified for effective collision avoidance.

A. *Dependability on weather and road conditions*
    In many developing countries like India, ADAS technology isn't fully implemented yet. Even taking urban scenario into consideration most of system failed because the environmental differences throughout the country. The unwanted factors like rain, storm, heat, road condition, improper traffic managements can degrade the performance of the system. Despite of numerous algorithms, protocols and researches, the reliable and accurate detection techniques are yet to be built. The roads of rural side are small and congested, the mid-range or distance region detection become difficult as full view of road is unavailable.

B. *False Positives*

VRU detection imposes many challenges to researchers such as, Small & partially occluded objects, reducing the rate of false positives and intention recognition of VRUs [42]. Movement of VRUs at an intersection convergence and leaning models for foreseeing VRU directions dependent on their present trajectory is recorded as per recent research reports. In the case of computer vision based VRU detection approaches, problems are faced as a result of false detections, which occur as a result of reflections and off-road wide advertising in the background of the scene, which feature people. Any false positives can be removed by taking advantage of established scenic geometry constraints (e.g., pedestrians or cyclists shall be detected on the ground plane). The rate of false positives can also be reduced by tracking objects between frames as reflections present in one frame may not be present in the next [43].

C. *Validation under different driving scenarios*

ADAS allows improved situation understanding and control to make driving easier and safer, ADAS technologies using FPGA/SoCs and advanced vehicle sensors may use local vehicle systems, such as vision/camera systems, sensor techniques, or smart, integrated networks such as vehicle-to-vehicle (V2V) or vehicle-to-vehicle (V2V) networks. However, the implementation of the computer vision algorithm reflects just one aspect of the overall design cycle of the product. One of the most difficult challenges is to validate entire system with a diverse range of driving situations. Usually, significant time and efforts are spent on evaluating and authorizing a specification as compared to designing an algorithm. A large number of long traffic videos are required to validate ADAS based on computer vision. OEMs also spend a great deal of time and resources in collection of recording of traffic which covers all possible scenarios in different weather and on urban roads.

For the reasons listed above, ADAS is emerging to adapt SoC designs to have embedded vision. These SoCs usually include a hardware-based on 32-bit MPU/MCU like ARM, advanced processors like GPU or DSP, and VLSI components like FPGA, CMOS. The low-level processing portion of the algorithm is performed in the FPGA/CMOS, GPU or DSP, and other processing is performed in the microprocessor, incorporating the advantages of the two architectures in order to produce better performance. In addition, since most of these components are physically contained inside the same chip, the overall power consumption is significantly lower than that of multiple separate chips [44, 45]. At present, solutions based on GPU tend to be the least desirable choice as they consume high power. In future, field of embedded vision for ADAS will prominently retain itself as an active area, and new technologies are expected to emerge along with smarter tools to validate them [46].

D. *Need of supporting infrastructure*

Urban concerns in developing countries are compounded by poor state of roads and a lack of traffic management strategies. The bad road conditions and changing appearance of the cars can have a significant impact on the technology performance. Some ADAS are using GPS as their tracking system. Global Positioning System (GPS) data could be unreliable. Using the cameras and GPU, the cost map can be created to calculate the geological properties of the roads [46, 47]. In the night, it is comparatively difficult to spot cars and roadside. Various lights, such as street-lights and reflectors, high-intensity focused beams of other vehicles can interfere in detecting road boundaries and monitoring edges. There should be proper depiction and verification of vast streams of information with radar, lidar, infrared, camera, ultrasonic and others [48].

V2X systems use on-board radio communication technique (short-range) to communicate vehicle's speed related safety messages, heading, braking status, size of the vehicle, and from these messages receives a similar information about surrounding vehicles. The V2X network can communicate over significantly longer distances by utilizing multi-hops to send messages through different nodes. This distance and capacity to "see" around corners or through surrounding vehicles helps V2X-enabled vehicles see a potential danger faster than sensors like cameras, lidar, radar, and caution their drivers appropriately [49]. If we provide adequate integrity and bandwidth of the network and data sources, all the vehicles and the infrastructural data, if accessible, could then be fused into a deep dynamic map [50].

7   Conclusion and Future Directions

In this paper, we have provided a review of advanced driver assistance methods with specific relevance to collision avoidance with obstacles and road users. With the discussion on various ADAS applications and their dependency on environmental and surrounding parameters, it can be therefore concluded that specification and the validation of the system is an essential aspect for the development of driver assistance systems. The ADAS modules from different manufacturers are becoming standard equipment and are now produced for luxury and semi-automated vehicles. Although the use of the ADAS has greatly improved road traffic safety so far, it is essential for drivers to adapt to the use of ADAS effectively. The ever-increasing role of sensors and ADAS modules additionally will facilitate the implementation of advanced intelligent inference systems. It will further allow the constructing vehicle driving systems totally based on AI (artificial intelligence), which is a step towards next levels of autonomous driving. The introduction of machine vision, vehicle-to-infrastructure and vehicle-to-pedestrian into the ADAS market is subject to a numerous challenge, associated to the adaptation of this powerful technology to the limitations imposed by the industry and surrounding real-world factors.

References

1. Alinda, A., et al.: Global Status Report on Road Safety 2018. World Health Organization, Geneva (2018)
2. Shantajit, T., Kumar, C.R., Zahiruddin, Q.S.: Road traffic accidents in India: an overview. Int. J. Clin. Biomed. Res. 4(4), 36–38 (2018)
3. Blanco, M., et al.: Human Factors Evaluation of Level 2 and Level 3 Automated Driving Concepts. (Report No. DOT HS 812 182), No. August, p. 300 (2015)
4. Bhushan, V.: An efficient automotive collision avoidance system for Indian traffic conditions. Int. J. Res. Eng. Technol. 05(04), 114–122 (2016)

5. Badue, C., et al.: Self-Driving Cars: A Survey (2019). http://arxiv.org/abs/1901.04407
6. Combs, T.S., Sandt, L.S., Clamann, M.P., McDonald, N.C.: Automated vehicles and pedestrian safety: exploring the promise and limits of pedestrian detection. Am. J. Prev. Med. **56**(1), 1–7 (2019)
7. Ziebinski, A., Cupek, R., Grzechca, D., Chruszczyk, L.: Review of advanced driver assistance systems (ADAS). AIP Conf. Proc. **1906** (2017)
8. Zhu, H., Yuen, K.V., Mihaylova, L., Leung, H.: Overview of environment perception for intelligent vehicles. IEEE Trans. Intell. Transp. Syst. **18**(10), 2584–2601 (2017)
9. Mosquet, X., Andersen, M., Arora, A.: A roadmap to safer driving through advanced driver assistance systems. Auto Tech Rev. **5**(7), 20–25 (2016)
10. Greguri, M., Mandžuka, S.: The use of cooperative approach in intelligent speed adaptation. In: 2018 26th Telecommunications Forum, pp. 1–4 (2018)
11. Abdi, L., Takrouni, W., Meddeb, A.: In-vehicle cooperative driver information systems. In: 2017 13th International Wireless Communications and Mobile Computing Conference IWCMC 2017, pp. 396–401 (2017)
12. Bengler, K., Dietmayer, K., Färber, B., Maurer, M., Stiller, C., Winner, H.: Three decades of driver assistance systems. In: XXVIII Encontro da Associação Nacional de Pós Graduação e Pesquisa em Administração, vol. 6, no. 4, pp. 1–9 (2004)
13. Ziebinski, A., Cupek, R., Erdogan, H., Waechter, S.: A survey of ADAS technologies for the future perspective. Int. Conf. Comput. Collect. Intell. **2**, 135–146 (2016)
14. Zhao, Z., Zhou, L., Zhu, Q., Luo, Y., Li, K.: A review of essential technologies for collision avoidance assistance systems. Adv. Mech. Eng. **9**(10), 1–15 (2017)
15. Kukkala, V.K., Tunnell, J., Pasricha, S., Bradley, T.: Advanced driver-assistance systems: a path toward autonomous vehicles. IEEE Consum. Electron. Mag. **7**(5), 18–25 (2018)
16. Andreone, L., Guarise, A., Lilli, F., Gavrila, D.M., Pieve, M.: Cooperative systems for vulnerable road users: the concept of the watch-over project. In: 13th World Congress Intelligence Transportation System Services, pp. 1–6 (2006)
17. Ahmed, S., Huda, M.N., Rajbhandari, S., Saha, C., Elshaw, M., Kanarachos, S.: Pedestrian and cyclist detection and intent estimation for autonomous vehicles: a survey. Appl. Sci. **9**(11), 1–38 (2019)
18. Rasouli, A., Tsotsos, J.K.: Autonomous vehicles that interact with pedestrians: a survey of theory and practice. IEEE Trans. Intell. Transp. Syst. **21**(3), 900–918 (2020)
19. Mars, F., Chevrel, P.: Modelling human control of steering for the design of advanced driver assistance systems. Annu. Rev. Control **44**, 292–302 (2017)
20. Rahman, M.M., Lesch, M.F., Horrey, W.J., Strawderman, L.: Assessing the utility of TAM, TPB, and UTAUT for advanced driver assistance systems. Accid. Anal. Prev. **108**, 361–373 (2017)
21. Ammoun, S., Nashashibi, F., Laurgeau, C.: Real-time crash avoidance system on crossroads based on 802.11 devices and GPS receivers. In: IEEE Conference on Intelligent Transportation Systems Proceedings, ITSC, pp. 1023–1028 (2006)
22. Schnelle, S., Wang, J., Jagacinski, R., Jun Su, H.: A feedforward and feedback integrated lateral and longitudinal driver model for personalized advanced driver assistance systems. Mechatronics **50**, 177–188 (2018)
23. Emirler, M.T., Wang, H., Güvenç, B.A.: Socially acceptable collision avoidance system for vulnerable road users. IFAC-PapersOnLine **49**(3), 436–441 (2016)
24. Jiménez, F., Naranjo, J.E., Anaya, J.J., García, F., Ponz, A., Armingol, J.M.: Advanced driver assistance system for road environments to improve safety and efficiency. Transp. Res. Procedia **14**, 2245–2254 (2016)
25. Korssen, T., Dolk, V., Van De Mortel-Fronczak, J., Reniers, M., Heemels, M.: Systematic model-based design and implementation of supervisors for advanced driver assistance systems. IEEE Trans. Intell. Transp. Syst. **19**(2), 533–544 (2018)

26. Hamid, U.Z.A., et al.: Multi-actuators vehicle collision avoidance system - experimental validation. IOP Conf. Ser. Mater. Sci. Eng. **342**, 012018 (2018)
27. Moon, S., Yi, K.: Human driving data-based design of a vehicle adaptive cruise control algorithm. Veh. Syst. Dyn. **46**(8), 661–690 (2008)
28. Yurtsever, E., Lambert, J., Carballo, A., Takeda, K.: A survey of autonomous driving: common practices and emerging technologies. IEEE Access **8**, 58443–58469 (2020)
29. Voulodimos, A., Doulamis, N., Doulamis, A., Protopapadakis, E.: Deep learning for computer vision: a brief review. Comput. Intell. Neurosci. **2018**, 1–13 (2018)
30. Sligar, A.P.: Machine learning-based radar perception for autonomous vehicles using full physics simulation. IEEE Access **8**, 51470–51476 (2020)
31. Shinar, D.: Safety and mobility of vulnerable road users: pedestrians, bicyclists, and motorcyclists. Accid. Anal. Prev. **44**(1), 1–2 (2012)
32. Janai, J., Güney, F., Behl, A., Geiger, A.: Computer Vision for Autonomous Vehicles: Problems, Datasets and State of the Art (2017)
33. Le, L., Festag, A., Baldessari, R., Zhang, W.: V2X communication and intersection safety. In: Meyer, G., Valldorf, J., Gessner, W. (eds.) Advanced Microsystems for Automotive Applications 2009, pp. 97–107. Springer, Heidelberg (2009). https://doi.org/10.1007/978-3-642-007 45-3_8
34. Goldhammer, M., Kohler, S., Zernetsch, S., Doll, K., Sick, B., Dietmayer, K.: Intentions of vulnerable road users-detection and forecasting by means of machine learning. IEEE Trans. Intell. Transp. Syst. **21**(7), 3035–3045 (2020)
35. Hurney, P., Waldron, P., Morgan, F., Jones, E., Glavin, M.: Review of pedestrian detection techniques in automotive far-infrared video. IET Intell. Transp. Syst. **9**(8), 824–832 (2015)
36. Cai, Y., Sun, X., Wang, H., Chen, L., Jiang, H.: Night-time vehicle detection algorithm based on visual saliency and deep learning. J. Sens. **2016**, 1–7 (2016)
37. Wang, H., Cai, Y., Chen, X., Chen, L.: Night-time vehicle sensing in far infrared image with deep learning. J. Sens. **2016**, 1–8 (2016)
38. Cai, Y., Liu, Z., Wang, H., Sun, X.: Saliency-based pedestrian detection in far infrared images. IEEE Access **5**, 5013–5019 (2017)
39. Jiménez, F., Aparicio, F., Páez, J.: Intelligent transport systems and effects on road traffic accidents: state of the art. IET Intell. Transp. Syst. **2**(2), 132 (2008)
40. Wiethoff, M., Oei, H.L., Penttinen, M., Anttila, V., Marchau, V.A.W.J.: Advanced driver assistance systems: An overview and actor position. IFAC Proc. Vol. **35**(1), 1–6 (2002)
41. Katteler, H.: Driver acceptance adaptation of mandatory intelligent speed (2005)
42. Mannion, P.: Vulnerable road user detection: state-of-the-art and open challenges, pp. 1–5 (2019). http://arxiv.org/abs/1902.03601
43. Wang, X., Wang, M., Li, W.: Scene-specific pedestrian detection for static video surveillance. IEEE Trans. Pattern Anal. Mach. Intell. **36**(2), 361–374 (2014)
44. Campmany, V., Silva, S., Espinosa, A., Moure, J.C., Vázquez, D., López, A.M.: GPU-based pedestrian detection for autonomous driving. Procedia Comput. Sci. **80**, 2377–2381 (2016)
45. Saponara, S.: Hardware accelerator IP cores for real time Radar and camera-based ADAS. J. Real-Time Image Proc. **16**(5), 1493–1510 (2016)
46. Borrego-Carazo, J., Castells-Rufas, D., Biempica, E., Carrabina, J.: Resource-constrained machine learning for ADAS: a systematic review. IEEE Access **8**, 40573–40598 (2020)
47. Chilukuri, D., Yi, S., Seong, Y.: Computer vision for vulnerable road users using machine learning. J. Mechatronics Robot. **3**(1), 33–41 (2019)
48. Velez, G., Otaegui, O.: Embedding vision-based advanced driver assistance systems: a survey. IET Intell. Transp. Syst. **11**(3), 103–112 (2017)

49. Chen, Y., Fisher, D., Ozguner, U.: T. A. S. Institute, C. I. S. U. T. Center, and R. and I. T. Administration. Pre-Crash Interactions between Pedestrians and Cyclists and Intelligent Vehicles (2018)
50. Bengler, K., Dietmayer, B., Maurer, M., Stiller, C., Winner, H.: Three decades of driver assistance systems. IEEE Intell. Transp. Syst. Mag. **6**(4), 6–22 (2014)

Segregation and User Interactive Visualization of Covid-19 Tweets Using Text Mining Techniques

Gauri Chaudhary$^{(\boxtimes)}$ ⓘ and Manali Kshirsagar

Department of Computer Technology, Yeshwantrao Chavan College of Engineering, Hingna Road, Wanadongri, Nagpur 441110, India

Abstract. One of the worst calamities the world is facing since early 2020 is corona virus or Covid-19 disease which has turned into a pandemic claiming millions of lives across the globe. Twitter sources huge number of tweets related to this disease from users globally. This research focuses on mining Covid-19 tweets using machine learning techniques. The tweets are first pre-processed and converted to a form suitable for applying clustering algorithms. Principal Components Analysis is used to separate most significant components. Similar tweets are categorized using Hierarchical agglomerative clustering. The segregated tweets are visualized on novel and interactive cluster plots, members of which can be identified on user interface interactively by user for easy interpretation. The implementation is done using R programming. Clusters of similar tweets can be used to analyze the response of people to the pandemic across countries, compare and adopt best practices across countries to address the pandemic based on people views, combat spread of rumors and other such applications.

Keywords: Text mining · Document clustering · Visualization · Covid-19 · Hierarchical clustering · Tweets

1 Introduction

With the increased usage of social media sites, Twitter has emerged one of the most popular platforms for disseminating and sharing information across the globe. It is one of the most liked mediums by communities at large to interact, communicate, share their ideas, views, news, knowledge etc. thus producing huge real-time data in the form of tweets which if mined effectively can provide truly valuable insights into the information. Previously published research in this area shows that tweets about a particular calamity provide better insights about the specific crisis by building, implementing and evaluating appropriate machine learning algorithms [1].

The candidate tweets chosen for this study are Covid-19 tweets. Covid-19 disease outbreak was observed in early 2020 with few countries to start with and over the time has spread rapidly across the world in an ugly fashion taking the form of a pandemic [2]. This disease which has claimed the life of millions and has created havoc is the most

© The Author(s), under exclusive license to Springer Nature Switzerland AG 2022
R. Misra et al. (Eds.): ICMLBDA 2021, LNNS 256, pp. 268–279, 2022.
https://doi.org/10.1007/978-3-030-82469-3_24

talked about subject of recent times. People have started much more increased usage of digital and social media platforms like twitter to communicate and share their views with respect to the pandemic for causes, response, numbers, measures taken, role of Governments, lockdowns, economic crisis, normalcy, comparison across countries and regions from various angles etc.

Covid-19 tweets for this study are obtained from Twitter using Twitter API and Python script for the period Feb 2020 to July 2020 and filtered with the hashtag "#covid19". The data is available in public domain [3]. In the rest of the paper, we showcase implementation and evaluation of text clustering done using R programming. This is an unsupervised machine learning approach to segregate the similar Covid-19 tweets [4]. We also showcase how these segregated tweets can be visualized interactively by user on interactive cluster plots and interactive dendograms designed using FactoExtra package in R and Plotly libraries.

2 Foundations

Information from social media platforms such as Twitter has largely been used in recent studies to showcase how fake information can be disseminated during pandemic causing damage and creating panic among people [5]. Application of clustering and an interactive dashboard for visualizing similar tweets can assist such researches in faster and easier analysis.

The tweets are text data that need to be first pre-processed using NLP techniques. Since Tweets largely differ from regular texts due to presence of hashtags, URLs, emoticons, hyperlinks etc. [6, 7], the tweets needs to be cleaned for the removal of these characteristics apart from regular text pre-processing like tokenization, stop word removal, white space removal and stemming [8, 9]. The tweets are converted to a document term matrix to show the frequency of each term in the corresponding documents [8].

Principal Components Analysis is a dimension reduction technique and especially useful and applicable in the context of text clustering as the document term matrix is generally sparse since the number of distinct words even after pre-processing is huge as compared to the number of documents [10, 11]. Principal components analysis is applied to reduce the data to most significant components indicating maximum variance in the data and less significant ones are dropped off from further analysis [12].

Clustering is a form of unsupervised learning since the number of clusters is not known in advance and we apply similarity metric to identify similarity between tweets so that similar tweets are placed in one cluster and tweets in different clusters are most dissimilar to each other [13, 14]. Hierarchical agglomerative clustering is a bottom-up clustering approach in which each tweet is placed in a separate cluster to start with. Then successively we go on finding pair-wise similarity between clusters and merge the most similar clusters [15, 16]. Cosine similarity measure is one of the most popular metrics to calculate similarity between documents [17, 18]. Data structure used to implement this pair-wise similarity and merging is distance matrix which is updated at each step to remove the similar clusters and merging them until only a single element is left. The output of hierarchical clustering is depicted using a tree-like structure called a dendogram. The dendogram is cut so as to get the desired number of clusters. Desired optimal number of clusters for which the dendogram is cut can be found using various available methods such as elbow method [19], average silhoutte, gap statistic, Calinski Harabasz and others.

3   Research Approach

This research involves cleaning or pre-processing of Covid-19 tweets data which is free flowing text embedded with lot of noise. Principal Components Analysis is applied to the cleaned tweets to identify most significant components capturing maximum spread in the data. Optimal clusters of tweets are obtained by application of Hierarchical Agglomerative Clustering on the most significant Principal Components. Clusters are plot using R and Plotly libraries on user interactive plot and various visualization techniques are facilitated. The approach is summarized in a flow chart depicted in Fig. 1.

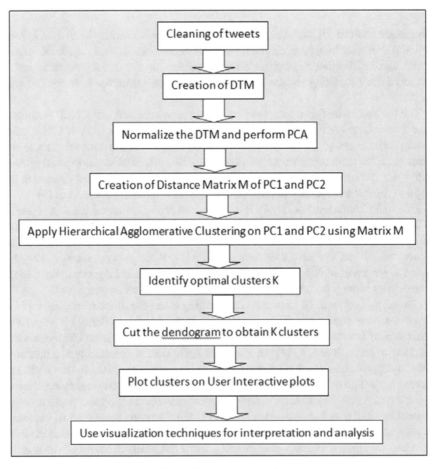

Fig. 1. Research approach

The research approach is detailed in the following algorithm:

Algorithm 1 Algorithm for segregation and visualization of Covid-19 tweets

Input: N (Number of tweets)
Output: K clusters showing cluster members on a user interactive plot
Method:
Segregation of similar tweets:
1. For each tweet i=1 to N do
2. Begin (Cleaning of tweets)
3. Tokenize based on whitespaces or line breaks
4. Eliminate special characters, stop words, hash tags, emoticons, URLs, excessive whitespaces
5. Apply stemming to replace each token with its root
6. Convert each token to lower case
7. End
8. Create Document Term Matrix(DTM) of the cleaned tweets
9. Normalize the DTM and perform Principal Components Analysis(PCA) to identify the most significant Principal Components PC1 and PC2
10. Identify optimal number of output clusters K using any of the available methods like elbow method, gap statistic, Average Silhoutte etc.
11. Construct Distance matrix (M) of PC1 and PC2 using cosine similarity measure
12. Apply Hierarchical Agglomerative Clustering on distance matrix M obtained in step 11 and cut the dendogram at K to obtain K clusters
Visualization of segregated tweets:
13. Plot clusters obtained in step 12 using R and Plotly libraries on user interactive plot
14. Use functions like box selection, zoom-in, zoom-out, pan, show data on hover, compare data on hover, download plot etc. for clear visualization and analysis.
15. For large dataset, double clicking on a particular cluster on legend can isolate one trace and thus can provide clear view of the members of that cluster.

4 Tweets Cleaning and Transformation

Document corpus for the purpose of this study is set of tweets filtered for "#covid19" hashtag from Twitter listed on kaggle website [3].

Portion of the original look of the data is depicted in Fig. 2 (Only "text" column from the dataset is displayed). As can be seen from the Fig. 2 the tweets contain special characters, hashtags, URLs and emoticons. The tweets are first pre-processed to remove all of these. Then the tweets are cleaned for stop words which really do not add much to the meaning of tweets such as conjunctions, articles etc. Whitespaces are removed and words are reduced to the root words. i.e. stemming so that we can actually identify the distinct words. In order that the vector sizes of the documents may be accommodated in the development environment we have used around 20% sampled documents for the experiments conducted. Figure 3 shows the most frequently used words found in the tweets corpus.

```
1  text
2  If I smelled the scent of hand sanitizers today on someone in the past, I would think they were so intoxicated thatâ€¦ https://t.co/QZvYb
3  Hey @Yankees @YankeesPR and @MLB - wouldn't it have made more sense to have the players pay their respects to the Aâ€¦ https://t.
4  @diane3443 @wdunlap @realDonaldTrump Trump never once claimed #COVID19 was a hoax. We all claim that this effort toâ€¦ https:/.
5  @brookbanktv The one gift #COVID19 has give me is an appreciation for the simple things that were always around meâ€¦ https://t.co/2
   25 July : Media Bulletin on Novel #CoronaVirusUpdates #COVID19
6  @kansalrohit69 @DrSyedSehrish @airnewsalerts @ANIâ€¦ https://t.co/MN0EEcsJHh
7  #coronavirus #covid19 deaths continue to rise. It's almost  as bad as it ever was.  Politicians and businesses wantâ€¦ https://t.co/hXMHo
8  How #COVID19 Will Change Work in General (and recruiting, specifically) via/ @ProactiveTalent #Recruitingâ€¦ https://t.co/bjZxzGPMbK
9  You now have to wear face coverings when out shopping - this includes a visit to your local Community Pharmacyâ€¦ https://t.co/OSu5Q
   Praying for good health and recovery of @ChouhanShivraj .
   #covid19
10 #covidPositive
```

Fig. 2. Portion of the original data view

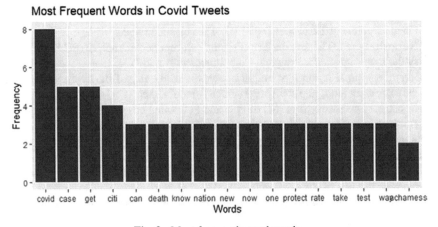

Fig. 3. Most frequently used words

```
> show(dtm)
<<DocumentTermMatrix (documents: 59, terms: 449)>>
Non-/sparse entries: 519/25972
Sparsity            : 98%
Maximal term length: 15
Weighting           : term frequency (tf)
```

Fig. 4. Meta data of the document term matrix

Figure 4 shows the Meta data of document term matrix and Fig. 5 shows a portion of the document term matrix showing documents as rows and words as columns.

In order to apply hierarchical clustering, we need to construct the distance matrix which is showing distance between each pair of documents using cosine similarity as the distance measure. Portion of the Distance Matrix is shown in Fig. 6.

Filter

	can	case	citi	covid	death	get	know	nation	new	now
consolid research hub free app built marklog ...	0	0	0	0	0	0	0	0	0	0
crisi dealt sever blow museum across world h...	0	0	0	0	0	0	0	0	0	0
juli ten lakh - aug twenti lakh covid case india...	0	1	0	1	0	0	0	0	0	1
march , gov. tate reev mississippi issu execut ...	0	0	0	0	0	0	0	0	0	0
retail protect improv guid nsai_standard outli...	1	0	0	0	0	0	0	0	0	0
run death rate usa juli . ' go pass days. even c...	0	1	0	0	2	0	0	0	0	0
sheeraf brianstelt facebook youtub china test ...	0	0	0	0	0	0	0	0	0	0
take bold step today, get plot plantat citi inve...	0	0	1	0	0	1	0	0	0	0
trump nation plan combat sinc begin worri ev...	0	0	0	0	0	0	0	1	0	0
world health organ warn thursday coronavirus...	0	0	0	0	0	0	0	0	0	0

Fig. 5. Portion of the Sparse Document term matrix

Filter   Cols: « <   1 - 50   > »

	govern respons loom crisi neither appropri support', say jason . kmpfqwvqir	trump nation plan combat sinc begin worri everi live person countri . jznsjhi	run death rate usa juli . ' go pass days. even case increasing, death rate . rklbvrsmi	theori wuhan huanan seafood wholesal market origin sourc animalhuman transmission, . riakitvk	crisi dealt sever blow museum across world however, war-torn yemen, compl . snrs	becom visibl global govern project.
govern respons loom crisi neither appropri su...	0.0000000	1.0000000	1.0000000	1	0.9087129	0.8585786
trump nation plan combat sinc begin worri ev...	1.0000000	0.0000000	1.0000000	1	1.0000000	1.0000000
run death rate usa juli . ' go pass days. even c...	1.0000000	1.0000000	0.0000000	1	1.0000000	1.0000000
theori wuhan huanan seafood wholesal marke...	1.0000000	1.0000000	1.0000000	0	1.0000000	1.0000000
crisi dealt sever blow museum across world h...	0.9087129	1.0000000	1.0000000	1	0.0000000	1.0000000
becom visibl global govern project.	0.8585786	1.0000000	1.0000000	1	1.0000000	0.0000000

Fig. 6. Portion of Distance matrix

5   Experimental Results

5.1   Implementation of Hierarchical Agglomerative Clustering

The dendogram obtained from application of HAC to the pre-processed tweets is shown in Fig. 7. The optimal number of clusters is found to be 5 using elbow method and dendogram is cut at level 5.

5.2   Interactive Visualization of Clusters Obtained from HAC

The clusters obtained from HAC can be viewed interactively using FactroExtra and Plotly libraries in R. Figure 8 shows one view of the interactive plot. The view has multiple interactive functions:

- Use functions like box selection, zoom-in, zoom-out, pan, show data on hover, compare data on hover, download plot etc. for clear interpretation of clusters and analysis.
- For large dataset, viewing all members is cumbersome as text seems superimposed on each other. Double clicking on a particular cluster on legend can isolate one trace and further zooming can provide clear view of each member of that cluster.

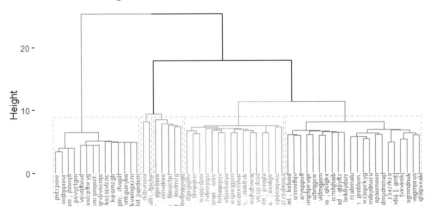

Fig. 7. Dendogram from application of HAC

Label being displayed in black box in Fig. 8 is displayed when mouse is hovered over the plot at a point in cluster 2. The text on the label reads –"10 lakhs cases in July, 20 lakhs cases in August in India, whether Prime Minister will speak about this". Figure 9 shows an interactive dendogram drawn using FactroExtra and Plotly libraries in R.

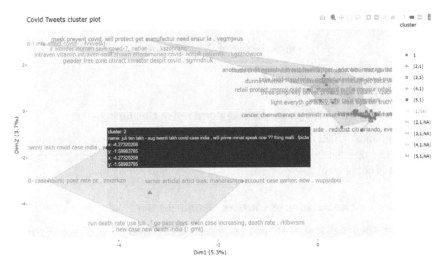

Fig. 8. Interactive cluster plot of Covid19 Tweets

The dendogram view supports multiple functions such as zooming, panning, scaling and showing members on hover. The members in the dendogram can be seen highlighted in green color when mouse is hovered at a particular node in the dendogram. Box selection and zooming can be used to clearly read the individual tweets. A portion of the tweets on using zoom out function is also showed in the Fig. 9.

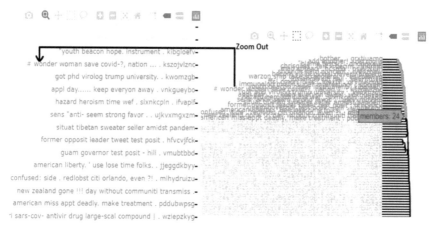

Fig. 9. Portion of interactive dendogram plot of Covid19 Tweets

Based on the sample words that can be observed from the plots, cluster members can be characterized and given labels as shown in Table 1.

Table 1. Indicative cluster labels

Cluser	Sample words	Cluster label
1	Improve, help, impact, policies, solution, transmission, good, dialogs, test	Tweets related to measures and policies set up by the Government for handling the pandemic
2, 5	Positive, covid, case, rate, death, increase, pass, lakh, prime minister, Trump, mess, Total, result, case, positive, test, user, recover, confirm, opposite, leader	Tweets related to statistics on number of positive cases, deaths, region and period and views about political leaders
3	Mask, wear, people, cure, detect, case, graph, analysis, update, manufacturer, protect, world, together	Tweets related to sharing preventive measures
4	Vitamin, intravenous, effect, research, data, scientist, alert, app, covid, trauma	Tweets related to research and solutions to disease cure
6	Report, fatal, new, case, congressman, Kazakhstan, Nigeria, compare, yesterday	Tweets related to comparative statistics such as number of positive cases and deaths

Example inferences can be derived out of the cluster labels such as:

- Cluster with label "Research and solutions to disease cure" – People's first hand feedback on solution to disease cure can be analyzed by location of the users posting tweets from this cluster. The solution adopted at majority location of positive tweet users can be a model for others to follow.
- With more data from early 2021 captured and if cluster with label "Discourage use of vaccine" is isolated, it can be analyzed for spread of misinformation about vaccine use and efficacy. Users and locations of the tweets from this cluster can be identified and appropriate action can be taken as required by Governments.

5.3 Implementation of PCA

Since the document term matrix is sparse, we use Principal Components Analysis for reduction of dimensions. Portion of terms in each of the Principal Component and their relative importance is shown as a sample in Fig. 10. Figure 11 shows plot view of the terms and their contributions in PC1 as a sample.

```
> tidied_pca
# A tibble: 4,490 x 3
      Tag          PC      Contribution
      <chr>        <chr>         <dbl>
 1 abilitylab   PC1         0.00664
 2 abl          PC1         0.00606
 3 accompani    PC1         0.0262
 4 account      PC1        -0.0438
 5 achamess     PC1         0.00664
 6 across       PC1         0.0386
 7 add          PC1         0.00193
 8 administr    PC1         0.0218
 9 ago          PC1         0.0115
10 agre         PC1         0.0180
# ... with 4,480 more rows
```

Fig. 10. Contribution of terms in each of the Principal components

5.4 Interactive Visualization of Clusters Obtained Using PCA and HAC

Hierarchical clustering on most significant principal components yields better cluster quality as compared to Hierarchical clustering on complete text dataset [20]. Distance Matrix of PC1 and PC2 is formed and HAC is applied using this matrix as per the approach specified in Algorithm 1. The resulting clusters plot is shown in Fig. 12. The black box in Fig. 12 appears when mouse is hovered on a point on cluster 4 on the plot.

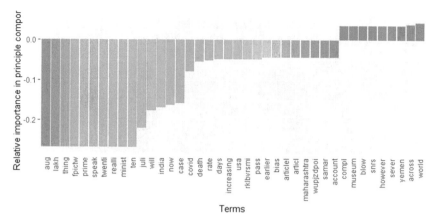

Fig. 11. Plot view for terms and their relative importance in PC1

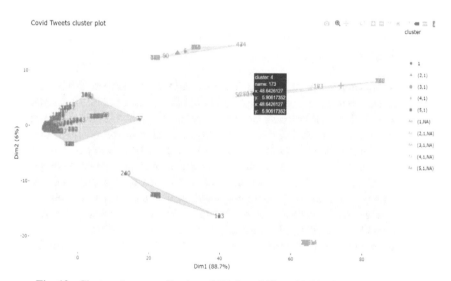

Fig. 12. Cluster plot on application of HAC on PC1 and PC2 of Covid19 Tweets

The box shows the coordinates and name of the point. Name of the point is actually a term from the Principal Component. (Example: name "173" stands for the term "India" in Principal Components).

Figure 13 shows the Average Silhoutte width of 0.85 for the clusters obtained in Fig. 12. Silhoutte value indicates the degree of cohesion i.e. similarity of points within a cluster as compared to degree of separation i.e. how dissimilar the points in a cluster than other clusters. Higher silhoutte value indicates good quality of clustering.

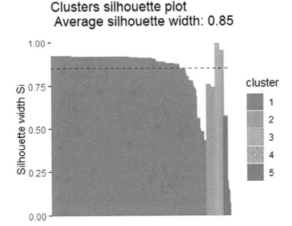

Fig. 13. Average silhouette width of clusters

6   Conclusion

Important applications of tweets mining in the scenario of covid times which has caused havoc could be using the tweets information to check people's views to plan response to the pandemic, check the behavior or sentiments of people and prevent dissemination of incorrect information causing damage to people at large in some way or the other. Clustering of Covid tweets for similar text can make any such analysis easier by limiting to documents/tweets in particular clusters which are of interest to us based on problem to be addressed at hand. The analysis can thus be focused on specific tweets clusters rather than the entire corpus and would thus yield results faster. The proposed approach used in this research yields 85% average silhoutte width of clusters which may further be improved with using different combination of clustering algorithms. To simplify the clusters interpretation, this research provides novel intuitive plots which can facilitate the user to interactively view the clusters and the member tweets within each cluster.

References

1. Lamsal, R.: Design and analysis of a large-scale COVID-19 tweets dataset. Appl. Intell. **51**(5), 2790–2804 (2020). https://doi.org/10.1007/s10489-020-02029-z
2. Shuja, J., Alanazi, E., Alasmary, W., Alashaikh, A.: COVID-19 open source data sets: a comprehensive survey. Appl. Intell. **51**(3), 1296–1325 (2020). https://doi.org/10.1007/s10 489-020-01862-6
3. Preda, G.: Covid19 Tweets, Tweets with the hashtag #covid19 (2020). https://www.kaggle.com/gpreda/covid19-tweets
4. Jiang, X., Shi, Y., Li, S.: Research of correction method in the feature space on text clustering. In: Proceedings of the 2012 International Conference on Computer Science and Service System, pp. 2030–2033 (2012)

5. Cruickshank, I.J., Carley, K.M.: Characterizing communities of hashtag usage on twitter during the 2020 COVID-19 pandemic by multi-view clustering. Appl. Netw. Sci. **5**(1), 1–40 (2020). https://doi.org/10.1007/s41109-020-00317-8
6. Bao, Y., Quan, C., Wang, L., Ren, F.: The role of pre-processing in twitter sentiment analysis. In: Huang, D.-S., Jo, K.-H., Wang, L. (eds.) ICIC 2014. LNCS (LNAI), vol. 8589, pp. 615–624. Springer, Cham (2014). https://doi.org/10.1007/978-3-319-09339-0_62
7. Anand, N., Goyal, D., Kumar, T.: Analyzing and preprocessing the twitter data for opinion mining. In: Tiwari, B., Tiwari, V., Das, K.C., Mishra, D.K., Bansal, J.C. (eds.) Proceedings of International Conference on Recent Advancement on Computer and Communication. LNNS, vol. 34, pp. 213–221. Springer, Singapore (2018). https://doi.org/10.1007/978-981-10-8198-9_22
8. Chaudhary, G., Kshirsagar, M.: Overview and application of text data pre-processing techniques for text mining on health news tweets. Helix **8**(5), 3764–3768 (2018)
9. Vijayarani, S., et al.: Preprocessing techniques for text mining – an overview. Int. J. Comput. Sci. Commun. Netw. **5**(1), 7–16 (2015)
10. Xiao, X., Zhou, Y.: Two-dimensional quaternion PCA and sparse PCA. IEEE Trans. Neural Netw. Learn. Syst. **30**(7), 2028–2042 (2019)
11. Han, X.: Nonnegative principal component analysis for cancer molecular pattern discovery. IEEE/ACM Trans. Comput. Biol. Bioinf. **7**(3), 537–549 (2010)
12. Yan, J., et al.: Trace-oriented feature analysis for large-scale text data dimension reduction. IEEE Trans. Knowl. Data Eng. **23**(7), 1103–1116 (2011)
13. Zhu, Y., Fung, B.C.M., Mu, D., Li, Y.: An efficient hybrid hierarchical document clustering method. In: Fifth International Conference on FuzzySystems and Knowledge Discovery, Shandong, China, pp. 395–399 (2008)
14. Carullo, M., Binaghi, E., Gallo, I., Lamberti, N.: Clustering of short commercial documents for the Web. In: Proceedings of the 2008 19th International Conference on Pattern Recognition, pp. 1–4 (2008)
15. Gascuel, O., McKenzie, A.: Performance analysis of hierarchical clustering algorithm. J. Classif. **21**(1), 3–18 (2004)
16. Jo, T.: String vector based AHC for text clustering. In: Proceedings of the 2017 19th IEEE International Conference on Advanced Communication Technology (ICACT), Bongpyeong, South Korea, pp. 673–678 (2017)
17. Yuan, S., Wenbin, G.: A text clustering algorithm based on simplified cluster hypothesis. In: Proceedings of the 2013 2nd International Symposium on Instrumentation and Measurement, Sensor Network and Automation (IMSNA), Toronto, ON, Canada, 23–24 December, 2013, pp. 412–415 (2013)
18. Li, Y., Luo, C., Chung, S.M.: Text clustering with feature selection by using statistical data. IEEE Trans. Knowl. Data Eng. **20**(5), 641–652 (2008)
19. Marutho, D., Handaka, S.H., Wijaya, E., Muljono: The determination of cluster number at k-mean using elbow method and purity evaluation on headline news. In: Proceedings of the 2018 IEEE International Seminar on Application for Technology of Information and Communication, Semarang, Indonesia, 29 November 2018, pp. 533–538 (2018)
20. Chaudhary, G., Kshirsagar, M.: Enhanced text clustering approach using hierarchical agglomerative clustering with principal components analysis to design document recommendation system. In: Advanced Research in Computer Engineering. Research Transcripts in Computer, Electrical and Electronics Engineering, vol. 2, pp. 1–18. Grinrey Publications (2021). ISBN: 978-81-948951-2-1

Sparse Representation Based Face Recognition Using VGGFace

Jitendra Madarkar[✉] and Poonam Sharma

Visvesvaraya National Institute of Technology, Nagpur 440010, India

Abstract. Face recognition is one of the most important applications of computer vision and it has been used in biometric, surveillance cameras and face tagging. Recently, the convolutional neural network (CNN) model has shown better performance on visual data and it attracted more attention due to automatic feature extraction. Sparse representation-based classification (SRC) has been shown tremendous success in the area of face recognition since the last decade. Convolution neural network is invariant to unconstrained variation whereas SRC is prone. SRC needs discriminative features to represent a test sample. This paper, exploits the significance of both the aforementioned methods and proposed a new CNN-SRC which alleviates the performance of unconstraint variation. The experimental results of the proposed method have shown better performance on different benchmark databases.

Keywords: Face recognition · Sparse representation based classification · ℓ^1-minimization · Biometric · Convolutional neural network

1 Introduction

The world lurks under the threat of terrorism and the need of the hour is a good system of human face recognition so that it recognizes the human face from the images and surveillance camera. Since the researchers have been working in the area of face recognition and they have achieved state-of-the-art results for constrained samples. Existing methods failed to tackle the issue of unconstrained conditions. Face recognition is mainly a two-step process: the first step is feature extraction such as principal component analysis (PCA) [1], local binary pattern LBP [2], Fisher discriminant analysis (FDA), independent component analysis (ICA), and locality preserving projections (LPP), curvelet [3] etc. and the second is classification such as nearest neighbor (NN) [5], nearest feature line (NFL), nearest feature plane (NFP), nearest feature subspace (NFS), SVM [4], linear regression-based classification (LRC), and Sparse representation based classification (SRC) [7]. Since the last decade, SRC has shown good performance overall existing classification methods [7–10]. The accuracy of face recognition immensely depends on features of training samples and SRC also needs discriminative features to represent a test sample. The sparsest representation of a test

© The Author(s), under exclusive license to Springer Nature Switzerland AG 2022
R. Misra et al. (Eds.): ICMLBDA 2021, LNNS 256, pp. 280–288, 2022.
https://doi.org/10.1007/978-3-030-82469-3_25

sample estimate from the dictionary which is constructed by all the training samples and classifies a test sample to the class which produces the less residual error. The sparse vector and dictionary play an important role in SRC and extended SRC methods have been developed by modifying these two factors. Deep learning models have shown state-of-the-art performance in the field of computer vision. The most popular three effective deep learning models developed till date are convolutional neural network(CNN), Boltzman family Deep belief network (DBN) and Deep Boltzman machine (DBM), and stack autoencoder. Recent studies show that CNN [6] based model is most effective on visual dataset among deep learning model. An architecture of CNN consists of a number of convolutional layers (commonly with subsampling step), fully connected, and softmax layer. Each convolution layer is followed by an activation layer such as ReLU, Sigmoid and it convoluted with the kernel or filter to produce the unique structure of a given image. A softmax layer is used for classification and uses probability distribution over different entities using Eq. 1. This function is not effective as other classification methods for face recognition.

$$Softmax(fc_i) = \frac{\exp(fc_i)}{\sum_i^K exp(fc_i)} \qquad (1)$$

where K number of classes, $Softmax(fc_i)$ and fc_i represents probability and features of the ith class.

In this paper, a new method CNN-SRC has been proposed which exploits the strength of CNN features and sparse representation classification. In [11] the author et al. have used only two layers and randomly generated weights. But CNN-SRC uses a number of convolutional layers and pretrained weights to improve the performance. The intention behind proposing this method is to alleviate the limitation of the SRC and CNN model. The experimental results of the CNN-SRC have shown better performance than the existing methods.

2   Proposed Method

The proposed CNN-SRC model is depicted in Fig. 1 and this model alleviates the significance of deep CNN and SRC. The architecture of the proposed CNN model consists of 13 convolutional layers followed by the pooling layer and 1 fully connected (fc) layer as shown in Fig. 1. This architecture uses RELU as an activation function and max-pooling layer. The CNN has gained more popularity due to its discriminative feature extraction technique. The training and testing features are extracted from the fc layer and the training features are encorporate into the dictionary and then apply minimization technique to estimate the sparse vector of a test sample. The Sparse coefficient value is estimated on the basis of similarity between train and test feature. The results from the last layer of CNN model (fc layer) are fed to the dictionary atoms which helps to represent a test image by linear Eq. 7 and it is finally optimized by ℓ^1-minimization. The flow of the proposed model, kernel size, and the number of kernels used in each convolutional layer have depicted in Fig. 1.

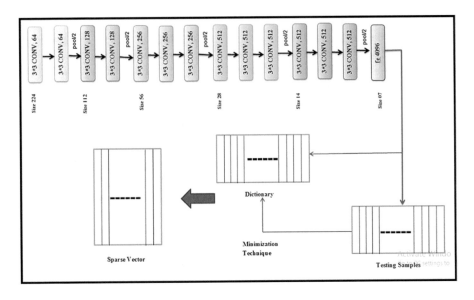

Fig. 1. Proposed model (CNN-SRC)

The convolution operation is performed by Eq. 2 and r convoluted feature map is created by Eq. 3.

$$h_{m,n} = k * I[m, n] \qquad (2)$$

where k is a filter, I is an input image and (m, n) are indices of the images and h is a convoluted image.

$$h_r^{l+1} = b_r^l + \sum_{k=1}^{m}\sum_{l=1}^{m} k_{r(k,l)}^l * h_{(i+l-2,j+k-2)}^l \qquad (3)$$

where h_r^{l+1} is a convoluted image using r^{th} kernel in $l+1$ layer, m is a size of the filter and r is a number feature map created in next layer. k_r^l is a kernel which convoluted with image of lth layer and produced r^{th} feature map in $l+1$ layer. b^l is bias in l^{th} layer.

After the convolutional layer, a large number of features are generated and are reduced by a subsampled operation that is done by Eq. 4. A max-pooling layer is used for subsampling where the maximum value is selected from the $m \times m$ values and it makes features robust against noise.

$$S_{ij}^{l+1} = \max_{0<k<m, 0<l<m} (h_{i^{l+1}*m+k, j^{l+1}*m+l}^l) \qquad (4)$$

where S^l subsampled of convoluted image after l layer, i^{l+1} is the indices of S^l subsampled image and m is a stride of the image.

In fully connected layer, each unit is connected to the all the units in the previous layer. This operation is done by Eq. 5.

$$fc[i] = \sum_{i}^{d} \sum_{j}^{r} S_j^l * w_j \qquad (5)$$

fc is a output of fully connected layer and $fc \in \Re^d$. r is a number of feature map in l^{th} layer.

Let $FC = [fc_1, fc_2, .., fc_K] \in \Re^{d*n}$ be the set of CNN features of all the training samples are called as dictionary. Each dictionary column is referred to as an atom or a basis. There are total n number of atoms in the dictionary. K represents number of classes and let $fc_i = [fc_{i,1}, fc_{i,2}, ..., fc_{i,N}]$ represent the set of samples that belongs to the ith class. N is number of samples in a class. $x \in \Re^n$ is a sparse vector which has very few non zero elements.

In ideal case, samples of same class lies in linear subspace and the test sample fc_y can be approximated with i^{th} class i.e. $fc_y = \alpha_{i,1} fc_{i,1} + \alpha_{i,2} fc_{i,2} + ... + \alpha_{i,n} fc_{i,n}$. A test sample fc_y can be linearly represented in terms of all training samples by using Eq. 6 [7].

$$fc_y = FC * x \quad \in \quad \Re^d \qquad (6)$$

where, $x = [0, 0, 0, \alpha_{i,1}, \alpha_{i,2}, ..., \alpha_{i,n}, 0, 0, 0, 0]^T$

Then Eq. 6 can be modified with the help of ℓ^1-minimization or convex relaxation methods [12] as shown in Eq. 7.

$$\hat{x} = arg \quad min_x \quad \|x\|_1 \quad s.t \quad fc_y = FC * x$$
$$or$$
$$\hat{x} = arg \quad min_x \quad \|x\|_1 \quad s.t \quad \|fc_y - FC * x\|_2 \le \varepsilon \qquad (7)$$

$\|.\|_1$ refer to as ℓ^1-norm which perform summation of all in x vector.

A test sample is assigned to the class that minimizes the residual error. The SRC assumes that the samples from the same class lie on the same subspace. Hence, test sample fc_y classified based on minimal residual or the class which has a more number of nonzero entries in sparse coefficient vector $x \in \Re^n$, the residual is computed by Eq. 8.

$$min \quad r_i(fc_y) = \|fc_y - FC * \delta_i(\hat{x})\|_2 \qquad (8)$$

Where, $r_i(fc_y)$ is a residual of test sample fc_y with respect to the ith class, $\delta_i(\hat{x})$ is a sparse coefficient of the ith class.

3 Experimental Results

In this paper, we have performed experimentation on different methods using five databases such as AR [16], Extended Yale B (EYB), LFW [17], CMU PIE [18] and GT. A CNN is implemented using Keras API. The experimentation of CNN-SRC use pretrained weight of vggface and compared the performance with other state-of-the-art methods such as SRC-PCA, ESRC, SSRC, SDR-SLR, SR-RLS and CNN-Softmax.

3.1 Face Databases

AR [15] face database contains 126 individuals having 26 color samples of each individual. Images are frontal views faces with different illumination conditions, occlusion, and facial expressions. Images were captured in the constraint condition. Extended yale B (EYB) face database contains 38 individuals and each individual having 9 different poses. Each pose has 64 different illumination samples. The LFW [17] face database images that are taken from the web of 1680 individuals. These images are varying in illumination, pose, expression, and different backgrounds. LFW database is a benchmark for face recognition applications. CMU PIE face database contains samples with different variations such as illumination, expression, and pose. The database samples are collected at Carnegie Mellon University in 2000. In GT face database contains images of 50 people and each individual has 15 samples. The images have taken in the Georgia Institute of technology for image processing. The images have a frontal, tilted views with different facial expressions and illumination effects.

Experimentation. The experimentation done on different databases

1. CMU PIE: Total 30 individuals, 1 individual has 20 samples, and choose randomly half samples for training and rest of the samples for testing.
2. EYB: Total 38 individuals, 1 individual has 20 samples, and choose randomly half samples for training and rest of the samples for testing.
3. LFW: Total 35 individuals, 1 individual has 9 samples, and choose randomly 5 samples for training and rest of the samples for testing.
4. AR: Total 100 individuals, 1 individual has 14 samples, and choose randomly half samples for training and rest of the samples for testing.
5. GT: Total 50 individuals, 1 individual has 15 samples, and choose randomly 8 samples for training and rest of the samples for testing.

The experimental results is shown in Table 1.

Result Analysis

1. The CNN-SRC is outperformed on all the databases.
2. Pretrained weights have given outstanding performance rather than randomly generated weights because for training it needs a number of training samples.
3. Randomly generated weights reduces the efficiency if the dataset has fewer samples.
4. Sparse coefficient of the proposed method has shown more sparsity.
5. Sparse coefficient value of the correct class is higher in CNN-SRC.

Table 1. Experimental results on CMU PIE, EYB, LFW, AR and GT database

Methods	CMU PIE	EYB	LFW	AR	GT
SRC-PCA [7]	98.1	94.0	48	94	76
ESRC [10]	98.5	94.4	52.5	94.4	77
SSRC [14]	97.6	94.1	51	94.1	78.5
SR-RLS [15]	99.0	94.7	43	94.7	74
SDR-SLR [13]	99.1	95.9	59.5	95.3	77.3
CNN-Softmax	99.5	96.1	98.3	96.1	91
CNN-SRC	100	96.66	99.3	96.7	94.3

6. Softmax function uses probability distribution to classify the test image but it can misclassify complex images such as the face. SRC uses linear representation to classify the image and it works better on fewer images.
7. The comparison of experimental results based on a different number of training samples is shown in Fig. 2. The proposed method has given better accuracy in a few training samples.

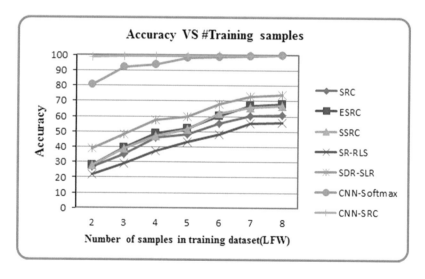

Fig. 2. The experimental results on different number of training samples

8. The sparse vector of a test sample of the LFW dataset is shown in Fig. 3. The dark brown lines indicate the coefficient value of the correct class and it is denser in the selected class whereas sparse in other classes. The coefficient value selected class is high as compared to the other classes. From Fig. 3 it is concluded that the proposed method has shown more sparsity.

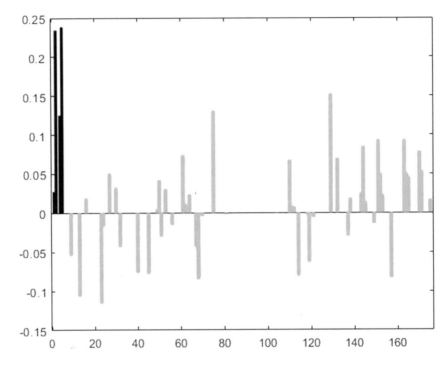

Fig. 3. Bar chart of Sparse vector (CNN-SRC)

9. Compared the experimental results using pre-trained and randomly generated weights on the LFW dataset. The proposed method has given better performance using both weights. From this experimentation, it is observed that randomly generated weights have given a poor performance than pre-trained weights. The training needs a large number of training samples to train the weights. The experimental results as shown in Table 2. In Table 2 R-indicate the randomly generated weights and P-indicate pre-trained weights.

Table 2. Experimental results using pretrained and random weights on LFW database

Methods	2	3	4	5	6	7
CNN-Softmax(R)	9.9	16.8	16.6	23.7	24.0	31.7
CNN-SRC(R)	18.7	23.7	30.0	32.8	33.9	37.3
CNN-Softmax(P)	80.7	92.3	93.7	98.3	98.7	99.3
CNN-SRC(P)	98.7	99.1	99.5	99.3	99.9	99.8

4   Conclusion

Face recognition is challenging for unconstrained images due to a different kind of variation. This paper proposed a CNN-SRC method that combines the strength of CNN features and sparse representation classification. The CNN features to maximize the discriminative power among the different individuals and robust to unconstraint variance. It uses 14 layers to extract the features from the image. SRC has shown more sparsity using CNN features than any other feature. The experimentation is performed on 5 open source databases such as LFW, AR, EYB, CMU, and GT databases and it has shown better performance.

Funding. This study was funded by the Ministry of Electronics and Information Technology (India) (Grant No.: MLA/MUM/GA/10(37)B).

Conflict of interest. The authors declare that there are no conflicts of interest regarding the publication of this paper.

References

1. Turk, M., Pentland, A.: Eigenfaces for recognition. J. Cogn. Neurosci. **3**, 71–86 (1991)
2. Ahonen, T., Hadid, A., Pietikäinen, M.: Face description with local binary patterns: application to face recognition. IEEE Trans. Pattern Anal. Mach. Intell. **28**, 2037–2041 (2006)
3. Sharma, P., Yadav, R.N., Arya, K.V.: Pose-invariant face recognition using curvelet neural network. IET Biometrics **3**, 128–138 (2014)
4. Jonathon, P.P.: Support vector machines applied to face recognition. In: Proceedings of the 1998 Conference on Advances in Neural Information Processing Systems II, pp. 803–809 (1999)
5. Cover, T., Hart, P.: Nearest neighbor pattern classification. IEEE Trans. Inf. Theory **13**, 21–27 (1967)
6. Parkhi, O.M., Vedaldi, A., Zisserman, A.: Deep face recognition. In: British Machine Vision Conference (2015)
7. Wright, J., Yang, A.Y., Ganesh, A., Sastry, S.S., Ma, Y.: Robust face recognition via sparse representation. IEEE Trans. Pattern Anal. Mach. Intell. **31**, 210–227 (2009)
8. Madarkar, J., Sharma, P., Singh, R.P.: Sparse representation for face recognition: a review paper. IET Image Process., 1–20 (2021). https://doi.org/10.1049/ipr2.12155
9. Madarkar, J., Sharma, P.: Occluded face recognition using noncoherent dictionary **1**, 6423–6435 (2020)
10. Deng, W., Hu, J., Guo, J.: Extended SRC: undersampled face recognition via intraclass variant dictionary. IEEE Trans. Pattern Anal. Mach. Intell. **34**, 1864–1870 (2012)
11. Cheng, E.-J., et al.: Deep learning based face recognition with sparse representation classification (2017)
12. Tropp, J.A.: Algorithms for simultaneous sparse approximation. Part II: convex relaxation. Sig. Process. **86**, 589–602 (2006)

13. Jiang, X., Lai, J.: Sparse and dense hybrid representation via dictionary decomposition for face recognition. IEEE Trans. Pattern Anal. Mach. Intell. **37**, 1067–1079 (2015)
14. Deng, W., Hu, J., Guo, J.: In defense of sparsity based face recognition. In: 2013 IEEE Conference on Computer Vision and Pattern Recognition, pp. 399–406 (2013)
15. Iliadis, M., Spinoulas, L., Berahas, A.S., Wang, H., Katsaggelos, A.K.: Sparse representation and least squares-based classification in face recognition, 2014 22nd European Signal Processing Conference (EUSIPCO), pp. 526–530 (2014)
16. Martinez, A.M., Benavente, R.: The AR face database. CVC Technical report (1998)
17. Huang, G.B., Ramesh, M., Berg, T., Learned-Miller, E.: Labeled faces in the wild: a database for studying face recognition in unconstrained environments, University of Massachusetts, Amherst, pp. 7–49 (2007)
18. Sim, T., Baker, S., Bsat, M.: The CMU pose, illumination, and expression (PIE) database. In: Proceedings of the 5th International Conference on Automatic Face and Gesture Recognition (2002)

Detection and Classification of Brain Tumor Using Convolutional Neural Network (CNN)

Smita Deshmukh and Divya Tiwari[✉]

Mumbai University, Terna Engineering College (TEC), Navi Mumbai, India
smitadeshmukh@ternaengg.ac.in

Abstract. An abnormal intracranial development of cells is called as Brain Tumor. This phenomenon occurs due to some abnormalities in the functioning of the body where the cells start to reproduce themselves continuously in an uncontrolled manner. Most research in the developed countries found that detecting Brain Tumor incorrectly, is the main cause of death of people suffering from it. It is important that the task of diagnosing Brain Tumor should be done quickly and accurately, as it is one of the deadliest diseases in the world. To produce images of the internal structure of the body, technologies like CT or MRI scan is used. It helps to detect the abnormalities in the internal organs, similarly it helps to identify the tumor if detected. To detect Brain Tumor accurately, segmentation of MR images is important. Classifying detected Brain Tumor through the segmented MR images, is a tedious task due to various characteristics of the tumor like its location, size, gray level intensities and shape. Nowadays due to increasing Tumor cases it is difficult to examine all the reports manually. Also, it sometimes becomes hazardous for a patient due to delay in the detection of the Tumor or the right time required for its surgery.

To eradicate this problem an intelligent system is required which can detect and classify the brain Tumor automatically. A method is proposed which can automatically consider MR images as an input and further analyze and process the input images and classify according to its presence and absence along with the type of Tumor detected. This is done using Convolutional Neural Network (CNN). It can perform analysis of data and extract the required features to understand the world as humans do. It stores the extracted features and knowledge in the knowledgebase called as weights, which is further used to detect or classify the object or image based on the extracted features stored in the knowledgebase.

Keywords: Brain tumor · MRI · Detection · Segmentation · Classification · CNN

1 Introduction

Human Body comprises of several types of cells. These cells have a precise function. The cells grow and divide in an orderly manner continuously forming the new cells to keep the body in good physical condition. Sometimes due to some abnormalities, few cells cease their capability to control their growth and they begin to grow in an improper and uncontrolled fashion which leads to the growth of extra cells resulting into

© The Author(s), under exclusive license to Springer Nature Switzerland AG 2022
R. Misra et al. (Eds.): ICMLBDA 2021, LNNS 256, pp. 289–303, 2022.
https://doi.org/10.1007/978-3-030-82469-3_26

the formation of a mass of tissue which is called a Tumor. The Tumor when formed within a brain is called "Brain Tumor." Brain Tumor is one of the most difficult types of Tumor to identify because of its vast variety of types. Till now, the different types of Brain Tumors known all over the world is more than 120. Primary or Secondary Tumors are the basic classification of Brain Tumor. If the cancerous cells originate in Brain then the case is classified as Primary Brain Tumor [40], whereas if the mass of cells develops anywhere in the body and then it spreads through the bloodstream to the brain in a process known as Metastasis [40] then it is classified as Secondary Brain Tumor. The most common secondary brain tumors arise from the lung, breast, kidney, colon and blood cells. Once the diagnosis is completed, medical experts try to understand whether the detected Tumor is within the central nervous system i.e. intra-axial or is located outside the central nervous system i.e. extra-axial and they generally classify it as either Benign or Malignant, even though this classification done is not always precise [40, 41] (Fig. 1).

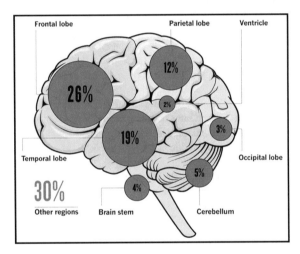

Fig. 1. Regions of brain affected by tumor

The diagram given above depicts the tumor affected regions of the Brain. Out of total, 26% of patients suffering from Brain Tumor are found to be suffering from the tumor growing in the Frontal lobe, 19% of the patients suffer from a tumor in the Temporal lobe, 12% of patients suffer from a tumor in the Parietal lobe, 5% of patients suffer from a tumor in Cerebellum, 4% of patients suffer from the tumor in Brain stem, 3% of patients suffer from a tumor in the Occipital lobe, 2% of patients suffer from a tumor in Ventricle and 30% of them suffer from the tumor in other regions of Brain. The count of Frontal lobe Tumor is the highest whereas in the Ventricle the count is lowest as compared to other regions of Brain. There is no scientific evidence about what causes Tumor to grow in different regions. The symptoms of all the patients are common like headache, nausea, fatigue, vomiting, loss of appetite and in some cases, poor vision. There are no distinguishing symptoms which can predict that the patient might be suffering from Brain Tumor. This shows the challenge faced by doctors to understand the disease the patient is suffering from. In most cases of the headache doctor directs patient for undergoing MRI scanning of Brain as this can provide internal images of Brain. Doctor reads and understands this report and after deep study concludes the type of Tumor the patient is suffering from. Abnormality in images can conclude that something is wrong in the Brain tissue, but the challenging task is to conclude the exact type of Tumor the patient is suffering from [44].

In some fatality case report, it was found that the patient lost its life because of delay in medication or surgery. This happens because sometimes doctors fail in deciding the appropriate time for the patient to undergo surgery or in some cases fail to understand the type of Tumor leading to delay in medication. This shows the need of an intelligent system in the medical domain which can assist doctors in gearing up the knowledge to treat patients. With the advancement in the field of science and technology efforts are being carried to develop an intelligent system which can assist doctors in Detecting and Classifying the Brain Tumor in less time along with the higher accuracy rate as compared to the time taken to do manually. It is expected that assistance of such systems in the future will help to reduce the fatality rate of patients caused due to delay in understanding of the reports, medication, or surgery.

Convolutional Neural Network (CNN)

Neurons with learnable weights and biases are used by a deep learning (DL) algorithm called Convolutional Neural Network (CNN) [45]. It consists of multiple hidden layers which are used to obtain required information from the input images [45]. CNN has four important layers given below:

1. Convolutional Layer.
2. ReLU Layer.
3. Pooling Layer.
4. Fully Connected Layer.

The first step in the process is Convolutional Layer. It is responsible for extracting valuable features from the images. It is made up of n-number of filters. These filters help to perform the convolution function. The input image, in this layer is taken as the matrix of pixel values. Once the extraction of feature maps is done, the next step is to move them to the next step i.e. in the ReLU layer. An element-wise function is performed by the ReLU layer. Non-linearity is introduced to the network as all the negative pixels are set to 0. It generates output which is called as the Rectified Feature Map. In CNN, every image is always scanned with n-number of convolutional layers and ReLU functions to locate and extract all the required features. Dimensionality of the feature map is reduced by using a down-sampling operation called Pooling. The Pooled Feature Map is generated by passing the Rectified Feature Map through Pooling Layer. All the detailed sections of the images like the edges, corners, area, etc. are identified using various filters by this layer. Before sending the Pooled Feature Map to Fully Connected (FC) Layer the Flattening process is done. Flattening process is used for creating a single continuous Linear Vector (LV). This is done by converting the resultant 2-Dimensional arrays from the Pooled Feature Map (PFM) [45]. This Flattened Matrix (FM) is then moved to the

Fully Connected (FC) Layer to carry out the process of classifying the image [45]. In short, CNN algorithm works as follows:

- The Convolution operation is performed at Convolutional Layer by considering the pixels from the image.
- This outcome of Convolutional Layer produces Convolved Map.
- ReLU function generates a Rectified Feature Map by taking the Convolved Map as an input.
- During the process, multiple Convolutional and ReLU Layers, processes the image for locating the features.
- For identifying the specific regions of the image various filters with different Pooling Layers are used.
- Flattening of the Pooled Feature Map is done first before it is moved for obtaining the final outcome to the Fully Connected (FC) Layer [45].

CNNs are basically used for the automation of models since it is capable of extracting features from datasets as well as training the model accordingly [10].

2   Literature Review

In the last decade, various methods have been developed to create an automated system to Detect and Classify Brain Tumor. Here various methods and techniques for medical diagnosis is presented. It presents a survey which shows different techniques such as hybrid clustering, segmentation, feature extraction, pattern recognition techniques used in the field, medical diagnosis, etc. The pros and cons of the methods observed are also pointed out. It is observed that most of them had difficulty in developing a model that can segment MR Brain images fully automatically. So, to extract the abnormality found in MRI Reports accurately and classify the type identified, classification and segmentation are very important steps to be carried out for clinical research, diagnosis and developing the applications which fulfill the requirement of robust, reliable, and adaptive techniques.

"Chirodip C., Chandrakanta M., Raghvendra K. & Brojo M. (2020)" proposed a method using CNN based model along with the DNN approach for classifying the detected Tumor. The outcome of the model is displayed as "Tumor Detected" or "Tumor Not Detected." The accuracy rate achieved by the model is 96.08% and f-score is 97.3%. CNN is implemented using 3 layers and in 35 epochs the result is produced [1]. The aim of their work is to enlist the efficacy of the Diagnostic ML Applications and the corresponding Predictive Treatment [1]. Detecting the brain tumour using the neutrosophical principles is their future work [1].

"Hajji T., Masrour T., Douzi Y., Serrhini S., Ouazzani M. & Jaara M. (2020)" proposed an architecture for the classification of Tumors in Brain, using the DL approach. Comparison and analysis of various famous convolutional architectures along with the AsilNet, which they proposed for concerning the classification of Tumors is done. They enlisted the convolution features of each type of Brain tumor and also developed a learning database. The classification rate on the local database is observed around 99%. This depicts the efficiency of the deep learning architecture for classifying Brain Tumors [2].

"Abhista B., Jarrad K. & Marc A. (2020)" investigated the role of CNN for the purpose of segmenting MR images of Brain Tumor. They proposed an automated segmentation method using Convolutional Neural Networks (CNNs) [3]. They explored the field of Radiomics as the future use of CNNs. This helped in examining the quantitative features of the brain tumor like its Shape, Texture, Location and Signal Intensity which can help for predicting results such as the survival rate of a patient and their response to the medication [3].

"Tonmoy H., Fairuz S., Mohsena A., MD Abdullah Al Nasim & Faisal S. (2019)" proposed a methodology that applies Fuzzy C-Means algorithm, for detecting Tumor from 2D MR images [5]. The same is also done using the traditional Classifiers and the Convolutional Neural Network. For tumor segmentation Fuzzy C-Means is used as it is known to predict tumor cells accurately. After tumor segmentation, classification is done using traditional Classifiers long with the Convolutional Neural Network. In traditional classifiers, they implemented and did comparative analysis of the results of different traditional classifiers. Among the traditional classifiers used, SVM classifier showed the max accuracy rate of 92.42%. To improve the results, they also implemented CNN and received the higher accuracy rate of 97.87% [5]. Achieving more efficiency rate in 3D brain images for brain tumor segmentation is the future work [5].

"Sunanda D., O.F.M. Riaz Rahman A. & Nishant N. (2019)" proposed a method to classify Brain Tumor. They developed a CNN classifier that classifies the tumor into three types namely glioma, meningioma and pituitary. Pre-processing of the image data is done first. The Gaussian Filter (GF) is used in pre-processing step for filtration of images [6]. The filtered images were also processed using the Histogram Equalization (HE) technique [6]. The system classified input by applying Convolutional Neural Network (CNN) algorithm. They observed the possibility of overfitting because the count of parameters used is very high as compared to the amount of dataset used for training. A regularization technique i.e. Dropout Regularization (DR) is applied for preventing the issue of overfitting [6]. The model achieved the accuracy rate of 94.39% and an average precision of 93.33% [6].

"Krishna P., Maheshkumar P., Nirali P., Dastagir M., Vandana S. & Bhaumik V. (2019)" proposed a methodology using CNN and Watershed algorithm for classifying Brain Tumor [7]. They performed the segmentation of Brain Tumor using Global Thresholding (GT) and Marker Based Watershed (MBW) algorithm [7]. They also proposed a method to calculate the area of a tumor. It is found that the Watershed algorithm represented better segmentation results. CNN classified the image into two types as Non-Tumorous Brain and Tumorous Brain. They achieved the training Accuracy rate of 98%. Modifying the algorithm for classifying the brain tumor into more types is future scope [7].

3   Problem Statement

Amongst various diseases Brain Tumor is considered as the 10th leading reason behind the loss of life for humans worldwide. The estimated data provided by the American Society of Clinical Oncology (ASCO) [43] represents that, about 18,020 adults i.e.10,190 men and 7,830 women will lose their lives due to Primary Cancerous Brain Tumor in the year 2020. The 5-year survival rate defines the % of people that will be alive for at least 5 years after they are diagnosed with tumor. The 5-year survival rate for the people that are diagnosed with a malignant brain tumor is almost 36% whereas, the 10-year survival rate of the same is almost 31%. Also, it is observed that the survival rate of a person decreases with aging. The 5-year survival rate for people that are diagnosed with a tumor and are aged less than 15 years is more than 74%, for the people that are diagnosed with a tumor and are aged between 15 to 39 years is about 71% and for people diagnosed with a tumor and aged 40 and over is only about 21% [43]. The survival rates differ widely and is dependent upon various factors, including the type of tumor detected. This shows the severity of the disease and urge of medical advancement in the domain [43] (Fig. 2).

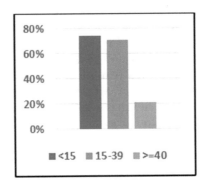

Fig. 2. 5-year survival rate of people diagnosed with tumor

As per data provided by the Cancer Research UK, detecting the Brain tumor in the early days is the only best possible opportunity to save lives from the disease. This tells that the only roadmap to increase the survival rate of the patient is the timely diagnosis of disease and its treatment. In many cases it is found that the patients lost their life due to delay in detecting and understanding the type of tumor the patient is diagnosed with. The severity of the reports shows the need of automation in the medical domain to make the process of detecting the severity of cases and understanding its insight faster by doctors so that the required treatment is given to patients on time. Doctors study the MRI scan reports which provides the insight of the brain in the form of images. This study is very critical as concluding the right type of Tumor is essential for them to further take the decisions about the medication required by the patient. With the increasing number of Brain Tumor cases the availability of doctors keeps on decreasing, as the ratio of patients to doctors is increasing at greater pace. It becomes difficult for doctors to study the reports of all the patients and hence the number of patients assigned to doctors get limited. This shows the need of automated assistance to the doctors for detecting and classifying the type of Tumor the patient is diagnosed with in less time an at a high accuracy rate so that the fatality case where the patient dies because of delay in treatment gets reduced and doctors can treat more patients on time [42].

4 Proposed Methodology

This section elaborates in detail about the method adopted to develop system that can detect and classify Brain Tumor using CNN. It includes detail about the structure of model, dataset, dataset preparation etc.

4.1 Structure of Model

The model depicted below shows the basic common approach used by researchers to develop an automated system using CNN algorithm for detecting and classifying Brain tumor (Fig. 3).

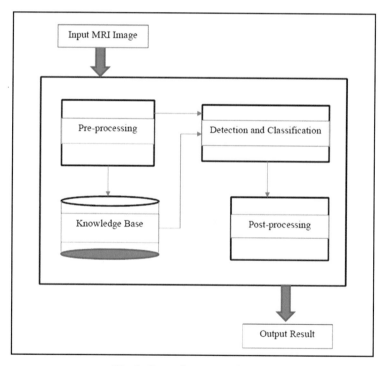

Fig. 3. Internal structure of model

Development of model consists of 3 basic steps:

1. Pre-processing
2. Processing (Detection & Classification)
3. Post-processing
1. *Pre-Processing:*
   This step generally focuses on collection and transformation of collected data into the usable format. In most of the cases, dataset is collected from the BraTS challenge of various years. After obtaining dataset work is carried to transform the data into the required format. Experiment with the discom, tiff format of dataset is most widely carried out. This process also includes removing of abnormalities in dataset. This consists of steps like filtration, skull stripping, etc. Once the data is ready to be used it is further processed in the processing step.
2. *Processing (Detection and Classification):*
   This is the main step of the model as it includes implementation of algorithm to carry out the desired task. Various algorithms like decision tree, support vector machine, k-means, convolutional neural network etc. as chosen is implemented in this step. Some cases also involve implementation of hybrid algorithms i.e., the combination of two or more algorithms as per desired. Storing of knowledge in knowledgebase is also carried out in this step.
3. *Post-Processing:*
   This step is the final step of the model. It consists of GUI, which accepts MR images and displays the outcome depending upon the processing of image and knowledge present in the knowledgebase. This step is essential as this is the part that a naïve user uses to obtain the outcome.

4.2 Dataset

Dataset is obtained from BRaTS 2018. Dataset obtained is of two types: Low-Grade Gliomas also known as Benign and High-Grade Gliomas also known as Malignant [37–39]. Dataset consists of 210 cases of HGG and 75 cases of LGG which will be used for training the model. Also, 67 cases of unknown data will be used for testing the trained model. The obtained dataset was in.nii format. To reduce the complexity while training the obtained 3D dataset is converted into 2D slices.

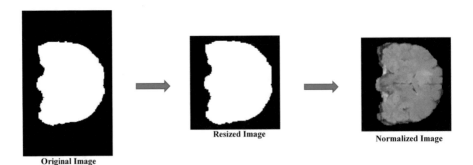

Resized Image

Normalized Image

Original Image

The dataset is converted into.tiff format. Tiff image format reduces the size of the image but retains the features and properties. After converting the volume into 2D slices, the size of image is reduced to 128 × 128 by cropping image from centre to obtain the Region of Interest (ROI) and eliminate the unwanted regions. Normalization is performed on the resized image to obtain the values in specific range which helps to reduce the calculation complexity of algorithm.

5   Experimental Results

V-net architecture of CNN algorithm is used to implement the detection part of the proposed methodology in which set of MR images are provided as an input and its corresponding segmentation mask which helps to train the model to find the Region of Interest (ROI). The masked images are used to understand the features of tumorous and non-tumorous regions.

Image   Segmented Mask   Detection with Ground Truth

Fig. 4. Performance of model on seen data

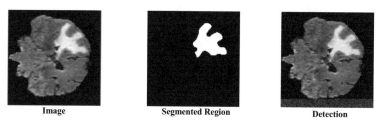

Image   Segmented Region   Detection

Fig. 5. Performance of model on unseen data

In Fig. 4 the ground truth is depicted using the contour of blue color. It is the segmented mask which is provided during training. The red contour shows the region predicted by model after training.

In Fig. 5 the region predicted by model as tumorous is shown using red contour. In this there is no ground truth displayed because the image provided was unknown to the model. The model predicted the outcome based on the features learnt during training.

Sequential Model of CNN algorithm is used to implement the classification part of the proposed methodology. Here the labels 0 and 1 are used to train the model to understand the features of HGG and LGG which are Malignant and Benign respectively (Figs. 6 and 7).

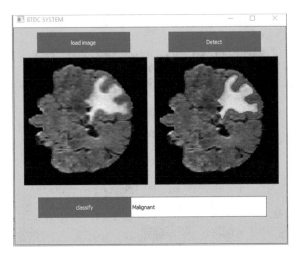

Fig. 6. Model classifying the detected tumor as malignant

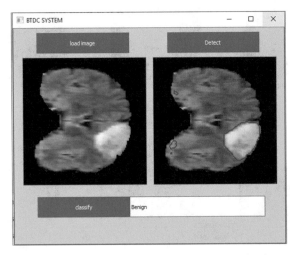

Fig. 7. Model classifying the detected tumor as benign

6 Performance Comparison

The model was trained for 10 epochs and it attained accuracy of 69.94% and 98% for detection and classification respectively. The graph generated by the model during training is depicted below (Figs. 8 and 9):

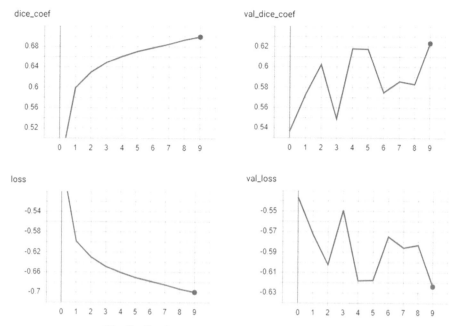

Fig. 8. Graph generated while training detection part

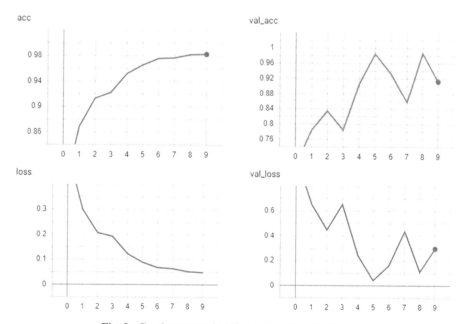

Fig. 9. Graph generated while training classification part

The performance of the model is compared with other models developed using CNN algorithm as shown below:

Paper title	Detection	Classification
"Towards an improved CNN Architecture for Brain Tumor Classification" [2]	–	Model attained accuracy of 99%
"Brain Tumor Detection using Convolutional Neural Network" [5]	Model attained accuracy of 92.98%	–
"Brain Tumor Classification using Convolutional Neural Network" [6]	–	Model attained accuracy of 94.39%
"Classification of Brain Tumor using Convolutional Neural Network" [7]	–	Model attained accuracy of 98%
"Brain Tumor Detection and Classification using CNN and Deep Neural Network" [1]	Overall accuracy of model is 96.08%	
"Detection and Classification of Brain Tumor using Convolutional Neural Network (CNN)."	Detection attained accuracy of 69.94%	Classification attained accuracy of 98%

7 Conclusion

The V-net architecture and Sequential Model provides good performance. Model achieved detection accuracy of 69.94% and classification accuracy of 98%. This is attained by training the model for 10 epochs. It is possible that if the model is trained for more epochs then the accuracy can be improved. By comparing with other CNN model it can be said that the performance of model implemented using V-net architecture and Sequential Model is better. Considering all sides of images to calculate the size of tumor to help medical practitioners to take decision on treatment and predicting the survival rate of patients diagnosed with tumor is the future scope.

References

1. Chirodip, C., Chandrakanta, M., Raghvendra, K., Brojo, M.: Brain Tumor Detection and Classification using Convolutional Neural Network and Deep Neural Network, 4 July 2020. IEEE (2020)
2. Tarik, H., Tawfik, M., Youssef, D., Simohammed, S., Mohammed, O.J., El Miloud, J.: Towards an Improved CNN architecture for brain tumor classification. In: Serrhini, M., Silva, C., Aljahdali, S. (eds.) Innovation in Information Systems and Technologies to Support Learning Research: Proceedings of EMENA-ISTL 2019, pp. 224–234. Springer International Publishing, Cham (2020). https://doi.org/10.1007/978-3-030-36778-7_24

3. Bhandari, A., Koppen, J., Agzarian, M.: Convolutional neural networks for brain tumor segmentation. Insights Imaging (2020)
4. Ozyurt, F., Sert, E., Acvi, D.: An expert system for brain tumor detection: fuzzy C-means with super resolution and convolutional neural network with extreme learning machine. Med. Hyp. (2020)
5. Hossain, T., Shadmani Shishir, F., Ashraf, M., Abdullah Al Nasim, M.D., Muhammad Shah, F.: Brain tumor detection using convolutional neural network. In: 1st International Conference on Advances in Science, Engineering and Robotics Technology 2019 (ICASERT). IEEE (2019)
6. Das, S., Riaz Rahman Aranya, O.F.M., Nayla Labiba, N.: Brain tumor classification using convolutional neural network. In: 1st International Conference on Advances in Science, Engineering and Robotics Technology (ICASERT). IEEE (2019)
7. Pathak, K., Pavthawala, M., Patel, N., Malek, D., Shah, V., Vaidya, B.: Classification of brain tumor using convolutional neural network. In: Proceedings of 3rd International Conference on Electronics Communication and Aerospace Technology (ICECA) (2019)
8. Deepak, S., Ameer, P.M.: Brain tumor classification using deep CNN features via transfer learning. Comput. Biol. Med. **111**, 103345 (2019)
9. Zhou, L., Zhang, Z., Chen, Y.C., Zhao, Z.Y., Yin, X.D., Jiang, H.B.: A deep learning-based radionics model for differentiating benign and malignant renal tumours. Transl. Oncol. **12**(2), 292–300 (2019)
10. Hemnath, G., Janardhan M., Sujihelen, L.: Design and implementing brain tumor detection using machine learning approach. In: 2019 3rd International Conference on Trends in Electronics and Informatics (ICOEI). IEEE (2019)
11. Swati, Z.N.K., et al.: Brain tumor classification for MR images using transfer learning and fine-tuning. Computer. Med. Imaging Graph. (2019)
12. Bernal, J.K.K., Asfaw, D.S., Valverde, S., Oliver, A., Marti, R., Llado, X.: Deep convolutional neural networks for brain image analysis on magnetic resonance imaging a review. Artif. Intell. Med. (2019)
13. Ostrom, Q.T., et al.: CBTRUS statistical report: primary brain and other central nervous system tumours diagnosed in the United States in 2012–2016. Neuro-oncology **21**, Supplement_5 (2019)
14. Global CEO Survey 2019 - Barometer of Corporate Opinion. PwC France Publications. https://www.pwc.fr/fr/publications/dirigeants-et-administrateurs/global-ceo-survey/22ndannual-global-ceo-survey.html
15. Tarik, H., Jamil, O.M.: Weather data for the prevention of agricultural production with convolutional neural networks. In: International Conference on Wireless Technologies, Embedded and Intelligent Systems, WITS (2019)
16. Park, A., Chute, C., Rajpurkar, P., et al.: Deep learning–assisted diagnosis of cerebral aneurysms using the HeadXNet Model. JAMA Netw. Open **2**(6), e195600 (2019). https://doi.org/10.1001/jamanetworkopen.2019.5600
17. Brain Tumor: Statistics, Cancer.Net Editorial Board, November 2017. Accessed 17 Jan 2019
18. General Information about Adult Brain Tumors. NCI. 14 April 2014. Archived from the original on 5 July 2014. Accessed 8 June 2014. Accessed 11 Jan 2019
19. cancer.org, 'Key Statistics for Brain and Spinal Cord Tumors', January 2019. https://www.cancer.org/cancer/brain-spinal-cord-tumors-adults/about/key-statistics.html. Accessed 9 Jan 2019
20. Training and Assessment Reform for Clinical Radiology. RANZCR ASM 2019 Conference, Auckland (2019)
21. Best, B., Nguyen, H.S., Doan, N.B., et al.: Causes of death in glioblastoma: insights from the SEER database. J. Neurosurg. Sci. **63**, 121–126 (2019)

22. Chang, K., Beers, A.L., Bai, H.X., et al.: Automatic assessment of glioma burden: a deep learning algorithm for fully automated volumetric and bidimensional measurement. Neuro Oncol. **21**, 1412–1422 (2019)
23. Sundararajan, R.S.S., Venkatesh, S., Jeya Pandian, M.: Convolutional neural network based medical image classifier. Int. J. Recent Technol. Eng. **8**(3), 4494–4499 (2019)
24. Yamashita, R., Nishio, M., Do, R.K.G., Togashi, K.: Convolutional neural networks: an overview and application in radiology. Insights Imaging **9**(4), 611–629 (2018)
25. mayoclinic.org. 'Brain tumor'. https://www.mayoclinic.org/diseases-conditions/brain-tumor/symptoms-causes/syc-20350084. Accessed 15 Dec 2018
26. Hasan, S.M.K., Linte, C.A.: A modified U-Net convolutional network featuring a nearest-neighbor re-sampling-based elastic-transformation for brain tissue characterization and segmentation. In: Proceedings of the IEEE West New York Image Signal Process Workshop (2018)
27. Mohsen, H., El-Dahshan, E.A., El-Horbaty, E.M., et al.: Classification using deep learning neural networks for brain tumors. Future Comput. Inform. J. **3**(1), 68–71 (2018)
28. Douzi, Y., Kannouf, N., Hajji, T., Boukhana, T., Benabdellah, M., Azizi, A.: Recognition textures of the tumors of the medical pictures by neural networks. J. Eng. Appl. Sci. **13**, 4020–4024 (2018)
29. Sobhaninia, Z., et al.: Brain tumor segmentation using deep learning by type specific sorting of images (2018)
30. Cui, S., Mao, L., Jiang, J., Liu, C., Xiong, S.: Automatic semantic segmentation of brain gliomas from MRI images using a deep cascaded neural network. J. Healthc. Eng. **2018**, 1–14 (2018)
31. Hajji, T., Itahriouan, Z., Jamil, M.O.: Securing digital images integrity using artificial neural networks. Conf. Ser. Mater. Sci. Eng. **353**, 12–16 (2018)
32. Wu, S., Zhong, S., Liu, Y.: Deep residual learning for image steganalysis. Multimedia Tools Appl. (2018)
33. Zaharchuk, G., Gong, E., Wintermark, M., Rubin, D., Langlotz, C.P.: Deep learning in neuroradiology. Am. J. Neuroradiol. **39**(10), 1776–1784 (2018)
34. Akkus, Z., Galimzianova, A., Hoogi, A., Rubin, D.L., Erickson, B.J.: Deep learning for brain MRI segmentation: state of the art and future directions. J. Digit. Imaging **30**(4), 449–459 (2017)
35. Hajji, T., Hassani, A.A., Jamil, M.O.: Incidents prediction in road junctions using artificial neural networks. IOP Conf. Ser. Mater. Sci. Eng. **353**, 012017 (2018). https://doi.org/10.1088/1757-899X/353/1/012017
36. Litjens, G., Kooi, T., Bejnordi, B.E., Setio, A.A., Ciompi, F., Ghafoorian, M., et al.: A survey on deep learning in medical image analysis. Med. Image Anal. **42**, 60–88 (2017)

Dataset

37. Menze, B.H., Jakab, A., Bauer, S., Kalpathy-Cramer, J., Farahani, K., Kirby, J., et al.: The multimodal brain tumor image segmentation benchmark (BRATS). IEEE Trans. Med. Imaging **34**(10), 1993–2024 (2015). https://doi.org/10.1109/TMI.2014.2377694
38. Bakas, S., Akbari, H., Sotiras, A., Bilello, M., Rozycki, M., Kirby, J.S., et al.: Advancing the cancer genome atlas glioma MRI collections with expert segmentation labels and radiomic features. Nat. Sci. Data **4**, 170117 (2017). https://doi.org/10.1038/sdata.2017.117
39. Bakas, S., Reyes, M., Jakab, A., Bauer, S., Rempfler, M., Crimi, A., et al.: Identifying the Best Machine Learning Algorithms for Brain Tumor Segmentation, Progression Assessment, and Overall Survival Prediction in the BRATS Challenge (2018). arXiv:1811.02629

Books

40. Navigating Life with a Brain Tumor. Lynne P. Taylor, By Alyx B. Porter Umphrey and With Diane Richard
41. Brain Tumors an Encyclopedia Approach 3rd Edition. By Andrew H. Kaye, Edward R. Laws. Jr

Websites

42. https://www.cancerresearchuk.org/about-us/cancer-news/press-release/2020-10-06-vital-ret hinking-in-cancer-early-detection-needed-to-save-lives
43. https://www.cancer.net/cancer-types/brain-tumor/statistics.
44. https://bestbraintumorcancer.blogspot.com/2018/10/brain-tumors-temporal-lobe.html
45. CNN simplilearn.com

Software Fault Prediction Using Data Mining Techniques on Software Metrics

Rakesh Kumar[(✉)] and Amrita Chaturvedi

Indian Institute of Technology (IIT-BHU), Varanasi, India
rakeshkumar.rs.cse18@iitbhu.ac.in
https://www.iitbhu.ac.in/

Abstract. Software industries have enormous demand for fault prediction of the faulty module and fault removal techniques. Many researchers have developed different fault prediction models to predict the fault at an early stage of the software development life cycle (SDLC). But the state-of-the-art model still suffers from the performance and generalize validation of the models. However, some researchers refer to data mining techniques, machine learning, and artificial intelligence play crucial roles in developing fault prediction models. A recent study stated that metric selection techniques also help to enhance the performance of models. Hence, to resolve the issue of improving the fault prediction model's performance and validation, we have used data mining, instance selection, metric selection, and ensemble methods to beat the state-of-the-art results. For the validation, we have collected the 22 software projects from the four different software repositories. We have implemented three machine learning algorithms and three ensemble methods with two metric selection methods on 22 datasets. The statistical evaluation of the implemented model performed using Wilcoxon signed-rank test and the Friedman test followed by the Nemenyi test to find the significant model. As a result, the Random forest algorithm produces the best result with an average median of 95.43% (accuracy) and 0.96 (f-measure) on 22 software projects. Based on the Nemenyi test, Random forest (RF) is performing better with 4.54 (accuracy mean score) and 4.41 (f-measure mean score) shown in the critical diagram. Experimental study shows that data mining techniques with PCA provide better accuracy and f-measure.

Keywords: Software fault prediction · Data mining · Ensemble methods · Software metrics · Nemneyi test

1 Introduction

Software engineering has become an attractive field for researchers in software-based industries [1]. Software industries strive for quality software to enhance fault prediction and removal in software projects. Human beings are depending more on application-based quality software [2]. On building the quality software, its results as increasing size and complexity of the software. So when processing

© The Author(s), under exclusive license to Springer Nature Switzerland AG 2022
R. Misra et al. (Eds.): ICMLBDA 2021, LNNS 256, pp. 304–313, 2022.
https://doi.org/10.1007/978-3-030-82469-3_27

a massive task with software, it can cause the software function failure. Such software failure occurs in real-time applications [3]. Then software tester analyses the software application to detect the faulty module [4].

Manual testing is not sufficient to find all faults and requires more effort and cost. Therefore, the software fault prediction models are widely used in the software industry for fault detection, effort estimation, risk analysis, and reliability prediction during the software development phase [5]. Software fault identification at an early stage of the software development life cycle (SDLC) can reduce the time, cost, human effort and improve software products' reliability. Many researchers proposed different approaches and techniques to build the software fault prediction models [6].

Software fault detection and locating the faulty module have become challenging issues. Machine learning and data mining techniques become promising methods to build prediction models. In data mining techniques, a machine learning algorithm is trained on the input dataset with the actual output to classify the module as faulty and non-faulty. Further, save the trained model and provide the unknown input to classify the module as faulty and non-faulty. Hence, data mining techniques such as Naïve Bayes [7], Support vector machine [8], K-nearest neighbor [9] bagging [10], boosting [11], etc., are used. These techniques provide better significant performance. But the prediction performance and complexity of the software architecture remain a challenging task. Data preprocessing and metric selection techniques resolve this challenge. It benefits by removing the irrelevant metrics from the dataset and reduce the software complexity [12] to improve the computational efficiency. Like principal component analysis (PCA), chi-square, correlation, entropy, etc., metric selection techniques are available.

The rest of the research paper is presented as follows: Sect. 2 presents the literature survey briefly for fault prediction using data mining techniques. Section 3 describes the used dataset, and Sect. 4 introduces ensemble methods. Section 5 shows the experimental setup and results. Next, Sect. 6 gives the conclusion and future scope of the research paper.

2   Literature Survey

This literature survey section presents a brief study of the recent techniques used for software fault prediction. As mentioned above that data mining-based machine learning algorithms have attracted more software researchers for building fault prediction models due to their significant performance. Data mining techniques are presented as promising techniques to improve the software fault prediction model for the software application system [13]. The available datasets are imbalanced that generated randomness in pattern recognition. So the researcher suggested that ensemble-based learning can help to build a fault prediction model. It helps to improve the performance of weak classifiers by combining the many single classifiers. Early, conventional techniques were used to remove the irrelevant metrics from the dataset [14]. This technique is implemented in two steps; firstly, they calculated maximum information from metrics and then applied machine learning algorithms but still degrading performance.

Recently, Duksan et al. infer that software fault prediction suffers from imbalanced datasets. This means a lower number of faulty instances are encountered; thus, the classification performance is degraded [15]. Therefore, to handle such an imbalance dataset, Abdi et al. [16] proposed an approach to maintain the classifier's train test ratio. Where k-nearest neighbor [9] used for instance selection and Naïve Bayes [17] used for knowledge learning to improve the performance of the classifier. Conventional techniques require more time and complexity instead of fails to provide precision results. Barajas et al. in 2015, present designs by comparing fuzzy regression techniques and statistical regression techniques to execute the software fault prediction in given time duration.

Shan et al. [8] used a Support vector machine with optimized constraints as ten-fold cross-validation and grid search techniques. The implemented result show that LLE-SVM provides better results. Yang et al. [18] introduce the radial basis function, neural network, and Bayesian methods to develop fault prediction models. By weight updating structure, the performance of RBF is improved. The artificial bee colony approach is used with the artificial neural network to update the optimal weight at the training phase, enhancing the performance of the fault prediction model [19].

Mejia et al. [20] also used the neural network to develop the fault prediction model. In this approach, the researcher used the Git hub repository to analyze the relationship between software code and fault. They found that the metric selection method produces an improved classification result. Hence, feature selection methods were proposed by Khoshgoftaar et al. for imbalanced software datasets [21]. These literature studies show that the metric selection approach with the machine learning algorithm implemented on software datasets can produce better results.

This brief related research summarizes metric selection methods, machine learning algorithms, optimization methods, instance selection techniques, and artificial intelligence to develop fault prediction models. These techniques can enhance performance and reduce complexity by using different data mining techniques and metric selection methods.

3   Used Dataset

3.1   Dataset Description

Early, fewer software projects have been used to design more efficient and generalize fault prediction models using data mining techniques. So, we have included more datasets from a different repository to generalize the validation. The 22 software projects collected from the four public software repositories named NASA [22], Eclipse [23], Elastic search [24], and Android [25] have been used in this research work. The description of the datasets is given in Table 1. It contains the name of the software repository, software project name, software code (SP code), number of metrics, number of modules, and faulty module percentage. More details about the software projects are publicly available.

Table 1. Software projects collected from four repository with the number of metrics and number of modules

Software repository	Software projects	SP code	No. of Metrics	No. of modules	Faulty module %	Software repository	Software projects	SP code	No. of metrics	No. of modules	Faulty module %
NASA	CM1	SP1	41	505	9.50	Eclipse	Eclipse_JDT	SP12	18	997	20.66
	KC3	SP2		458	9.38		Eclipse_PDE	SP13		1497	13.96
	KC4	SP3		125	48.80		Equinox	SP14		324	39.50
	MC2	SP4		161	32.29		Lucene	SP15		691	9.26
	MW1	SP5		403	7.69		Mylyn	SP16		1862	13.05
	PC1	SP6		1107	6.86	Android	Android_2013_1	SP17	107	73	27.39
	PC2	SP7		5589	0.41		Android_2013_2	SP18		98	17.34
	PC3	SP8		1563	10.23		Android_2013_3	SP19		109	1.83
	PC4	SP9		1458	12.20		Android_2014_1	SP20		116	6.03
Elastic	AR1	SP10	30	121	7.43		Android_2014_2	SP21		119	1.68
	AR6	SP11		101	14.85		Android_2015_1	SP22		124	0.00

3.2 Dataset Preprocessing

After collecting the raw dataset from the software repositories, we have performed the necessary wrangling on the dataset. Firstly, we got some metrics that have missing values. The missing values are replaced with the mean values of the metrics. Some datasets are enormous and found that some metrics have the same value, e.g., 0 or 1. So such types of metrics are removed with the consideration that they can not affect the results. The insight behind these removals is that the variance of such metrics is zero. These metrics don't make any pattern and significance.

3.3 Dataset Outlier Detection and Iterative Instance Removal/Selection

After cleaning the datasets, the interquartile range (IQR is the difference between the 75th (Q1) and the 25th percentiles (Q3)) method has been used to detect the outliers in metrics. The cutoff for outliers is 1.5 times the IQR. The lower and upper limit is defined by subtracting the cutoff from the 25th percentile and adding it to the 75th percentile. The outlier defined as the data point x_i is outside of the lower and upper limit. After detecting all the outliers, count the number of outliers in each instances. There is no rule to remove the definite number of instances based on outlier detection. So, we have started removing instances one by one, starting from the ones that have the most number of outliers. After removing the first instance, the result is stored and then removed next two instances with the maximum number of outliers and results updated. Repeat these steps similarly. Stop removing instances either no improvement in results or maximum n/12 instances removed, where n is the total number of instances.

3.4 Iterative Metric Removal/Selection

A software metric is a more useful parameter to build the fault prediction model using data mining techniques. Metrics having higher metric values create more complexity, and some metrics are not significant to build the fault prediction model. So, choosing the optimal and adequate sub metric is a data mining task. The following two metric selection/removal methods have been used in this research after removing irrelevant instances.

1. Select KBest(SKB): It is a data mining technique based on Chi-Square, which assigns a unique score to each metric based on their significance to predict the fault. The metric with the least score works as the least important metric. So, we removed the least essential k-metrics with the minimum score. There is no fixed threshold to select the optimal value of k. We have analyzed all the possible cases. E.g., Firstly, remove the least important metrics, then calculate the result. Then remove the two metrics with the least score and update the results. Repeat these steps similarly. Stop removing the metric either no improvement in results or maximum m/5 metric removed, where m is the total number of metrics in a dataset.
2. PCA: Principal component analysis is a metric extraction technique. PCA generates a new subset of PCA components based on the existing metrics values. We removed the top k principal components based on the variances. Remove the first PCA component with the least variance and calculate the results. Then remove two PCA components with the least variance and update the results. Repeat these steps similarly. Stop removing components either no improvements in results or maximum m/5 components removed, where m is the total number of PCA components.

Hence at last, after preprocessing (cleaning), instance removal, and metric removal, we have obtained the refine and effective dataset to trained the fault prediction model.

4   Ensemble Methods

Ensemble methods are work as the concept of strength in unity. It is work to hypothesize that combining base/single classifiers with some rules can provide better results. There are three types of ensemble methods: Bagging, boosting, and stacking.

Base Classifier: The hypothesis is that a single weak classifier works as a building model to design the prediction model. It is also found that a single classifier suffers from high bias or low variance. As a result, compared to single classifiers, ensemble models achieving better results. In this research, three base classifiers Naïve Bayes (NB) [7], Support Vector Machine(SVM) [26], and K-nearest neighbor (KNN) [9], have been used.

Combining the base classifier (Ensembling): There are two rules to combine the result of base classifiers: Homogeneous and Heterogeneous. In Homogeneous methods, base classifiers are the same, while in heterogeneous, base classifiers are different. Bagging and Boosting are Homogeneous, and the majority voting (MV) method is Heterogeneous. The homogeneous process classifies into two types: parallel and sequential. Bagging behaves as parallel (e.g., Random forest) and boosting as a sequential classifier (e.g., Adaboost). Majority voting is a heterogeneous method in which the three base classifier outcomes are combined based on majority voting methods. In this experimental work, Random forest (RF), Adaboost (AB), and Majority voting (MV) are implemented as ensemble methods.

5 Experimental Setup and Results

5.1 Experimental Setup

In this experiment, six data mining techniques were implemented on 22 software projects (Table 1). After preprocessing, Outliers detection using IQR method and instance removal based on the most number of outliers are performed. Then, two metric selection techniques (SKB and PCA) were used to select the significant metric. The holdout validation method (means training on 80% dataset and tested on 20% dataset) was used to compare fault prediction models' performance. The performance parameter accuracy and f-measure are evaluated using the Eq. 1 and 3.

The statistical evaluation paired Wilcoxon signed-rank test (non-parametric) has been used to compare the classifiers with mean difference values and p-value. Post-hoc Nemenyi test was conducted to compare many classifiers implemented on many datasets. To perform the Nemneyi test, firstly performed the Friedman test to validate that at least one classifier performs significantly differently and then used the Nemenyi test to group the classifiers by similar performance. The performance of the two classifiers is significantly different, if their mean score differs by more than some critical distance. The value of critical distance (1.112) depends on the number of classifiers, datasets, and the critical value (at a given significant level (p-value)). We can see in CD; the p-value is 0.000 (Friedman test) represent that at least one classifiers performance is significantly different. It is found that the ensemble method (RF) producing the best result compared to the state of art techniques [4]. Table 2 shows each method's accuracy and f-measure with the two metric selection techniques. The diagrammatic performance of each classifier and each metric selection methods show in Fig. 1 as Box Plot and critical diagram (CD).

$$Accuracy = \frac{TN + TP}{TN + FP + FN + TP} \tag{1}$$

$$Precision = \frac{TP}{TP + FP}; Recall = \frac{TP}{TP + FN} \tag{2}$$

$$F - measure = \frac{2 * Precision * Recall}{Precision + Recall} \tag{3}$$

Table 2. Accuracy and f-measure performance of different classifier using SKB and PCA feature selection methods

SP Code	Accuracy (%)												F-measure											
	NB		SVM		KNN		RF		AB		MV		NB		SVM		KNN		RF		AB		MV	
	SKB	PCA	SKB	PCA	SKB	PCA	SKB	PCA	SKB	PCA	SKB	PCA	SKB	PCA	SKB	PCA	SKB	PCA	SKB	PCA	SKB	PCA	SKB	PCA
SP1	90.91	94.74	95.79	94.74	95.79	94.74	95.79	97.87	93.68	94.74	95.79	95.79	0.91	0.95	0.96	0.95	0.96	0.95	0.96	0.98	0.94	0.95	0.96	0.96
SP2	89.13	94.57	95.29	97.65	94.12	95.35	95.51	95.35	94.44	97.65	95.29	97.65	0.89	0.95	0.95	0.98	0.94	0.95	0.96	0.98	0.94	0.98	0.96	0.98
SP3	80.00	83.33	76.00	79.17	76.00	79.17	80.00	83.33	84.00	70.83	80.00	87.50	0.8	0.83	0.76	0.79	0.76	0.79	0.8	0.83	0.84	0.71	0.8	0.88
SP4	75.00	87.10	80.00	84.38	77.42	74.19	83.33	83.33	80.00	83.33	86.67	83.87	0.75	0.87	0.8	0.84	0.77	0.74	0.83	0.83	0.8	0.83	0.9	0.84
SP5	86.42	94.81	97.40	96.25	97.40	96.25	96.30	96.25	97.40	96.25	97.44	96.25	0.86	0.95	0.97	0.96	0.97	0.96	0.96	0.96	0.97	0.96	0.97	0.96
SP6	89.19	95.71	96.80	96.19	96.35	96.19	98.17	98.10	96.35	96.19	97.72	97.14	0.89	0.96	0.97	0.96	0.96	0.96	0.98	0.98	0.96	0.96	0.98	0.98
SP7	97.50	99.81	99.90	100.00	99.90	100.00	99.90	100.00	99.90	100.00	99.90	100.00	0.97	1.00	1.00	1.00	1.00	1.00	1.00	1.00	1.00	1.00	1.00	1.00
SP8	86.80	93.90	92.98	94.59	93.23	94.58	93.56	95.95	92.98	94.24	93.56	95.61	0.87	0.94	0.93	0.95	0.93	0.95	0.94	0.96	0.93	0.94	0.94	0.96
SP9	89.13	91.61	91.90	94.81	91.90	93.70	93.66	94.44	89.78	92.96	91.55	94.44	0.89	0.92	0.92	0.95	0.92	0.94	0.94	0.94	0.9	0.93	0.92	0.94
SP10	46.67	100.00	66.67	92.86	66.67	92.86	85.71	100.00	66.67	92.86	71.43	100.00	0.47	1.00	0.67	0.93	0.67	0.93	0.86	1.00	0.67	0.93	0.71	1.00
SP11	80.00	90.00	89.47	85.00	89.47	85.00	89.47	95.00	84.21	85.00	89.47	85.00	0.80	0.90	0.89	0.85	0.89	0.85	0.89	0.95	0.84	0.85	0.89	0.85
SP12	100.00	100.00	100.00	100.00	100.00	100.00	100.00	100.00	100.00	100.00	100.00	100.00	1.00	1.00	1.00	1.00	1.00	1.00	1.00	1.00	1.00	1.00	1.00	1.00
SP13	90.91	100.00	95.45	100.00	95.45	100.00	95.65	100.00	95.45	100.00	95.45	100.00	0.91	1.00	0.95	1.00	0.95	1.00	0.96	1.00	0.95	1.00	0.95	1.00
SP14	100.00	100.00	100.00	100.00	100.00	100.00	100.00	100.00	100.00	100.00	100.00	100.00	1.00	1.00	1.00	1.00	1.00	1.00	1.00	1.00	1.00	1.00	1.00	1.00
SP15	100.00	100.00	100.00	100.00	100.00	100.00	100.00	100.00	100.00	100.00	100.00	100.00	1.00	1.00	1.00	1.00	1.00	1.00	1.00	1.00	1.00	1.00	1.00	1.00
SP16	79.90	82.99	85.19	83.92	84.66	84.02	86.02	86.56	84.66	83.92	84.13	84.41	0.80	0.83	0.85	0.84	0.85	0.84	0.86	0.87	0.85	0.84	0.85	0.84
SP17	70.23	88.63	90.22	90.64	90.22	90.64	89.86	89.76	90.22	90.64	89.86	90.64	0.70	0.89	0.90	0.91	0.90	0.91	0.90	0.90	0.90	0.91	0.90	0.91
SP18	58.46	70.49	67.74	70.77	68.25	73.85	68.85	75.00	63.93	70.77	68.85	72.58	0.58	0.70	0.68	0.71	0.68	0.74	0.69	0.75	0.64	0.71	0.70	0.72
SP19	86.33	95.38	93.23	96.15	93.38	96.15	93.23	96.24	93.98	96.15	93.98	96.15	0.86	0.95	0.93	0.96	0.93	0.96	0.93	0.96	0.94	0.96	0.94	0.96
SP20	78.81	89.78	88.22	89.92	89.67	90.88	89.22	90.48	88.79	89.92	88.22	90.06	0.79	0.90	0.88	0.90	0.90	0.91	0.89	0.90	0.89	0.90	0.89	0.90
SP21	91.67	100.00	95.83	100.00	95.83	100.00	100.00	100.00	95.83	100.00	95.83	100.00	0.92	1.00	0.96	1.00	0.96	1.00	1.00	1.00	0.96	1.00	0.96	1.00
SP22	95.00	89.47	95.00	84.21	95.00	84.21	95.00	89.47	94.74	84.21	95.00	84.21	0.95	0.89	0.95	0.84	0.95	0.84	0.95	0.89	0.95	0.84	0.95	0.89

5.2 Results and Discussion

Statistical evaluation: In this section, we are providing the following results of implemented data mining techniques. 1. From the Table 2, it is found that the Random forest (RF) with PCA is performing maximum average accuracy 93.96%, then Majority voting (MV) as 93.24%, and in term of f-measure, both RF and MV with PCA obtained 0.94 average value on 22 datasets. 2. The box-plot comparison of the data mining techniques shown in Fig. 1a (Accuracy) and Fig. 1c (f-measure). It can be seen that the median accuracy value (95.43%) of RF is maximum among all, RF and MV performing same in term of median f-measure value (0.96). 3. The critical diagram of the Post-hoc Nemenyi test shown in Fig. 1e and 1f, showing that RF with mean score 4.54 (accuracy) and 4.41 (f-measure) is performing better among all classifiers. The shaded background used to differentiate the significant group. So NB is the least performing classifier. KNN, SVM, RF, and MV are performing better and belong to the same group. AB performance is at the border; it can not be decided to which group it will belong. 3. When comparing the metric selection techniques, PCA performs better than the SelectKbest algorithm with an average mean difference of 2.74% shown in Table 4. The boxplot performance of both is shown in Fig. 1b and 1d. It can easily be seen that PCA is performing better in terms of accuracy and f-measure. 4. Table 3 shows the mean difference and p-values among the classifiers in term of accuracy and f-measure. P-value is calculated using the paired Wilcoxon signed-rank test. It shows that the accuracy of the random forest is 4.36% (maximum) greater than other classifiers. If the p-value less than 0.05 shows that the classifiers are significantly different (Y) else significantly not different (N). Suppose the row classifier is greater than column classifier performance then represented as G else less as (L). In Table 3, almost all the p-value

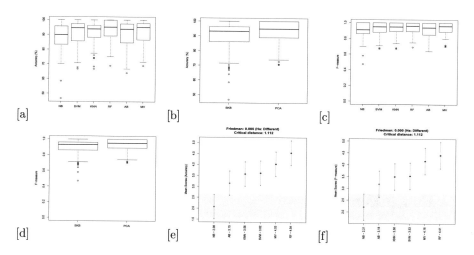

Fig. 1. Box plot and critical diagram performance on testing dataset). [a] Comparative Accuracy of classifiers. [b] Comparative accuracy of PCA and SKB. [c] Comparative f-measure of classifiers. [d] Comparative f-measure of PCA and SKB. [e] Comparing classifiers statistically using mean score (Accuracy). [f] Comparing classifiers statistically using mean score (f-measure) test.

Table 3. Wilcoxon signed rank test: mean difference (Greater (G)/Less (L)), p-value ($p < 0.05$) with statistical significance difference (Yes (Y)/ NO (N)) between different classifiers

(a) Accuracy (%)

Mean difference (Row-Column)							p-value, G/L, Y/N					
	NB	SVM	KNN	RF	AB	MV		NB	SVM	KNN	RF	AB	MV
NB	0.00	−2.73	−2.45	−4.36	−2.32	−3.57	NB	–	0.00	0.01	0.00	0.00	0.00
SVM	2.73	0.00	0.27	−1.64	0.40	−0.84	SVM	G, Y	–	0.45	0.00	0.05	0.00
KNN	2.45	−0.27	0.00	−1.91	0.13	−1.12	KNN	G, Y	L, N	–	0.00	0.37	0.01
RF	4.36	1.64	1.91	0.00	2.04	0.79	RF	G, Y	G, Y	G, Y	–	0.00	0.10
AB	2.32	−0.40	−0.13	−2.04	0.00	−1.24	AB	G, Y	L, N	L, Y	L, N	–	0.00
MV	3.57	0.84	1.12	−0.79	1.24	0.00	MV	G, Y	G, Y	G, Y	G, Y	G, Y	–

(b) F-measure

Mean difference (Row-Coloumn)							p-value, G/L, Y/N					
	NB	SVM	KNN	RF	AB	MV		NB	SVM	KNN	RF	AB	MV
NB	0.00	−0.03	−0.02	−0.04	−0.02	−0.04	NB	–	0.00	0.01	0.00	0.01	0.00
SVM	0.03	0.00	0.00	−0.02	0.00	−0.01	SVM	G, Y	–	0.37	0.00	0.05	0.00
KNN	0.02	0.00	0.00	−0.02	0.00	−0.01	KNN	G, Y	E, N	–	0.00	0.49	0.00
RF	0.04	0.02	0.02	0.00	0.02	0.01	RF	G, Y	G, Y	G, Y	–	0.00	0.26
AB	0.02	0.00	0.00	−0.02	0.00	−0.02	AB	G, Y	E, N	E, N	L, N	–	0.00
MV	0.04	0.01	0.01	−0.01	0.02	0.00	MV	G, Y	G, Y	G, Y	L, Y	G, Y	–

Table 4. Wilcoxon signed rank test: mean difference (Greater (G)/Less(L), p-value, and Significance difference (Yes (Y)/No (N)) between feature selection methods

Parameter	Mean (PCA-SKB)	p-value	Significance
Accuracy (%)	2.74	0.00	G, Y
F-measure	0.03	0.00	G, Y

is near to 0.00, which means each classifier is significantly different and have different predictive power (Y).

6   Conclusion

This research work focused on software fault prediction methods using machine learning-based data mining techniques and ensemble methods. It is concluded that cleaning the dataset, removing instances based on outliers, and removing metrics helpful in enhancing the classifiers performances. This work has increased the fault prediction model's performance using data mining techniques and ensemble methods implemented on 22 datasets collected from four repositories. The random forest with PCA produced the best result as 95.43% (median accuracy) and 0.96 (median f-measure). The statistical analysis concluded that the best performing model is 4.36% greater than the low-performing methods (NB). P-value (using Wilcoxon signed-rank) among the classifiers concludes that these have significantly different predictive power. The post hoc Nemenyi test (applied after the Friedman test) was used to compare the multiple data mining techniques on multiple datasets, and it concluded that RF was performing better with a mean score of 4.45 (accuracy) and 4.41 (f-measure). Overall, it can be concluded that data mining techniques with machine learning algorithms were helpful to build the prediction model on many imbalanced datasets with high dimensions. In the future, this work is extended by developing a more robust and generalized fault prediction model for unlabeled datasets and private datasets.

References

1. Jayanthi, R., Florence, L.: Software defect prediction techniques using metrics based on neural network classifier. Cluster Comput. **22**(1), 77–88 (2018). https://doi.org/10.1007/s10586-018-1730-1
2. Tian, J.: Software Quality Engineering: Testing, Quality Assurance, and Quantifiable Improvement. Wiley, Hoboken (2005)
3. Salfner, F., Lenk, M., Malek, M.: A survey of online failure prediction methods. ACM Comput. Surv. **42**(3), 1–42 (2010)
4. Canaparo, M., Ronchieri, E.: Data mining techniques for software quality prediction in open source software: an initial assessment. In: CHEP, vol. 2018, pp. 1–8 (2019)
5. Chauhan, N.S., Saxena, A.: A green software development life cycle for cloud computing. IT Prof. **15**(1), 28–34 (2013)

6. Kumar, L., Sripada, S.K., Sureka, A., Rath, S.K.: Effective fault prediction model developed using Least Square Support Vector Machine (LSSVM). J. Syst. Softw. **137**, 686–712 (2018)

7. Okutan, A., Yıldız, O.T.: Software defect prediction using Bayesian networks. Empirical Softw. Eng. **19**(1), 154–181 (2012). https://doi.org/10.1007/s10664-012-9218-8

8. Shan, C., Chen, B., Hu, C., Xue, J., Li, N.: Software defect prediction model based on LLE and SVM. In: IET Conference Publications, vol. 2014, no. CP653 (2014)

9. Cover, T., Hart, P.: Nearest neighbor pattern classification. IEEE Trans. Inf. Theor. **13**(1), 21–27 (1967)

10. Kuncheva, L.I., Skurichina, M., Duin, R.P.W.: An experimental study on diversity for bagging and boosting with linear classifiers. Inf. Fusion **3**(4), 245–258 (2002)

11. Aljamaan, H.I., Elish, M.O.: An empirical study of bagging and boosting ensembles for identifying faulty classes in object-oriented software. In: 2009 IEEE Symposium on Computational Intelligence and Data Mining, CIDM 2009 - Proceedings, pp. 187–194 (2009)

12. Mccabe, T.J.: A Complexity. IEEE Trans. Softw. Eng. **2**(4), 308–320 (1976)

13. Wang, T., Zhang, Z., Jing, X., Zhang, L.: Multiple kernel ensemble learning for software defect prediction. Autom. Softw. Eng. **23**(4), 569–590 (2016)

14. Xu, Z., Xuan, J., Liu, J., Cui, X.: MICHAC: defect prediction via feature selection based on maximal information coefficient with hierarchical agglomerative clustering. In: 2016 IEEE 23rd International Conference on Software Analysis, Evolution, and Reengineering, SANER 2016, January 2016, pp. 370–381 (2016)

15. Ryu, D., Baik, J.: Effective multi-objective naïve Bayes learning for cross-project defect prediction. Appl. Soft Comput. J. **49**, 1062–1077 (2016)

16. Abdi, Y., Parsa, S., Seyfari, Y.: A hybrid one-class rule learning approach based on swarm intelligence for software fault prediction. Innov. Syst. Softw. Eng. **11**(4), 289–301 (2015). https://doi.org/10.1007/s11334-015-0258-2

17. Taheri, S., Mammadov, M.: Learning the Naive Bayes classifier with optimization models. Int. J. Appl. Math. Comput. Sci. **23**(4), 787–795 (2013)

18. Yang, Z.R.: A novel radial basis function neural network for discriminant analysis. IEEE Trans. Neural Netw. **17**(3), 604–612 (2006)

19. Arar, Ö.F., Ayan, K.: Software defect prediction using cost-sensitive neural network. Appl. Soft Comput. J. **33**, 263–277 (2015)

20. Mejia, J., Muñoz, M., Rocha, Á., Calvo-Manzano, J. (eds.): Trends and Applications in Software Engineering. AISC, vol. 405. Springer, Cham (2016). https://doi.org/10.1007/978-3-319-26285-7

21. Khoshgoftaar, T.M., Gao, K.: Feature selection with imbalanced data for software defect prediction. In: 8th International Conference on Machine Learning and Applications, ICMLA 2009, pp. 235–240 (2009)

22. NASA Dataset. https://github.com/klainfo/NASADefectDataset

23. Eclipse Dataset. http://bug.inf.usi.ch/download.php

24. Elastic Search Dataset. http://www.inf.uszeged.hu/ferenc/papers/UnifiedBug DataSet/

25. Android Dataset. http://www.inf.uszeged.hu/~ferenc/papers/UnifiedBugData Set/

26. Singh, Y., Kaur, A., Malhotra, R.: Software fault proneness prediction using support vector machines. In: Lecture Notes in Engineering and Computer Science, vol. 2176, no. 1, pp. 240–245 (2009)

MMAP: A Multi-Modal Automated Online Proctor

Aumkar Gadekar$^{(\boxtimes)}$, Shreya Oak, Abhishek Revadekar, and Anant V. Nimkar

Department of Computer Engineering, Sardar Patel Institute of Technology,
Mumbai, India
{aumkar.gadekar,shreya.oak,abhishek.revadekar,anant_nimkar}@spit.ac.in

Abstract. With the surge in online education, more universities have shifted classes online. The growing popularity of MOOC courses and the changing education landscape could mean more and more people switching to online education. A primary drawback is the difficulty in monitoring of students during an online examination which leads to a lot of malpractices used by candidates. This paper explores computer vision based techniques to propose a five-fold proctoring mechanism for online tests. The features incorporated are authentication, head movement, eye motion tracking, speech detection and object detection. The solution has an overall accuracy of 91% accuracy.

Keywords: Online education · Automated proctor · Image processing · Authentication · Object detection · Eye tracking · Computer vision

1 Introduction

There has been a gradual shift from the traditional teaching methods to technology-based forms of education. Online learning makes education available for those who cannot travel to schools and colleges. The 2020 COVID-19 pandemic highlighted this need as numerous educational institutions were suspended. However it also exposed loopholes in the system.

The primary problems associated with distance education is a lack of supervision. This results in widespread plagiarism in assignments and tests conducted on online platform. Often such degrees are not valued due to the ease with malpractice can occur in tests and courses given from home. D. L. King et al. [12] found that incidence of cheating in online exams was as high as 73%. Dale Varble [23] reported that 58% percent of students surveyed felt inclined to cheat in online tests that led to difference in performance.

A variety of measures have been employed for conducting tests on online platforms. Currently online live proctoring is used for many examinations, which involves the need for a human to be behind the screen and monitor the students. Such a solution is expensive and difficult to implement in case of hundreds of thousands of candidates. Therefore, AI-powered detection of malpractice is the

© The Author(s), under exclusive license to Springer Nature Switzerland AG 2022
R. Misra et al. (Eds.): ICMLBDA 2021, LNNS 256, pp. 314–325, 2022.
https://doi.org/10.1007/978-3-030-82469-3_28

viable alternative. The objective of this research work is to conduct a thorough review of malpractices in online tests to identify various unethical methods employed by candidates, and design a proctoring mechanism. The ideal solution would be easy to implement with no additional hardware, as automated as possible to minimize the need for human proctoring and would use a combination approaches to identify malpractice.

This research posits a five-fold approach using computer vision to detect use of unethical means in online tests. The candidate's laptop is the only necessity and no extra hardware is needed. The live video feed from the webcam is assessed frame by frame by the automated proctor, which looks for the following activities: head movement, eye gaze and mouth movements indicative of speech. It also carries out regular face authentication in the duration of the test.

The rest of the paper is organized as follows. Section 2 covers the literature survey relevant to the topic. Section 3 comprises of the methodology that this study implements. The results and accuracy obtained by the proposed method have been evaluated against existing data-sets and presented in Sect. 4 The paper then concludes by in Sect. 5.

2 Literature Review

A survey of various online resources reveals the different commonly used unethical tactics by students in online exams, and the previous work on preventing this malpractice [7].

2.1 Review of Cheating Strategies and Counter Measures

The following are the most commonly undertaken malpractices by test takers:

Fake Identity: It is possible for a person to pretend to be the intended candidate and take the test on their behalf. Traditional methods of password protection have been replaced by biometric authentication [7]. Biometric authentication is comparing the user's fingerprints, voice signals, facial geometry, etc. In [14], the authors use keystroke dynamics for user authentication in online examinations. Their proof of concept system measures the time between subsequent keystrokes and also uses Steinhaus' technique to implement cosine correlation to perform user authentication by using keystroke dynamics. Face recognition is also a popular method as cited in the papers [20,22]. Speech recognition, retina or iris scanning, hand geometry are described in [15] with their relative accuracy and error incidence. A balance between accuracy and ease of use for users should be maintained. Face recognition has a better trade-off and is suitable since only the laptop's webcam is used and other methods require special hardware.

Malpractices by Communication Between Candidates: Candidates attempt to discuss answers even in supervised exams, and are more likely to do so in unsupervised examinations. They could communicate with people around them, or contact via phones. The work proposed in [2], uses a webcam based approach to detect phones and objects with which students can communicate. Question bank randomization [5,21] ensures that different students get differing questions, or in a random order. The works in [2,17] also implement sound based detection to detect human voice and speech, in which the candidate is communicating with someone else in the room.

Malpractice from Online/Offline Resources: Students can access offline and online resources. Accessing the internet from the laptop/device on which the person is taking a test can be prevented using Active-Window [2,5,21] detection and tab locking. They record how many times the user switches out of the tab during the test and reports the test taker or shuts down the test [11]. In [2], the authors give a method that is a multimedia analytics system. The candidate wears a headcam which detects text, objects and people around the person. A Arinaldi et al. [1] use gesture tracking to catch forbidden actions like use of cheating material. Eye and gaze tracking is used to see if a person is looking outside of their laptop screen [2,4,17].

2.2 Review of Specific Techniques to Detect Cheating

Many of the previous solutions required complicated hardware and configurations. Keeping that in mind, the following techniques were reviewed:

User Verification with Face Authentication. Face authentication using a Viola Jones face detector and histogram of gradients technique is proposed in [7]. The authors of [8,10] made use of the Eigen-face method to authenticate a user with face detection. H. Xia and C. Li proposed a face recognition algorithm trained on TV actors, using convolutional neural networks in [24]. The variety of high performing algorithms, coupled with the fact a webcam is sufficient for capturing the candidate's face, makes face-authentication a viable option. There exist libraries such as OpenFace and Facenet which analyse the features of a face and find similarities between two faces.

Eye Tracking. High quality eye tracking involves methods like infrared imagery and video oculography [13]. These methods are able to detect with high accuracy, but require expensive hardware and headgear. The authors reviewed webcam based eye trackers as entailed in the work of Höffner [8], and A. Kar et al. [10]. C. Meng et al. [15] outline a CNN-based approach for iris-tracking using webcams. Alexandra Papoutsaki et al. introduced webgazer, which employs a calibrated approach to fit the users pupil position to points on the screen and trains a regression model [16]. In his research, Imamuchi et al. [9] combines head pose detection along with calibration of eye gaze to successfully build a eye tracking model that was used as a driver alert system. A survey of all of these papers reveals that webcams have similar accuracy to that of infrared cameras. Under decent lighting conditions, they can be used for eye tracking.

Gaze Tracking Using Head Posture. Calculating the orientation of the head can give a good idea of where a person is looking, and thus detect suspicious behaviour during a test. The work proposed in [2] implements head gaze calculation using the field of view of the wearable headcam. In the work by Höffner [8], various methods for 3-D modeling of head gaze using the webcam are discussed. It uses webcam imagery, applies an affine transform followed by Perspective-n-Point (PnP) processing to calculate head gaze relative to the device screen. The authors of [17] have used a simple but effective trigonometric approach to calculate the head orientation via the yaw angle of the face-points.

Object Detection. For the detection of objects such as phones, ipods, tablets, books or other such materials, we studied various detection frameworks. YOLOv3 is a good choice as it has been trained on millions of images, has various inbuilt classes, and works fast, something needed in an online test environment where the video is being processed frame by frame [12].

3   Multi-modal Automated Proctor

3.1   Overview

The multi-modal automated proctor (MMAP) proposed in the study assumes that the test is being conducted in locked mode (i.e. tab switching is not allowed). This can either be implemented by using a browser, e.g. Respondus, or by keeping a track of active window and shutting down the test in case of a switch. Now, since the candidate cannot cheat using the same device that s/he is giving the test on, the proposed model will aim to detect cheating activity outside of the system.

Based on common malpractices identified, the MMAP system consists of five modules:

1. Face Authentication
2. Head-pose Estimation
3. Eye Tracking
4. Speech Detection
5. Detection of proscribed objects

Each of these five modules are implemented using Computer Vision techniques. The system requires a laptop with a working camera, from which the live feed is processed frame by frame. See Fig. 1 for the overview of our approach.

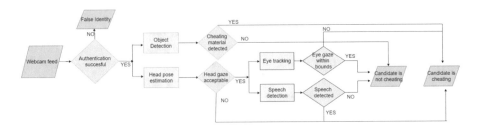

Fig. 1. An overview of our solution

3.2 Relevant Theory

Facial Landmark Extraction with DLIB: The study used Dlib's (an open source library) face detector, from the OpenCV library. It uses a histogram of gradients (HOG) detector coupled with support vector machine to extract 68 facial landmarks. Histogram of gradients computes the distribution of intensity gradients within an image. The image is broken down into pixel grids. The direction and magnitude of variation in the intensities of pixels is used to calculate gradients of the pixels. The magnitude and direction of gradient for each pixel is calculated using the formulae:

$$g = \sqrt{g_x^2 + g_y^2} \tag{1}$$

$$\theta = \arctan(\frac{g_y}{g_x}) \tag{2}$$

The HOG Feature descriptors obtained are then used to train an SVM-Model, which can thus detect faces and return co-ordinates of the facial landmarks.

OpenFace for Facial Feature Vector: OpenFace is an opensource library that analyses facial features and structure. [3] It can be used to find a similarity between two pictures, for authentication and verification purposes. The landmarks obtained from dlib are passed through OpenFace which returns a 128 dimension vector describing the various features of the face. If two faces look alike, their vectors are close together. Therefore, a comparison of 2 faces can be found using the difference between their two vectors.

Perspective N Point Problem: The orientation and tilt of a head can be expressed using the three Tait-Bryan angles: the yaw, pitch and roll, as show in Fig. 3.

To compute these angles, the research work used PnP (Perspective N Point) problem. The PnP problem is used when we know the position of an object in a three- dimensional coordinate system, and also know its co-ordinates on two-dimensional image [18]. The image of the object can be used to obtain its rotational and translational shift using the following equation.

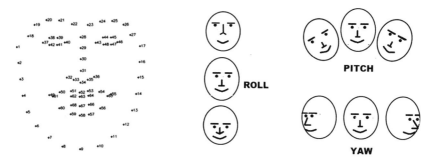

Fig. 2. Facial detection landmarks   **Fig. 3.** The three angles of rotation for head movement

$$m = A[R|t]M \tag{3}$$

Which is expanded as:

$$\begin{bmatrix} u \\ v \\ 1 \end{bmatrix} = \begin{bmatrix} f_x & 0 & c_x \\ 0 & f_y & c_y \\ 0 & 0 & 1 \end{bmatrix} \begin{bmatrix} r_{11} & r_{12} & r_{13} & t_1 \\ r_{21} & r_{22} & r_{23} & t_2 \\ r_{31} & r_{32} & r_{33} & t_3 \end{bmatrix} \begin{bmatrix} X \\ Y \\ Z \\ 1 \end{bmatrix} \tag{4}$$

Where,

- (u, v) are the coordinates of the projection point in pixels, on the 2D image, representing the matrix in Eq. 3.
- A is the calibration matrix of intrinsic parameters of the camera.
- (cx, cy) a principal point, taken as centre of image
- fx, fy are the focal lengths expressed in pixel units
- R|t in Eq. 3 is the rotation and translation matrix, respectively
- (X, Y, Z) are co-ordinates in the 3-D co-ordinate system, denoted by M in Eq. 3

The approximate focal lengths in pixels can be calculated as:

$$f = \frac{w}{2} \cot(\frac{\alpha}{2}) \tag{5}$$

Where f = focal length, α = angle of view of camera and w is pixel width/height of image.

The focal lengths and centre of image values are used in the calibration matrix. The following 6 points are taken as the 3-D co-ordinate points: the chin, tip of nose, corner of both eyes, and corners of the mouth. These are given predefined 3-D co-ordinates assigned by OpenCV's module, with the tip of nose as origin (0,0,0).

Thus, using the 2-D points of the facial landmarks, the 3-D model and the camera calibration parameters like focal length, the PnP problem can be solved to calculate the rotational and transformation matrix, which can be decomposed to give the yaw, pitch and roll value, and thus tell where the person is looking using the head tilt.

3.3 Experimental Setup and Implementation

This subsection discusses the overall flow of the application and the experimental set up. The candidate is expected to give the test in a reasonably well lit area. The input feed from the image is first sent through the dlib facial extraction module, which gets facial landmarks, refer Fig. 2. For authentication, detection of head-pose, eye tracking and speech detection these landmarks will be needed.

Authentication: The landmarks extracted are processed using OpenFace to generate the 128 dimensional vector. The distance between the vector and that of the mugshot of the candidate present in the system is calculated. A threshold distance of 0.38 was found to work the best for the test data, which gives minimum false positives.

Head Pose Estimation: Once the identity is confirmed, the head tilt is identified. As explained in the previous subsection, using the PnP problem, the yaw and pitch of the candidate's head tilt are computed. From Eq. 4, it is evident that the camera intrinsic parameters such as field of view are needed. However, most webcams have a field of view, α, between 50–65°. Thus α is assumed as 55 and substituted in Eq. 5. Using experimental results, the threshold value for yaw was set to $\pm$ 15°, to detect if the candidate is looking sideways. To check if the candidate is looking down for extended periods, perhaps at a hidden phone or book, the pitch threshold value was set as $\pm 10°$, below which it is flagged as suspicious.

Eye Tracking: If the person's face is centred, the eyes are tracked to see if the candidate is staring out of the screen. Using the facial landmarks that have been extracted as shown in Fig. 2, points 37 to 42 give the left eye while the points 43 to 48 mark the right. The eye is gray-scaled as shown in Fig. 4a.

Unlike infrared imagery, visible light images cannot clearly differentiate between the iris and the pupil. Therefore, assuming the centres of the iris and pupil are detected, the entire iris is detected. The iris can be deformed in the image due to light reflections [6]. Therefore these reflections have to be removed, or else it can lead to false detection of boundary [8,16]. The image is converted into black and white pixels only, by applying binary thresholding. Threshold values between 25 and 70 gave good results. The noise in the image is reduced by applying three operations. The first two are morphological operations - dilation and erosion. This is followed by median blur which is an averaging operation where a central pixel is assigned the median value of its surrounding pixels.

(a) Grayscaled   (b) Threshold   (c) Erosion   (d) Dilation   (e) Median-Blur

Fig. 4. Various stages of processing for pupil extraction

Using OpenCV's contour function, the largest circle detected is identified as the iris. The centre of this circle denotes the position of the pupil. Then the centre of the blob (pupil) thus obtained is compared with the centre of eye, previously calculated using dlib's shape predictor. The distance between the two is compared with the width of the eye.

$$W = (x2 - x1)/2 \tag{6}$$

$$dist > W/2 \tag{7}$$

If Eq. 7 holds true,it is assumed that the person is not staring into the screen. Since a candidate cannot be expected to stare straight ahead without any eye movement at all, a threshold number of 10 frames is set, to allow for brief eye movements. If the number of frames for which the candidate is found staring sideways exceeds the threshold, the activity is marked as suspicious.

Speech Detection: Alongside the eyes, the mouth is monitored for lip movements. A quantity called the mouth aspect ratio (M.A.R.) is computed. It is a measure to find how wide open the mouth is extended from driver-drowsiness systems where it was used for detecting yawns [19].

The mouth aspect ratio is the ratio of the vertical distance between the points p1 and p2 and the horizontal distance between points p3 and p4.

In dlib, these points are represented by landmark numbers 62, 67, 60 and 64, as shown in Fig. 2. When the mouth is shut, the points p1 and p2 are very close, hence the ratio is near zero. When a person talks, there is a sharp increase, followed by multiple dips and increases. This is because while talking, the mouth repeatedly opens and shuts (Fig. 5).

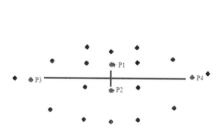

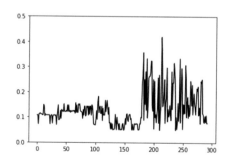

Fig. 5. Extracting keypoints for lip detection

Fig. 6. M.A.R plot showing mouth movement

We can easily distinguish between the region where the person is quiet versus the second when there is significant mouth movement, refer Fig. 6. A threshold of 0.1 is set as the limit for determining a shut mouth. However, a single frame, where it crosses the threshold, or a brief movement is not recorded as suspicious. If a sustained sequence of dips and rises is observed, the activity is flagged as suspicious. This to ensure that a normal activity like yawning or momentarily moving your lips doesn't get misidentified as speech.

YOLO for Object Detection: Alongside tracking facial features, for object detection, YOLO, which is one of the fastest deep learning object detectors, is used. It is computationally faster than two stage detectors and can spot objects like books, phones and other cheating material, as well as other people in the background. The pre-trained yolo was trained to detect objects like headphones, electronic devices.

Each frame is resized to 256*256 and is passed through the convolutional neural net, to get bounding boxes and labels. YOLO detects location and the label of the object. The output is the coordinates of the bounding box and the predicted class label. If objects such as books or phone is detected, a suspicious flag is issued.

The system generates a report at the end with a flag denoting each suspicious activity and the time at which it has occured.

4   Results and Discussion

There were no data-sets of students or subjects in video lectures available in open source. Moreover, it is difficult to give a binary classification as to whether a person is cheating in the duration of test, as multiple instances of cheating behaviour can be detected which might include false positives as well. As such, instead of giving a binary verdict, it is better to generate a report with every suspicious instance noted. The examination committee can then view the report for the student and take a call. To test the system as a whole, as well as it's constituent modules, a database of 50 videos was created with 15 subjects. The video clips were of 5 min each had subjects simulating the behavior of a candidate in front of a camera, and engaging in cheating as well as non-cheating behavior.

The total running time of the 50 videos was 4 h 10 min. The subjects engaged in a total of 258 instances of head movements, 290 eye movements, spoke a total of 220 times, and used banned objects such as phones 200 times.

Authentication Results. For face authentication,the 15 subjects who had participated in the study provided 12 images per person for testing. Then each of 180 images was compared with every other image. If the distance between the two images compared was less than the set threshold, the system would return an 'authenticated' verdict. The accuracy was calculated based on whether two images of the same person were correctly identified as being identical, and vice versa for images of different subjects. Given below is a plot of the accuracy and F1 score vs threshold (Fig. 7).

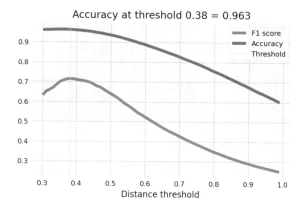

Fig. 7. Optimum threshold for face authentication

The best threshold is found to be 0.38, which gives an F1 score 0.72 and accuracy of 96.3%. Thus threshold 0.35-0.4 can be used to correctly determine authenticity of the test taker.

System Results. The database of 50 videos was tested using the system. Out of 258 instances of head movements, 244 were correctly identified by the proctor, with 94.57% accuracy. Additionally, on only 8 occasions, a false flag for head movement was issued.

Out of 290 instances of eye movements, 262 were correctly flagged by the proctor as staring out of the screen with 90.34% accuracy. However, there were a total of 28 false positives being issued across the 50 videos.

There were 220 instances of the subjects speaking, from which 205 were correctly identified, representing an accuracy of 93.18%. No false positives were identified.

Finally, for the object detection module, YOLO was able to correctly identify 171 out of 200 banned objects in use giving an accuracy of 85.5% (Fig. 8 and Table 1).

Table 1. Results acquired from the Test Dataset

Attribute	Total	Detected	Accuracy%
Head movements	258	244	94
Eye movement	290	262	90
Speech	220	205	93
Banned objects	200	171	85

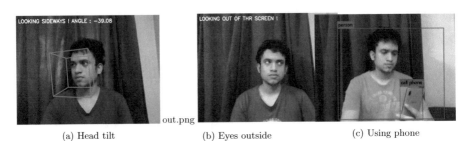

(a) Head tilt   (b) Eyes outside   (c) Using phone

Fig. 8. Results

5   Conclusion

The system proposed is a result of an extensive survey. Since there are numerous opportunities for malpractice in an environment, this study incorporates multiple detection techniques. The results obtained, though with limitations, are promising and show a clear advantage over manual proctoring, with an overall accuracy of 91%. There is no hardware needed besides a laptop with a functional camera, which makes it cost-efficient and easy to implement. The use of multiple features ensures that if even if one module fails, some other part of the system will ensure that the unethical activity is detected. The proposed system can be used by itself, or as an accessory to live proctoring.

References

1. Arinaldi, A., Fanany, M.I.: Cheating video description based on sequences of gestures. In: 2017 5th International Conference on Information and Communication Technology (ICoIC7), pp. 1–6. IEEE (2017)
2. Atoum, Y., Chen, L., Liu, A.X., Hsu, S.D., Liu, X.: Automated online exam proctoring. IEEE Trans. Multimed. **19**(7), 1609–1624 (2017)
3. Baltrusaitis, T., Zadeh, A., Lim, Y.C., Morency, L.: Openface 2.0: Facial behavior analysis toolkit. In: 2018 13th IEEE International Conference on Automatic Face Gesture Recognition (FG 2018), pp. 59–66 (2018)
4. Bawarith, R., Abdullah, D., Anas, D., Dr, P.: E-exam cheating detection system. Int. J. Adv. Comput. Sci. Appl. **8** (2017)
5. Chua, S.S., Bondad, J.B., Lumapas, Z.R., Garcia, J.D.: Online examination system with cheating prevention using question bank randomization and tab locking. In: 2019 4th International Conference on Information Technology (InCIT), pp. 126–131. IEEE (2019)
6. Daugman, J.: New methods in iris recognition. IEEE Trans. Syst. Man Cybern. Part B (Cybern.) **37**(5), 1167–1175 (2007)
7. Flior, E., Kowalski, K.: Continuous biometric user authentication in online examinations. In: 2010 Seventh International Conference on Information Technology: New Generations, pp. 488–492. IEEE (2010)
8. Höffner, S.: Gaze tracking using common webcams. Institute of Cognitive Science Biologically oriented Computer Vision (2018)

9. Imabuchi, T., Prima, O.D.A., Ito, H.: Visible spectrum eye tracking for safety driving assistance. In: Fujita, H., Ali, M., Selamat, A., Sasaki, J., Kurematsu, M. (eds.) IEA/AIE 2016. LNCS (LNAI), vol. 9799, pp. 428–434. Springer, Cham (2016). https://doi.org/10.1007/978-3-319-42007-3_37

10. Kar, A., Corcoran, P.: A review and analysis of eye-gaze estimation systems, algorithms and performance evaluation methods in consumer platforms. IEEE Access **5**, 16495–16519 (2017)

11. Kasliwal, G.: Cheating detection in online examinations. San Jose State University Master's thesis, p. 399, May 2015

12. King, D.L., Case, C.J.: E-cheating: incidence and trends among college students. Issues Inf. Syst. **15**(1), 20–27 (2014)

13. Li, X., Chang, K.M., Yuan, Y., Hauptmann, A.: Massive open online proctor: Protecting the credibility of moocs certificates. In: Proceedings of the 18th ACM Conference on Computer Supported Cooperative Work & Social Computing, pp. 1129–1137 (2015)

14. McGinity, M.: Let your fingers do the talking. Commun. ACM **48**(1), 21–23 (2005)

15. Meng, C., Zhao, X.: Webcam-based eye movement analysis using CNN. IEEE Access **5**, 19581–19587 (2017)

16. Papoutsaki, A., Sangkloy, P., Laskey, J., Daskalova, N., Huang, J., Hays, J.: Webgazer: Scalable webcam eye tracking using user interactions. In: Proceedings of the Twenty-Fifth International Joint Conference on Artificial Intelligence-IJCAI 2016 (2016)

17. Prathish, S., Bijlani, K., et al.: An intelligent system for online exam monitoring. In: 2016 International Conference on Information Science (ICIS), pp. 138–143. IEEE (2016)

18. Rocca, F., Mancas, M., Gosselin, B.: Head pose estimation by perspective-n-point solution based on 2D markerless face tracking. In: Reidsma, D., Choi, I., Bargar, R. (eds.) INTETAIN 2014. LNICST, vol. 136, pp. 67–76. Springer, Cham (2014). https://doi.org/10.1007/978-3-319-08189-2_8

19. Savaş, B.K., Becerikli, Y.: Real time driver fatigue detection based on SVM algorithm. In: 2018 6th International Conference on Control Engineering & Information Technology (CEIT), pp. 1–4. IEEE (2018)

20. Sawhney, S., Kacker, K., Jain, S., Singh, S.N., Garg, R.: Real-time smart attendance system using face recognition techniques. In: 2019 9th International Conference on Cloud Computing, Data Science & Engineering (Confluence), pp. 522–525. IEEE (2019)

21. Siyao, L., Qianrang, G.: The research on anti-cheating strategy of online examination system. In: 2011 2nd International Conference on Artificial Intelligence, Management Science and Electronic Commerce (AIMSEC), pp. 1738–1741. IEEE (2011)

22. Sukmandhani, A.A., Sutedja, I.: Face recognition method for online exams. In: 2019 International Conference on Information Management and Technology (ICIMTech), vol. 1, pp. 175–179. IEEE (2019)

23. Varble, D.: Reducing cheating opportunities in online test. Atlantic Mark. J. **3**(3), 9 (2014)

24. Xia, H., Li, C.: Face recognition and application of film and television actors based on DLIB. In: 2019 12th International Congress on Image and Signal Processing, BioMedical Engineering and Informatics (CISP-BMEI), pp. 1–6 (2019)

A Survey on Representation Learning in Visual Question Answering

Manish Sahani[(✉)], Priyadarshan Singh, Sachin Jangpangi, and Shailender Kumar

Delhi Technological University, Delhi, India
{manishsahani_2k17co179,priyadarshansingh_2k17co242,
sachinjangpangi_2k17co286}@dtu.ac.in, shailenderkumar@dce.ac.in

Abstract. Visual question answering stands among the most researched computer vision problems, pattern recognition, and natural language processing. VQA extends the computer vision world's challenges and directs us toward developing some basic reasonings on visual scenes to answer questions on the specific elements, actions, and relationships between different objects in the image. Developing reasonings on the image has always been popular among computer vision and natural language processing researchers. It is directly dependent on the expressivity of the representations learned from the datasets. In the past decade, with advancements in computing machinery, neural networks, and the introduction of highly optimized and efficient software, a substantial amount of research has been done to solve VQA efficiently. In this survey, we present an in-depth examination of representation learning of state-of-the-art methods proposed in the literature of VQA and compare them to discuss the future directions in the field.

Keywords: Computer vision · Visual Question Answering · Natural language processing · Representation learning

1 Introduction

One of the ultimate goals of computer vision is to understand the various dynamics of the scene and understand it as a whole [1], which requires a system to capture all kinds of information and relationships among them on numerous semantics levels. Visual question answering plays an essential role in achieving that goal as it serves as a niche to those researchers who are invested in developing ways to generate these reasonings. The task of visual question-answering in the most common form is provided with an image and a textual question related to the image. The machine is supposed to determine the best answer composed of a single or few words. The tasks combine both artificial intelligence fields: computer vision and natural language processing and have always attracted researchers from both communities. Over time the computer vision community has seen similar tasks that require machines to form some reasoning; these are mainly image captioning [2–7], visual grounding [8], visual madlibs [9], etc. VQA challenge [10] differs from these because it is a more open-ended and complex problem. It requires more than generating reasoning from the question on image and often requires a common

© The Author(s), under exclusive license to Springer Nature Switzerland AG 2022
R. Misra et al. (Eds.): ICMLBDA 2021, LNNS 256, pp. 326–336, 2022.
https://doi.org/10.1007/978-3-030-82469-3_29

sense of the specific domain, which introduces a new set of challenges and makes VQA a truly AI-complete problem [10]. This way, we can consider visual question answering as a proxy for evaluating and answering an important question - "how far we are in advanced reasoning and AI capable systems for image understanding." This has been a great source of motivation for researchers in the area to approach these problems with various perspectives with the same ultimate goal. In the past decade, the advancements in both computer vision and natural processing language have produced a lot of literature. The objective of this survey is to provide a compressive survey to upcoming researchers of the fields, covering common approaches, their representation learning techniques, available datasets, models and suggest a promising future direction.

In the first section of this survey, we examine the available datasets for visual question answering training and discuss their level of complexity, short-comings, and improvements. In the second part of this survey, we present an in-depth examination of the various methods proposed in the field. We classify these methods based on the nature of processing and the modalities to learn representation from them. We start our examination with the most common approaches in VQA for learning representation, which are based on creating a joint embedding with their shortcomings. We continue our discussion in the direction of approaches with stronger and expressive representation learning, like methods that used attention, transformers, encoders, etc. finally, we compare the performance of discussed models and present our conclusion and talk about the future directions of the fields.

2 Datasets

In recent decades, a number of datasets have emerged for training and evaluating the visual question answering models. The most common structure or the minimum requirement for a VQA dataset is the availability of triples composed of Image, question, and answer. Each of these datasets was proposed to solve a specific problem. Widely vary from each other in terms of the level of reasoning required, the complexity of datasets, size, and types of images and questions. These additional data may be present in annotations to the pairs, such as support for question-specific image regions, image captions, and multiple-choice candidates. And, newer datasets like VQA already provide this extra information within the dataset in multiple-choice annotations. There are also other datasets used by VQA researchers that can act as the potential source of this extra information required, ex - image captioning generating [11–14] and understanding [15, 16]. The first effort in curating a VQA dataset was made by Geman *et al.* [17] and Tu *et al.* [18]; these were of smaller size and had limited settings. The following datasets discussed are relatively larger, composed of real-world, and address the limited settings restriction.

2.1 DAQUAR

This was the first large real-world images dataset proposed by Malinowski *et al.* [19] to solve the VQA tasks. The authors of DAQUAR build the dataset over the NYU-Depth V2 image dataset [20]. It has 1,449 images and 12,468 total question-answers (QA) pairs, out of which 2,483 are unique questions. The test or validation split contains 5,674 question-answer pairs, while training is done on 6,794 QA pairs. The QA pairs can be categorised in synthetic and human. The first type of QA pairs has humans as the main object, and the synthetic QA pair are based on templates.

2.2 COCO-QA

The COCO-QA [21] is also a real-world images dataset, composed of images taken from a popular image real-world image dataset: Common Objects in Context data curated by Microsoft [13] and has 123,287 real-world images. The authors developed this dataset in order to increase the amount of training data for the models. Each image in the dataset has a question-answer pair available that has an answer to one of the following types: object, color, number, and location.

2.3 VQA-Real

VQA dataset is among the most important and widely used dataset for visual question answering. This dataset has played a vital role in the history of VQA problem-solving. The dataset consists of two parts based on the type of images - real and abstract. In the VQA-real, the 82,783 train images, 40,504 validation images, and 81,434 test images are also taken from Microsoft's COCO dataset. The early version of VQA (now known as VQA 1.0 [22]) consists of 2,483,490 question answers for training, 1,215,120 question answers pairs for validation, and 244,304 questions for the testing. For every image, the dataset has three questions based on human-subjects. In addition to this, ten different subjects provide answers to those questions. Furthermore, each question has 18 candidate responses, which were created for the tasks of multiple-choice VQA. The major problem with the VQA 1.0 dataset was in its inherent bias. In addition to this, the baseline methods, which only used short-term memory units like LSTM for the representation of the question, and which did not have the image as input, achieved an overall accuracy of 48.76%. This clearly states a bias in the dataset, and this huge impact of the language priors on the answers motivated the design of a more stable version of the VQA dataset.

The VQA 2.0 [23] dataset was created to address the biasing in the dataset. While the authors were developing the workaround for the bias, they noticed that almost all of the models ignored the visual information. The images have structural information that is extremely hard to relate with the natural language question and learn the resultant features via traditional ways. The images in VQA 2.0 are the same as VQA 1.0; however, there is a significant increase in the number of questions, VQA 2.0 has almost doubled the number of VQA 1.0 questions i.e., 443,757 training questions in VQA 2.0. This forces the learning network to use both visual and textual information for answering the question. This is how the bias on VQA 1.0 was addressed and led to the rise of a more stable with balanced question-answer pairs dataset VQA 2.0.

3   Approaches

In this section of the paper, we examine the different approaches introduced in visual question answering and discuss their method of representation learning and the level of reasoning they generate on images. We start our examination with a discussion on different modalities involved in the VQA and proceed in the direction of approaches with stronger representation learning and better reasoning. Visual question-answering in the most common form can be formalized as: *Given an image I, and a textual question q related to the image represented in natural language. The intelligent system is supposed to predict the best natural answer from the set of all possible candidate answers.*

$$\hat{a} = argmax_{a \in \Omega} p(a|I, q, \theta) \tag{1}$$

Where in Eq. (1)Ω is the set of all possible answers and is the model parameter vector.

There are at least two modalities involved in the VQA task: visual and textual modality. Both these modalities are the foundation for the model's reasoning generation capabilities and to the success of the VQA task. In almost all of the VQA approaches the image feature vector I is generated by processing the image using a CNN modal like Resnet [24], VGGnet [25], etc. which are pre-trained on a big visual dataset [26] and q is generated using one-hot encoding, or by using sophisticated NLP techniques like Glove [27]. we categories the models on the basis of their method of processing these modalities to form some reasoning on the image.

Simpler Joint Embeddings: Concatenation or Inner Product. Early work in VQA for example, the combination of Bayesian methods and semantic text parsing by Malinowski *et al.* was restricted to limited settings and does not do well with the open-ended nature of the problem. The motivation for generating basic reasoning on the image was the advancement of deep learning techniques like CNN, RNN, LSTM, etc., and the researcher's aim was to learn features from a common feature space of the image and textual questions. The image modality was created by processing an image using CNN and textual embeddings were created by processing word mapping using RNN, where words were mapped to space and distance reflects the similarities in semantics. The joint embedding is the result of operations on these two modalities, like - concatenation, elementwise product sum.

Methods: Neural-Image QA [28] was one of the early models in VQA proposed by Malinowski *et al.* to use joint embeddings, It has an RNN created with LSTM [29] cells. Image features from CNN along with questions are passed into a first "encoder" LSTM followed by a "decoder" LSTM to produce answers of different lengths recurrently, single word per iteration until the "END", a special symbol isn't discovered. Later, a classifier was introduced by Ren *et al.* [21] instead of using decoders, making a "VIS+LSTM" model and features from the LSTM encoder were directly fed into the classifier. The above method uses both inputs to create a single modality, another slightly different method was introduced by Gao *et al.*, "Multimodal QA" [30] where it also has an LSTM and CNN but both these are used to learn different parameters. CNN was used to create image modality and then fed into the LSTM encoder at every step. This way of treating

VQA as a classification problem gained popularity and similar classifiers can be seen at the end of most of these approaches [31, 32] which may have methods of processing modalities.

Expressive Joint Embeddings: Approximate Outer Product. This method creating simpler joint embeddings with operations like concatenation or inner product of embeddings do not preserve the spatial information about the image and weakly relate the question with the image. To cope with the loss of structural and spatial information, the researcher came up with different ways to learn more expressive representations. The motivation for this approach was to learn more parameters by approximating the outer product of the modalities.

Methods: MCB [8] is one of the fine demonstrations of efficiently learning more parameters, which showed a fast and efficient way of computing a compact bilinear pooling. The motivation for this research was a known fact that bilinear pooling or outer product works great for fine-grain vision tasks. But the challenge was the parameter explosion, as the final parameter of the outer product of the image and question would result in a higher dimension around 12.5 billion ($2048 \times 2048 \times 3000$), which is not feasible to compute. Therefore, the authors ofCB proposed a method called multimodal compact bilinear pooling that uses count sketch and computes outer product indirectly by convoluting the count sketches of both modalities, they learn only up to 16 K parameters. Similar to this method, the authors of DPPnet [31] prospered a dynamic parameter prediction layer in their VGG+GRU+CNN Classifier network, to allow having multiplicative interactions between visual and textual modalities. They employed a known hashing trick for learning the parameters of the CNN classifier from LSTM encodings. For the input of network's dynamic parameter layer, its output vector is denoted by f^o which is given by:

$$f^o = W_d(q)f^i + b \tag{2}$$

In Eq. (2), $W_d(q) \in R^{M \times N}$ denotes matrix constructed using the dynamic parameter prediction network and b denotes bias

$$w^d_{mn} = p_{\psi(m,n)} \cdot \xi(m, n) \tag{3}$$

where in Eq. (3) w^d_{mn} be the element at (m, n) in $W_d(q)$, which represents weight between mth and nth neuron and $\xi(m, n) : N \times N \rightarrow \{+1, -1\}$.

Attention-Based Learnings. The above methods use image-wide global features for representation of the visual scene, which introduces a significant amount of noise in the visual input which may be irrelevant to the final answer and the models are unable to learn the silent region of the images that have some weightage in the answer. This acted as the motivation for researchers to use soft attention in V + L tasks like image captioning, visual grounding, and VQA. The attention component emphasizes the region of interest and where the models should look for the right answer.

Methods: Vahid Kazemi *et al.* [38] employed the stack attention network, "SAN" [33] in a simple CNN+LSTM model to mimic the working of the human brain in their paper titled "Show, Ask, Attend and Answer". They outperformed the state of the art methods of that time by carefully choosing the choices like activation function, initializers, optimizers, and regularization functions. A similar approach and more powerful approach, named Region Selection [34] proposed a method for generating a relevant region, a common latent space created by the projection of potential sub-regions of the image along with the question and followed by an inner product. The method first projects the image features into a shared N-dimensional space and an inner product is computed between textual and these project image features and this inner product is used to compute relevance weighting of the region. The attention mechanism can help models to learn salient features of the images that are contributing to the answers and they have been really popular among computer vision and VQA researchers and in the recent year.

Multimodal Fusion-Based Learnings. The Attention-based approaches only focus on the visual region relevant to the questions and do not generalize well with the questions requiring more complex reasoning. Therefore, recent research work has been more focused on the feature-rich representation of images with textual questions.

Methods: A recent approach MUREL [35] introduced a new method of rich feature representation of image regions merged with textual representation and introduced a multimodal relational network to learn over images using a single unit called MURAL Cell. To scale the expressivity of the region's relation between using pairwise combinations of between questions and image regions. These units further combine to refine the region and question interaction and create a MURAL network. The authors employ Tucker Decomposition [36] of third-order tensors to model these interactions with a lower number of parameters efficiently. In the next step, each representation receives information based on the relation with neighboring representation to achieve semantic and spatial context awareness and this context representation contains an aggregation of the messages provided by the neighbors. Finally, the output of the MUREL cell is computed using residual functions and is used to answer the question. The advantage of this method is that it overcomes the linguistic bias in VQA datasets by relying on visual information. Later, a more recent approach by Zhou Yu *et al.* [39] showed the effect of attention mechanism on MUREL networks. They have deep embedded co-attention layers including components for self attention "SA" of question and image and guided attention "GA" to image by question. These two stacked co-attention layers act as an encoder-decoder and then Multimodal fusion is performed and passed to an output classifier.

Transformers and Encoders Based Learning. Other methods like LXMERT [37] emphasises on learning visual-and-language connections using transformers and encoders. The authors of LXMERT aim to learn the relationship between two modalities using a transformer model containing three encoders: Object relationship model, Language encoder, and Cross-modality encoder. To enable the model to connect visual and language semantics, they pre-trained their model with a dataset of image and sentence pairs via five pre-training tasks: Masked cross-modality language modeling, Masked object prediction via ROI feature regression, Masked object prediction via detected-label classification, Cross Modality matching, Image question answering. The input layers convert the inputs into two feature sequences: Word-level sentence embeddings: Using the WordPiece Tokenizer, a sentence is split into words, each word and its index is then projected to vectors followed by addition to index aware word embeddings Object-Level Image Embeddings: It takes detected objects as image embeddings. Each object is defined by its position vector and ROI feature vector, by adding these two it gets the position aware embedding. Two single modality encoders, language and object-relationship encoder are then applied on the embedding layers. Each encoder layer contains self-attention and feed-forward layer. The cross-modality encoder contains two self-attention and one bi-directional cross-attentional layer. The query and context vectors are the outputs from the second last layer of this encoder. The cross-attention sublayer exchanges the information between two modalities.

4   Comparison

In this section, we present a comparison of above approaches on the basis of methods of features processing, extracting, influence of attention, and performance on various available datasets. Table 1 shows a comparative analysis of feature extraction and processing. The simpler methods like [21] using concatenation and inner product of modalities are easy to implement but result in longer training time due to exponential growth of features as a result of inner product and do not preserve information about relation between image and questions. Approximating the outer product shows good results by reducing the parameters as done by MCB [8]. But above methods use image as a global feature, resulting in addition of noise. The attention-based methods SAN [33], SAAA [38] and region-selection [34] reduce the noise in regions of interest via self attention. Attention based methods only focus on image features and fail to generalize on complex questions. Multimodal fusion based learning methods like MUREL [35] emphasize on using both modalities together and provide better results on datasets that involve reasoning from both image and textual questions as compared to attention based models. Current transformer based models like LXMERT [37] perform better than attention based models on both splits of VQA as they learn the relationships between cross and self modalities using transformers. Table 2 presents characteristics and performances of different models respectively. the performance of these models on the various available datasets.

Table 1. Comparison of discussed approaches on the basis of image features, textual features attention mechanism and representation learning

Model	Image features	Textual features	Attention	Representation learning
Neural Image-QA [28]	CNN	LSTM	No	Simpler Joint Embeddings
VIS+LSTM [21]	CovNet	LSTM	No	Simpler Joint Embeddings
Multimodal QA [30]	CNN	LSTM	No	Simpler Joint Embeddings
MCB [8]	Resnet	LSTM	Yes	Compact bilinear pooling
DPPnet [31]	VGG	GRU	No	Dynamic hashing
SAN [33]	VGG	LSTM	Yes	Attention
SAAA [38]	Resnet	LSTM	Yes	Concatenation + attention
Region-Selection [34]	VGG	LSTM	Yes	Inner product + attention
MCAN [39]	F-RCNN	LSTM	Yes	Multimodal Fusion + attention
MUREL [35]	F-RCNN	GRU	No	Multimodal Fusion
LXMERT [37]	F-RCNN	GRU	No	Transformer and Encoders

Table 2. Comparison of discussed approaches on the basis of performances on available datasets

Model	DAQUAR ACC (%)	VQA test-dev open-ended	VQA test-standard open-ended
Neural Image-QA [28]	19.43	–	–
VIS + LSTM [21]	34.41 (reduced)	–	–
MCB [8]	–	66.7	66.5
DPPnet [31]	28.98	57.2	57.3
SAN [33]	29.30	58.7	58.9
SAAA [38]	–	–	59.7
Region-selection [34]	–	62.4	62.4
MCAN [39]	–	70.6	70.9
MUREL [35]	–	68.0	68.4
LXMERT [37]	–	69.9	72.5

5   Conclusion

In this survey, we presented a review of the VQA literature from the narrative of features or representations learning and a comparison of the representations proposed state-of-the-art methodologies. We discussed the basic approaches and the improvements based on these methods like attention layers, joint-embedding, transformers and pre-training methods. We feel that the field's main goals are now more focused on capturing the fine-grain relationships between images and questions. Encoders and transformers techniques provide a good starting point for the researchers that aim to generate reasonings on the images. This survey can help researchers by providing them a different way of looking at the problem — which has emphasis on the relationships between image and questions. The methods we discussed above are a good baseline in their respective methodologies to solve VQA.

References

1. Yao, S.F., Urtasun, R.: Describing the scene as a whole: joint object detection, scene classification and se-mantic segmentation. In: CVPR, p. 1 (2012)
2. Donahue, J., et al.: Long-term recurrent convolutional networks for visual recognition and description. In: Proceedings of the IEEE Conference on Computer Vision and Pattern Recognition (2015)
3. Karpathy, A., Joulin, A., Li, F.F.: Deep fragment embeddings for bidirectional image sentence mapping. In: Proceedings of the Advances in Neural Information Processing Systems (2014)
4. Mao, J., Xu, W., Yang, Y., Wang, J., Yuille, A.: Deep captioning with multimodal Recurrent Neural Networks (m-RNN). In: Proceedings of the International Conference on Learning Representations (2015)
5. Vinyals, O., Toshev, A., Bengio, S., Erhan, D.: Show and tell: a neural image caption generator. In: Proceedings of the IEEE Conference on Computer Vision and Pattern Recognition (2014)
6. Yao, L., et al.: Describing videos by exploiting temporal structure. In: Proceedings of the IEEE International Conference on Computer Vision (2015)
7. Wu, Q., Shen, C., Hengel, A.V.D., Liu, L., Dick, A.: What value do explicit high level concepts have in vision to language problems? In: Proceedings of the IEEE Conference on Computer Vision and Pattern Recognition (2016)
8. Fukui, A., Park, D.H., Yang, D., Rohrbach, A., Darrell, T., Rohrbach, M.: Multimodal compact bilinear pooling for visual question answering and visual grounding. In: Proceedings of Empirical Methods in Natural Language Processing, EMNLP, pp. 457–468 (2016)
9. Yu, L., Park, E., Berg, A.C., Berg, T.L.: Visual madlibs: fill in the blank image generation and question answering. In: Proceedings of the IEEE International Conference on Computer Vision (2015)
10. Antol, S., et al.: VQA: visual question answering. In: Proceedings of the IEEE International Conference on Computer Vision (2015)
11. Chen, X., et al.: Microsoft COCO captions: data collection and evaluation server (2015). arXiv:1504.00325
12. Hodosh, M., Young, P., Hockenmaier, J.: Framing image description as a ranking task: data, models and evaluation metrics. In: JAIR, pp. 853–899 (2013)
13. Lin, T.-Y., et al.: Microsoft COCO: common objects in context. In: Fleet, D., Pajdla, T., Schiele, B., Tuytelaars, T. (eds.) Computer Vision – ECCV 2014: 13th European Conference, Zurich, Switzerland, September 6-12, 2014, Proceedings, Part V, pp. 740–755. Springer International Publishing, Cham (2014). https://doi.org/10.1007/978-3-319-10602-1_48

14. Young, P., Lai, A., Hodosh, M., Hockenmaier, J.: From image descriptions to visual denotations: new similarity metrics for semantic inference over event descriptions. In: Proceedings of the Conference on Association for Computational Linguistics, p. 2 (2014)
15. Kazemzadeh, S., Ordonez, V., Matten, M., Berg, T.L.: Referitgame: referring to objects in photographs of natural scenes. In: Conference on Empirical Methods in Natural Language Processing, pp. 787–798 (2014)
16. Hu, R., Xu, H., Rohrbach, M., Feng, J., Saenko, K., Darrell, T.: Natural language object retrieval. In: Proceedings of the IEEE Conference on Computer Vision and Pattern Recognition (2016)
17. Geman, D., Geman, S., Hallonquist, N., Younes, L.: Visual turing test for computer vision systems. Proc. Natl Acad. Sci. **112**(12), 3618–3623 (2015)
18. Tu, K., Meng, M., Lee, M.W., Choe, T.E., Zhu, S.-C.: Joint video and text parsing for understanding events and answering queries. IEEE Trans. Multimedia **21**(2), 42–70 (2014)
19. Malinowski, M., Fritz, M.: A multi-world approach to question answering about real-world scenes based on uncertain input. In: Proceedings of the Advances in Neural Information Processing Systems, pp. 1682–1690 (2014)
20. Silberman, N., Hoiem, D., Kohli, P., Fergus, R.: Indoor segmentation and support inference from RGBD images. In: Fitzgibbon, A., Lazebnik, S., Perona, P., Sato, Y., Schmid, C. (eds.) ECCV 2012. LNCS, vol. 7576, pp. 746–760. Springer, Heidelberg (2012). https://doi.org/10.1007/978-3-642-33715-4_54
21. Ren, M., Kiros, R., Zemel, R.: Image question answering: a visual semantic embedding model and a new dataset. In: Proceedings of the Advances in Neural Information Processing Systems (2015)
22. Antol, S., et al.: VQA: visual question answering. In: International Conference on Computer Vision (ICCV) (2015)
23. Goyal, Y., Khot, T., Agrawal, A., Summers-Stay, D., Batra, D., Parikh, D.: Making the V in VQA matter: elevating the role of image understanding in visual question answering. Int. J. Comput. Vis. **127**(4), 398–414 (2018)
24. He, K., Zhang, X., Ren, S., Sun, J.: Deep residual learning for image recognition. In: Proceedings of the IEEE Conference on Computer Vision and Pattern Recognition (2016)
25. Simonyan, K., Zisserman, A.: Very deep convolutional networks for large-scale image recognition. In: ICLR (2015)
26. Deng, J., Dong, W., Socher, R., Li, L.-J., Li, K., Fei-Fei, L.: Imagenet: a large-scale hierarchical image database. In: Proceedings of the IEEE Conference on Computer Vision and Pattern Recognition (2009)
27. Pennington, J., Socher, R., Manning, C.: Glove: global vectors forword representation. In: Conference on Empirical Methods in Natural Language Processing, Doha, Qatar (2014)
28. Malinowski, M., Rohrbach, M., Fritz, M.: Ask your neurons: a neural-based approach to answering questions about images. In Proceedings of the IEEE International Conference on Computer Vision (2015)
29. Hochreiter, S., Schmidhuber, J.: Long short-term memory. Neural Comput. **9**(8), 1735–1780 (1997)
30. Gao, H., Mao, J., Zhou, J., Huang, Z., Wang, L., Xu, W.: Are you talking to a machine? Dataset and methods for multilingual image question answering. In: NIPS 2015: Proceedings of the 28th International Conference on Neural Information Processing Systems, vol. 2, pp. 2296–2304 (2015)
31. Noh, H., Seo, P.H., Han, B.: Image question answering using convolutional neural network with dynamic parameter prediction. In: Proceedings of the IEEE Conference on Computer Vision and Pattern Recognition (2016)

32. Yang, Z., He, X., Gao, J., Deng, L., Smola, A.: Stacked attention networks for image question answering. In: Proceedings of the IEEE Conference on Computer Vision and Pattern Recognition (2016)
33. Yang, Z., He, X., Gao, J., Deng, L., Smola, A.: Stacked attention networks for image question answering. In: CVPR (2016)
34. Shih, K.J., Singh, S., Hoiem, D.: Where to look: focus regions for visual question answering. In: CVPR (2016)
35. Cadene, R., Ben-Younes, H., Cord, M., Thome, N.: Multimodal relational reasoning for visual question answering. In: CVPR (2019)
36. Ben-Younes, H., Cadene, R., Thome, N., Cord, M.: Mutan: multimodal tucker fusion for visual question answering. In: ICCV (2017)
37. Tan, H., Bansal, M.: Lxmert: learning cross-modality encoder representations from transformers. In: EMNLP (2019)
38. Kazemi, V., Elqursh, A.: Show, ask, attend, and answer: A strong baseline for visual question answering (2017). arXiv:1704.03162
39. Yu, Z., Yu, J., Cui, Y., Tao, D., Tian, Q.: Deep modular co-attention networks for visual question answering. In: Proceedings of the IEEE Conference on Computer Vision andPattern Recognition, pp. 6281–6290 (2019)

Evidence Management System Using Blockchain and Distributed File System (IPFS)

Shritesh Jamulkar[1], Preeti Chandrakar[1(✉)], Rifaqat Ali[2], Aman Agrawal[1], and Kartik Tiwari[1]

[1] Department of Computer Science and Engineering, NIT Raipur, Raipur, India
pchandrakar.cs@nitrr.ac.in
[2] Department of Mathematics and Scientific Computing, NIT Hamirpur, Hamirpur, India

Abstract. Evidence gathering is at the core of every analysis process. The ability to verify the results and have appropriate paperwork, especially if a case lasts for several years, is vital. In later periods, information gathered at the outset of a prosecution may become crucial. If the documentation is handled by a system, the judicial authority can find the important facts in the appropriate time quicker. This system provides knowledge about electronic evidence collecting, processing, transportation and handling. Maintaining documents based on paper may be a tedious job that is exposed to human interference by mistakes and modifications. Security issues emerge from stored electronic records in a centralized consolidated archive. Any evidence obtained five or six years ago is very difficult to preserve in a paper-based evidence storage system. We propose a secure evidence management system to store evidence in a secure, distributed peer-to-peer (p2p) file storage network(IPFS) using blockchain technology. The system is designed with a custom transaction family on Hyperledger Sawtooth to document every transaction from the moment the evidence is collected, ensuring that only approved individuals can access or possess evidence. Our proposed framework provides a safe compromise between different stakeholders such as law enforcement agencies, attorneys, and forensic professionals that protects the integrity and permissibility of evidence.

Keywords: Digital evidences · Digital forensics · Distributed file systems · Consortium blockchain · Hyperledger sawtooth

1 Introduction

Since the medieval times, the process of judiciary and law - enforcement has been done based on evidences collected on any criminal or civil case. The justice system works at its finest when all of the material evidence and case records are presented with utmost accuracy. The diligence required to maintain such a track record is exemplary. Especially in recent times, that is from the mid-19th

© The Author(s), under exclusive license to Springer Nature Switzerland AG 2022
R. Misra et al. (Eds.): ICMLBDA 2021, LNNS 256, pp. 337–359, 2022.
https://doi.org/10.1007/978-3-030-82469-3_30

century, the justice system has seen some massive reforms and have been shaped into a structure that stands solidly as the foundation of any and every country's government in the world. It takes life-altering decisions that can even have an impact on a global scale. For such a system to work smoothly and efficiently, the authorities have been using various methods to collect and store these decision-changing evidences which have proved to be of little success unless very properly handled, secured and maintained.

One of those methods is maintaining paper-based records in files which is still currently being used at a very large scale. This paper-based recording system is slowly but surely becoming very naive in this ever-changing modern world of digitization and it is prone to a lot of human error as well as external intervention. In this era of the Internet, when everything is on the fingertips of apparently all of the world's population, the maintenance of paper-based records is not only worn out but is also becoming a little unnecessary. The evidence collection, management, transfer and maintenance can all now be done digitally. This digital revolution is also proving to be very beneficial in other aspects of the judiciary especially in case of a severe pandemic.

Another point to be noted is that the internet application has lately moved from hostcentric to content-centric. The key prerequisite for Internet users is the publishing and storage of contents. The material can be a system or a file, such as web pages, images, audios or videos, that is transported across the internet. As a result, content protection is an important aspect of cyber security and focuses primarily on three security features: anonymity, transparency and non-repudiation. Not all contents are, sadly, sufficiently covered. Due to network threats or other purposes, certain content objects are interfere with. This kind of file modification has many adverse consequences. For instance, web site exploitation can be used for phishing attacks or for the dissemination of illicit material. The intruder may inject malicious files into executable files to monitor the user actions or to enter private data illegally. It is also important to review the corresponding digital evidence for file tampering for scientific, financial and legal reasons.

1.1 Objectives of the System

The handling of paper based evidence and records has already proved to be fatal to many cases in the past. The first-responders and the authorities collecting the evidence may misrepresent and mishandle the evidence, sometimes intentionally as well, to hurt or favour a case. The records are also prone to human intervention as is evident from the past. There can be unwanted external changes to records on paper that cannot be tracked down which can make the evidence inadmissible or in some cases, false convictions. This poses life-threatening situations for various innocents and law-abiding citizens. This problem can be solved to some extent with the use of a digital system that handles and maintains the evidence and track records which are almost fool-proof to external changes. Hence, the digital evidence management system is the need of the hour.

Digital evidence gains rising significance in today's ever-changing digital world, with the exponential growth of cybercrimes, since it is used to prove factual information or to prosecute convicts. Therefore, the quality of digital evidence in any criminal inquiry is highly critical in its life-cycle. The collection and handling of digital evidences has picked up some interest world-wide looking at the current scenario. When digital evidences are collected, they are almost always submitted directly to the management of third parties or stored in local computers. The storage, handling and dissemination of evidence is all focused on these systems. The protection of automated evidence systems has been overlooked however. There may be vulnerabilities in the automated evidence storage and collection method. The manipulation of these vulnerabilities will lead to (1) data that can be misrepresented, removing or traceability of evidence, and (2) privacy leaks. Private information can be leaked, such as evidence material, suppliers of evidence, and other information. It is worth learning how to maintain the security of digital data.

Centralized designs are often followed in stable digital evidence management schemes, like those mentioned in [7,20]. They have deceptive frameworks with protected applications, stable hardware, physical isolation or hybrid techniques on a single computer or a central structure. The core architecture faces certain challenges: (1) a single point of weakness that can override the system; (2) a problem of scalability that occurs when the volume of evidences is too high to store. The main objective of this white paper is to build a secure and easily accessible digital system to maintain and handle evidence efficiently.

2   Motivation

When storing physical evidence there are a lot of challenges and things to be careful about. From the collection of the evidence right to its presentation in a court of law, evidence may have to be stored for years and years. Adding to this, their tracking, maintaining and recording is all done on paper which is prone to human intervention and requires a lot of effort. To reduce some of these complications, the idea of a digital evidence management system comes to mind. But digital evidence comes with their own unique troubles.

The challenge with digital evidence usually occurs with the chain of custody (CoC) [9]. CoC plays an important role in a digital forensic inquiry, and during its cross-over across various layers of hierarchy, it tracks every minute information about digital data, i.e. the first respondent to senior authorities responsible for the investigation into cybercrime. CoC gathers information such as how, when, where, and who came in contact with data that was obtained, processed and stored for production, etc. However, if data is not preserved and maintained during the life-cycle of digitally recorded evidence, Forensic CoC is liable to compromise, making it unacceptable to prove any situation in connection with cybercrime in the court of law.

Hence, a suitable decentralised framework which is immune to changes is needed for this scenario. Blockchain is just the right technology that fits this description. This white paper aims to secure the evidence properly so that they can be used correctly to convict the right felons, and to make the work of the authorities easier. Adding to that, the presented study focuses on the use of Hyperledger Sawtooth as the framework for implementing blockchain as other frameworks such as Ethereum provide services on a payment basis. Hence, Hyperledger Sawtooth proves to be cost effective and just as efficient.

3   Related Work

Many industries have adopted it and are revolutionising the technology of the future due to the extensive growth of blockchain in the past decade. The use of the distributive ledger makes it possible for secure, immutable, controllable and transparent data to be registered and transmitted [30]. Blockchain has a revolutionising potential in the industry that makes the enterprise secure, effective, open, decentralised, making it much easier to track data.

Many industries have emerged and applied this technology to their respective fields in the past decade and have dramatically improved [11,24]. New studies emerging in the blockchain area are seen by Anjum et al. [3]. There are primarily 3 types of public, private and hybrid blockchain implementations which is also known as consortium. In this white paper we are focusing on the use of blockchain in the digital forensic domain. The comparison between existing systems is described in the Table 1.

Tian et al. [25] made a "Secure digital evidence framework using blockchain: Block-DEF". It is made by custom blockchain networking running on PBFT [5] consensus algorithm as its working mechanism. The paper also discussed evidence retrieval and verification. The limitation of the paper is, it does not include the details of file storage on decentralized file systems.

Lone et al. [18] made "Forensic-chain: Blockchain based digital forensics chain of custody with PoC in Hyperledger Composer". It is based on Hyperledger Composer which uses PBFT [5] consensus algorithm as its working mechanism. The limitation of paper is Hyperledger Composer gets deprecated recently, and there is no official support from the Hyperledger Organization in this project.

Ahmad et al. [1] developed "Blockchain-based chain of custody". It is built on Ethereum blockchain [28] which is fully a public blockchain network and uses PoS [26] consensus algorithm. The limitation of the paper is it is built on the Public Ethereum blockchain network [28], on which making state changes can cost the user some money.

Jeong et al. [16], made a "Digital Evidence model on Hyperledger Fabric". It is made using Hyperledger Fabric which is running on PBFT [5] consensus algorithm as its working mechanism. The limitations of the paper is it is very specific to the Korean Evidence Management System, it cannot be implemented in any other country.

Wang et al. [27] made "Lightweight and Manageable Digital Evidence Preservation System on Bitcoin". It is built on Bitcoin [19] network which uses PoW [15] consensus algorithm which requires a significant amount of computation to achieve consensus. The limitation of this system is thus it requires a significant amount of computation just to submit a evidence into the network.

Yunianto et al. [29] made "B-DEC: Digital Evidence Cabinet based on Blockchain". It is also built on Ethereum blockchain [28] which is fully a public blockchain network and uses PoS [26] consensus algorithm. The limitation of the paper is it is built on the Public Ethereum blockchain network [28], on which making state changes can cost the user some money.

4   Proposed Model

The proposed model is built on top of Hyperledger Sawtooth [21] to mange the evidences and Distributed File Storage Network (IPFS) [4] to store the evidences, with Web client connected with the Hyperledger Sawtooth's node using REST-APIs and IPFS node. The proposed system architecture is described in Sect. 4.1. The reason for choosing the framework is it is highly scalable and it comes with two groups of permissioning [23]. The reason for using IPFS is it supports the permanent Web so that a file once added into the network then it cannot be removed from the network.

4.1   Architecture of Evidence Management System

The architecture of the proposed system comprises the following 3 components, See Fig. 1 to know about how they connect with each other.

- **Hyperledger Sawtooth Distributed Blockchain Network:** It is the component which is responsible for storing the information related to the user and evidence.
- **IPFS Distributed file storage Network:** It is the component responsible for storing different evidence in a distributed network. Each file in the network is represented by a unique multihash to address the file [17]. This hash string of file is stored in the Hyperledger Sawtooth Network.
- **The Web Client:** The client is responsible for registering new users, submitting evidence, retrieving evidence, from the IPFS Network and Sawtooth Network.

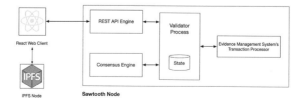

Fig. 1. High level evidence management system's architecture

The Evidence Management System ensures security and transparency by utilizing Hyperledger Sawtooth's Permissioning System [23]. The Evidence Management System's permissioning is designed so that it runs on a consortium blockchain format to allow only specific nodes to join the network and allow all "Evidence Providers" to submit evidence and allow all the "Evidence Requester" to access different evidences.

The evidence is stored in IPFS(Interplanetary File System) which is a distributed peer to peer file storage network. This evidence can be managed and accessed using Hyperledger Sawtooth Distributed Ledger Technology (DLT). The Evidence Management System uses 70 characters to address different evidence's data and IPFS's Content Identifier (CID) to access the file from the IPFS. This CID is a 46 characters long multihash address of a file in a IPFS and it is used to address different files in the IPFS network [17].

IPFS provides users with access to physically distributed systems that enable them to share their data and resources through a common file system. This set of workstations and mainframes is a Distributed File System Configuration itself, interconnected by a Local Area Network (LAN). As part of the operating system, a Distributed File System is executed where a namespace is generated and this mechanism is made clear for the clients. IPFS is a distributed file system for storing and accessing files, websites, applications, and data. A protocol for peer-to-peer hyper-media intended to make the internet quicker, simpler, and more accessible. It stands for Interplanetary System of Files. It also describes how data travel around a network, rendering it, just like BitTorrent, a distributed file system. IPFS creates a modern permanent network by integrating these two properties and improves the way we utilise current internet protocols such as HTTP. Figure 2 describes the working of IPFS.

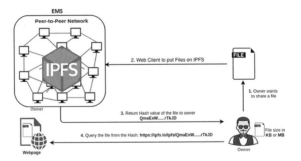

Fig. 2. Working of IPFS

4.2   The Evidence Management System

The high level architecture of the Evidence management System is demonstrated in Fig. 1. According to official Hyperledger Sawtooth documentation, application in Hyperledger Sawtooth is made up of Transaction Family. So, the Evidence

Management System is built on top of Hyperledger Sawtooth using Evidence Management System Transaction Family [23]. The system is accessed using a Reactjs [12] based Web GUI to register different users and to access and submit different evidence to the blockchain network. There are two types of users in a network: "Evidence Requester" and "Evidence Provider". The "Evidence Requester" can access the certain evidence in a network through Web Interface. The "Evidence Provider" can submit the evidence through the Web Interface and he/she can access evidence submitted by him or any other "Evidence Provider".

Table 1. Comparison between various existing systems

Paper title	Authors	Year	Parameters of judgement	Advantages	Limitations
Block-DEF: a secure digital evidence framework using blockchain [25]	Tian et al.	2019	Block rate	Explained about system in detail with two types of users	No explanation of how to store files
Blockchain based digital forensics chain of custody with PoC in Hyperledger composer [18]	Lone et al.	2019	Block rate Throughput	Explained about the working of system in hyperledger composer	Hyperledger composer is deprecated recently and there is no official support from the hyperledger organization
Blockchain-based chain of custody: towards real-time tamper-proof evidence management [1]	Ahmad et al.	2020	Gas Price block rate	Built CoC on top of Ethereum blockchain network	If user wants to store the evidence, it requires to pay certain amount of money to make their evidence admissible
Design and implementation of a digital evidence Management model based on hyperledger fabric [16]	Jeong et al.	2020	Block rate	Built digital evidence management system on Hyperledger fabric	The system is very specific to Korean evidence management system. It cannot be implemented in other countries
Lightweight and manageable digital evidence preservation system on bitcoin [27]	Wang et al.	2018	Block rate	Built digital evidence management system on Bitcoin network	It is built on top of the Bitcoin network which uses PoW consensus which requires high amount computation
B-DEC: digital evidence Cabinet based on blockchain for evidence management [29]	Yunianto et al.	2019	Gas price block rate	Built CoC on top of Ethereum blockchain network with different types of users	If user wants to store the evidence, it requires to pay certain amount of money to make their evidence admissible

As the Hyperledger Sawtooth stores data in Key-Value pairs, with values as buffers, we have used a standard predefined JSON structure to store data in buffers which deserializes to JSON and vice-versa to store data in JSON which serializes to buffer. See Sect. 4.2.6 to know about this predefined JSON structure.

4.2.1 The Evidence Management System's Transaction Family

The EMS Transaction family has three major components, a Web Client, a Validator, and a custom Transaction Processor nammed EMS Transaction Processor. According to Hyperledger Sawtooth official documentation, each Transaction Family comprises a transaction format and a namespace prefix. The EMS Transaction Family consists of three components: a web client, a validator from Sawtooth's Node and a custom transaction processor nammed EMS Transaction Processor [23].

According to Hyperledger Sawtooth official documentation, we have to register the Transaction Family with the validator with Transaction Family

Name and its version, In the case of EMS Transaction Family it is simply **evidence_management_system** with version v1.0 and namespace prefix as **d23299**. The namespace prefix of the Transaction Family is **evidence_management_system** with version v1.0. The Namespace prefix of the EMS Transaction family can be generated by the Algorithm 1.

Algorithm 1: Algorithm to Generate Namespace of Evidence Management System

```
H = SHA512("evidence_management_system")

N = H.slice(0, 6)
```

In the Algorithm 1, we first take the SHA512 hash of the string **evidence_management_system**, which will generate a 64 characters long hashed string, we then take its first six characters as our Namespace Prefix, which is **d23299**. All the address and incoming requests of transactions starts from this string.

4.2.2 The Web Client

The web client is responsible for submitting evidence and creating users. It is designed to create these two types of requests by creating two types of transactions. These transactions have a specific payload design to serialize the request for further processing at the EMS Transaction Processor's side. Figure 3 describes about the serializing algorithm of these payloads.

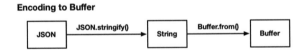

Fig. 3. Serializing to buffer

The Web client gets the serialized data from the REST-API engine of Hyperledger Sawtooth while querying the state using EMS's addressing scheme (described in Sect. 4.2.6), on which it runs a deserialization algorithm to get the data in JSON format. Figure 4 describes about the deserializing algorithm to get the data in JSON format. See Sect. 4.2.3 to know more about the payload design.

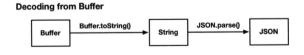

Fig. 4. Deserialization from buffer

The client is always connected to an IPFS Node, hence it is responsible for submitting and retrieval of files form the IPFS Node. According to the official IPFS documentation, each file in IPFS node gets a unique Content Identifier(CID) of 46 characters which is used to identify the file in the IPFS network [17]. The algorithm of submission and retrieval of files to the IPFS node is demonstrated in Fig. 5.

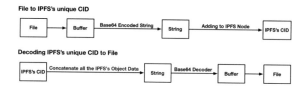

Fig. 5. File encoding and decoding algorithms

4.2.3   The Payload Design

Payloads are the predefined data format used to perform a specific action on the given data like, in this system to create a person or to create an evidence. The Web client is responsible for creating these payloads. All these payloads are in JSON format with three major keys: **action**, **data** and **timestamps**, these are further serialized to buffer and sent with the transaction, see next section to know more about the transaction design. The action defines what action it is to be performed and the data on which the action is performed and the current UNIX timestamps which is the moment of creation of the payload. The Evidence Management System is designed to handle two types of actions:

- **CREATE_PERSON.** This is the payload used to create a person in the network, it takes two parameters of the person, name of the person and their email. See Algorithm 2 for a sample *CREATE_PERSON* payload.
- **CREATE_EVIDENCE.** This is the payload used to create evidence in the network, it takes three parameters regarding the evidence: the Content Identifier (CID) from the IPFS, title of the evidence and application file format of the evidence. See Algorithm 3 for a sample CREATE_EVIDENCE payload.

Algorithm 2: CREATE_PERSON payload JSON structure

```
{
  "action": "CREATE PERSON",
  "timestamp": 1604080782322,
  "data": {
    "name": "test",
    "email": "test@gmail.com"
  }
}
```

Algorithm 3: CREATE_EVIDENCE payload JSON structure

```
{
  "action": "CREATE EVIDENCE",
  "timestamp": 1603996424136,
  "data": {
    "cid" :"QmYejFQ3SdYNKTaXMXpZ9EGofn9nGAfTzwNeWc
      1UKWTbA9",
    "title": "Resume",
    "mimeType": "application/pdf"
  }
}
```

4.2.4 Batches and Transactions

The Web client then packs the payload JSON into transactions by creating buffers by the algorithm described in Fig. 3. All transactions are created in a specific format explained in the official Sawtooth documentation [23]. Transactions are created using the following Algorithm 4.

Algorithm 4: Algorithm to create Transaction

```
const encodedPayload = Utils.encode(payload);
  // JSON payload to buffer

const transactionHeaderBytes = TransactionHeader
  .encode({
        familyName: "evidence_management_system",
        familyVersion: "1.0",
        inputs: [ "d23299"],
        outputs: [ "d23299"],
        signerPublicKey: keys.publicKey,
          // signer's public key
        batcherPublicKey: keys.publicKey,
          // signer's public key
        nonce: "a9sr2bsf3bej",
          // random 12 chars nonce
        dependencies: [],
        payloadSha512: SHA512(encodedPayload),
          // SHA512 of payload
}).finish();

const transaction = Transaction.create({
        header: transactionHeaderBytes,
        headerSignature:
          sign(signer, transactionHeaderBytes),
            // Digital Secp256k1 ECDSA Signature
        payload: encodedPayload,
});

return transaction;
```

These transactions are then packed into a single batch. Batches are created using the following Algorithm 5. All these parameters are explained in official Sawtooth's documentation [23].

Algorithm 5: Algorithm to create Batch

```
const batchHeaderBytes = BatchHeader.encode({
  signerPublicKey: keys.publicKey,
    // signer's public key
  transactionIds:
    transactions.map((tx) => tx.headerSignature),
  // list of transaction signatures
}).finish();

const batch = Batch.create({
        header: batchHeaderBytes,
        headerSignature:
          sign(signer, batchHeaderBytes),
            // Digital Secp256k1 ECDSA signature
        transactions: transactions,
          // array of transactions
});

return batch;
```

These batches are sent with the REST-API request as octet streams, which gets unpacked at the validator to generate Transaction receipt.

4.2.5   The EMS Transaction Processor

The EMS Transaction Processor has the actual business logic to store the data into the state and to retrieve data from the state using the APIs given by the validator [23]. The validator generates the transaction receipts from the above transaction requests on successful validation of the transactions. These transactions are handled by the EMS Transaction Handler, which is designed to handle the two payloads mentioned in Sect. 4.2.3. Figure 6 describes the Transaction flow of Evidence Management System.

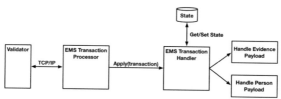

Fig. 6. EMS's transaction flow

The algorithms to handle these payloads contain interaction with the state, and a addressing scheme (Explained in Sect. 4.2.6). Algorithm 6 handles the *CRE-ATE_PERSON* payload and Algorithm 7 handles the *CREATE_EVIDENCE* payload.

Algorithm 6: Algorithm to handle CREATE_PERSON payload

```
const header = transaction.header;

const publicKey = header.signerPublicKey;
    // Signer's Public key

const payload = new EMSPayload(transaction.payload);
    // Deserialized Transaction payload object

const state =
  new EMSState(context, payload.getTimestamp());
    // State at timestamp

if (payload.getAction() === "CREATE_PERSON"){

    // Check if the Person Already exist
    const address =
      Utils.getPersonAddress(publicKey);
        // convert signer's public key to addressing
        // scheme to read state's value

    state.getState([address])
      .then((entries) => {
            const entry = entries[address];
          if (!entry || entry.length === 0) return null;
            throw new InvalidTransaction(
                      `Person with the public key
                        ${publicKey} already exists`);
    });

    // Create Person
    const person = {
      name: payload.data.name,
      email: payload.data.email,
      timestamp: payload.getTimestamp(),
      evidences: [],
      publicKey: signer,
    };

    state.setState(
      { [address]: Utils.encode(person) }
        // person object to buffer
    );
}
```

Algorithm 7: Algorithm to handle CREATE_EVIDENCE payload

```
const header = transaction.header;

const publicKey = header.signerPublicKey;
    // Signer's Public key

const payload = new EMSPayload(transaction.payload);
    // Deserialized Transaction payload object

const state =
  new EMSState(context, payload.getTimestamp());
    // State at timestamp

if (payload.getAction() === "CREATE_EVIDENCE"){

  // Check if the Person Exist
  const pAddress = Utils.getPersonAddress(publicKey);

  // Convert signer's public key to addressing scheme
  // to read state's value
  const person = state.getState([address])
    .then((entries) => {
      const entry = entries[address];
      if (!entry || entry.length === 0)
        throw new InvalidTransaction(
          `Person with the public key
            ${publicKey} does not exists`);
          return Utils.decode(entry);
            // person object from buffer
  });

  // Create evidence
  const evidence = {
    title: payload.data.title,
    timestamp: payload.getTimestamp(),
    owner: publicKey,
    cid: payload.data.cid,
      mimeType: payload.data.mimeType,
  };

  const eAddress =
    Utils.getEvidenceAddress(payload.data.cid);
        // convert evidence's cid to addressing scheme
        //  to read state's value

  person.evidences.push(eAddress);
  // push the evidence address to array of person's
    list of evidences

  state.setState({
    [pAddress]: Utils.encode(person),
      // person object to buffer
    [eAddress]: Utils.encode(evidence)
      // evidence object to buffer
  });
}
```

4.2.6   Addressing Scheme and State Storage Model

Checkout official Hyperledger Sawtooth's documentation to know more about the addressing structure [23]. The addressing of person and evidence is composed of three segments. The address of a person can be generated using the public key of the person by taking the first 62 characters of SHA512 hash of public key, appending it with **d23299** (Namespace of EMS Transaction Family) and **01**. The address of an evidence can be generated using the content Identifier (CID from IPFS) of the file by taking the first 62 characters of SHA512 hash of content Identifier (CID from IPFS), appending it with **d23299** (Namespace of EMS Transaction Family) and **00**. Figure describes 7 the addressing scheme of Person. Figure 8 the addressing scheme of Evidence.

Fig. 7. Addressing scheme of person

Fig. 8. Addressing scheme of evidence

These addresses represent different leaf nodes of the Merkle-Radix tree. These leaf nodes store the data in buffers which can be deserialized using the algorithm described in Fig. 4. Algorithm 8 describes what data is stored at the leaf of this Merkle-Radix tree for the person and Algorithm 9 describes what data is stored at the leaf of this Merkle-Radix tree for the evidence. This state can be queried from the Web Client to REST-API engine of Sawtooth's node which gives the buffered response of the state, it will be deserialized to known readable JSON format.

Algorithm 8: Deserialized buffer data of Person at leaf of Merkle-Radix tree

```
{
    "address": "d232990190a785b2de3bc8f0079ce3e3f4919b39
      19b47e5cd6a72F9Fcb9495dc442bcFf",
    "data": {
      "name": "test",
      "email": "test@gmail.com",
      "timestamp": 1603916360338,
      "evidences": [
        "d2329900825dbb85799d5741d0214649675d63632de5da6
          4cd709b01529b9992827049" ,
        "d23299008ed814d361885b2754fc4b79 f98abad396F77d
          c0eecbcc3f7ba6F6299b081",
        "d23299007300fec6de47eca0d9172c9804ed829ca72f7ee
          4d4Fe862651165674ac3817",
        "d2329900460216e737e886a16T5f979096bad56fa26686c
          67f195e0dd3e772al1e057f7"
      ],
      "publicKey": "0244cfa7605b0f0094cdcOb90ce3c08Fcal4
        372a57f5acb2876bd584b8427813c7"
    }
}
```

Algorithm 9: Deserialized buffer data of Evidence at leaf of Merkle-Radix tree

```
{
    "address": "d23299007300fec6de47eca0d912c9804ed829ca
      2F7ee4d4fe862651165674ac3817",
    "data": {
      "title": "Resume",
      "timestamp": 1603996424136,
      "owner": "Q244cfa7605b0f0094cdcOb90ce3cO8Ffcal4372
        a57f5acb2876bd584b8427813c7",
      "cid": "QmYejFQ3SdYNKTaXMXpZ9EGofn9nGAfTzwNeWc1UKW
        TbA9",
      "mimeType": "application/pdf"
    }
}
```

4.3 Workflow of the Proposed Model

Figure 9 describes the operational workflow of the Evidence Management System. It describes that can be multiple users connected to the Front-End web interface, a user can get or submit the Digital Evidence.

4.3.1 Web Interface Workflow and Sawtooth Explorer GUI

The web interface is built using ReactJs which makes the transaction requests to Hyperledger Sawtooth network. The REST-API engine of Sawtooth responds with appropriate response. The Web interface has 5 screens which are described below. Figure 10 Describes the Web Interface flow of the Evidence Management System. The Sawtooth Explorer has a GUI to querying and viewing the ledger in Real Time [22]. It connect with the REST-API engine of Sawtooth's Node for querying the ledger.

Login Screen: Registered users can login with this screen. It takes the user's private key and generates the user's public key. The details of the user are fetched from the network, and then the user is redirected to the List of Evidences Screen. If the user does not exist then it redirects the user to the Register Screen.

Register Screen: New users can register with this screen. It takes the user's name and email and generates the unique user's public key and private key. The interface creates a new Transaction of *CREATE_PERSON* payload with parameters of user's name and user's email. On successfully receiving the response from the REST-API engine, this page displays the user's generated keys and on clicking the **List of Evidences** button, the user is redirected to the to the List of Evidences Screen. Currently we are just displaying the user's public key and private key, we can also send them in his email as it is sensitive information.

List of Evidences Screen: It displays the list of evidence to the logged in users in tabular format. Users can sort and filter from the list of evidence. The list of evidence is fetched in real time from the network. It has a **Submit Evidence** button, which redirects the user to the Submit Evidence screen. It also has the **Logout** button which logs out the current user. Each evidence has a **Download** button, which lets the user to download that evidence.

Submit Evidence Screen: This screen allows the logged in user to submit evidence into the network. It takes the evidence file and the title of the evidence. The file is then uploaded to the IPFS network to get the unique Content Identifier (CID) of the file [17]. The interface creates a new Transaction of *CREATE_EVIDENCE* payload with parameters of this CID, title of the evidence and file type of the evidence. On successfully receiving the response from the REST-API engine, this page redirects the user to the List of Evidence screen.

5 Performance and Evaluation

The Evidence Management System's performance metrics are collected on a real time basis from the InfluxDB database [13]. InfluxDB [13] is an open-source time series database developed by InfluxData. These metrics can be viewed from a tool named Grafana [10]. Grafana [10] is the open source analytics and monitoring solution for every database. The Grafana [10] interface measures many metrics in real time. InfluxDB [13] uses a tool named Telegraf [14] to collect system's metrics. These measures show characteristics such as traffic in the network, efficiency of the network etc. Figure 11 describes the testing architecture of Hyperledger Sawtooth Node.

5.1   Experimental Setup

The system is running in Docker Containers [8] with four Sawtooth Nodes. The containers are running on top of Ubuntu OS 18.0 host system with 16 GB RAM and with Intel's core i9 processor. The system can also be runned on AWS EC2 [2] Ubuntu running containers with dedicated 1 GB RAM for each node.

The Evidence Management System is running on PBFT consensus [5], which is suitable for small scale consortium networks. The system can also be implemented using PoET [6] consensus as the Network grows.

A sample image file of 200 KB is taken for this experimentation purpose which is stored in the IPFS and this file can be retrieve using it's multihash generated, by querying the file from IPFS Network from URL https://ipfs.io/ipfs/¡multihash of the file¿.

5.2   Performance Criteria and Evaluation

The proposed system's performance is measured using the InfluxDB [13] and Grafana [10]. The Grafana [10] interface provides some main metrics namely

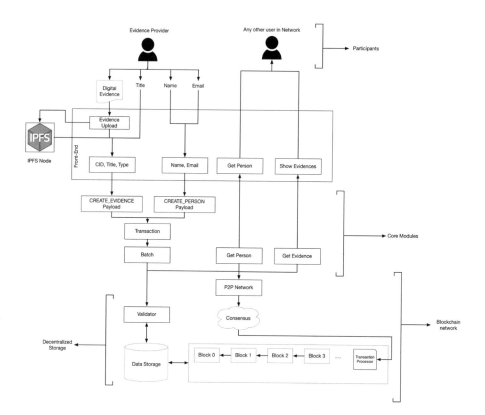

Fig. 9. Operational workflow diagram

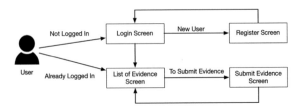

Fig. 10. WEB interface UI Workflow

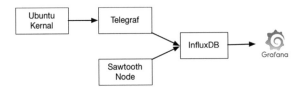

Fig. 11. Testing architecture of hyperledger Sawtooth's node using InfluxDB and Grafana

Block rate and transaction rate, etc. in real time using database InfluxDB [13]. These measures show characteristics such as traffic in the network, efficiency of the network etc. Some of the metrics are explained below.

5.2.1 Block Rate

Block rate is defined as the number of blocks which are added in unit time to the blockchain. If the unit time is taken as a minute the Block Rate can be specified as blocks per minute. If the value of the parameter is high this shows that the network has high traffic and is active. Figure 12 shows the Block Rate of Evidence Management System. Right side of the graph shows the Docker Container [8] IDs of different validators on Sawtooth Network and their number of blocks.

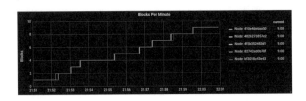

Fig. 12. Blocks per minute

5.2.2 Block Publication Rate

Block Publication Rate of different nodes of the Sawtooth Network. If the unit time is taken as a minute the Block Publication Rate can be specified as blocks per minute. If the value of the parameter is high this shows that the network has high traffic and is active. Figure 13 shows the Block Publication Rate of Evidence Management System by different validators of Hyperledger Sawtooth in PBFT [5] consensus. Right side of the graph shows the Docker [8] Container IDs of different validators on Sawtooth Network and their number of blocks published.

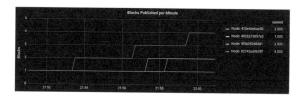

Fig. 13. Block publication rate of different nodes in Sawtooth network

5.2.3   Committed Transaction Rate

It is also a performance measure to evaluate the performance of Sawtooth Network, by measuring the no of transactions committed by each individual node in Sawtooth. If the value of the parameter is high this shows that the network has high traffic and is active. If the unit time is taken as a minute the Committed Transaction rate can be specified as Committed Transactions per minute. Figure 14 shows the Committed Transaction Rate of Evidence Management System of different validators of Hyperledger Sawtooth in PBFT [5] consensus. Right side of the graph shows the Docker [8] Container IDs of different validators on Sawtooth Network and their number of committed transactions.

5.2.4   Transaction Execution Rate

Transaction Execution rate is a performance measure of Sawtooth Network, in which rate of processing of a transaction at the Sawtooth's Node is measured. If the value of the parameter is high this shows that the network has high traffic and is active. Figure 15 shows the Transaction Execution Rate of different Sawtooth Nodes.

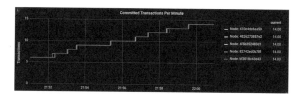

Fig. 14. Committed transactions per minute

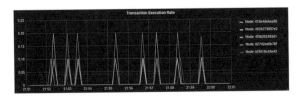

Fig. 15. Transaction execution rate of different Sawtooth nodes

5.2.5   REST-API Block Publication Rate

It is the batch submission per minute rate of different Sawtooth Nodes. Since the application is running locally, only one Node is responsible for submission of batches. If the unit time is taken as a minute the number REST-API Batch submission rate can be specified as number of REST-API Batches per minute. Figure 16 shows the REST-API batch submission rate of different Sawtooth Nodes. Right side of the graph shows the Docker Container IDs of different validators on Sawtooth Network and their REST-API Batch Submission rate.

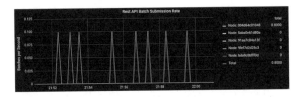

Fig. 16. REST-API batch submission rate of different Sawtooth nodes

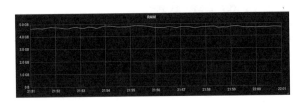

Fig. 17. RAM usage of evidence management system with 4 nodes

5.2.6   System Matrices: RAM Usage

Hyperledger Sawtooth's performance is also measured by how much RAM it is consuming. See Fig. 17 for RAM usage graph for 4 nodes running on PBFT [5] consensus. It can be said from the graph the 4 nodes use almost 4 GB of RAM apart from all the other application processes to deliver optimized performance.

5.2.7   System Matrices: CPU Utilization

Hyperledger Sawtooth's performance is also measured by how much CPU it is utilizing. See Fig. 18 for CPU utilization graph for 4 nodes running on PBFT [5] consensus.

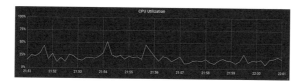

Fig. 18. CPU utilization of evidence management system with 4 nodes

6   Conclusion and Future Scope

We proposed a digital security forensic model that can exchange and control data by accessing a network in a distributed environment with only approved participants. No user can alter and extract the data from the digital forensic until written in the block and has the benefit of improving accountability and security, as they are used by all blockchain members. The digital forensic data can be used reliably from anywhere in the world with ease of access. Using Hyperledger Sawtooth framework for the blockchain in addition to the Inter-Planetary File System (IPFS) for decentralized storage of files proved to be beneficial in various ways. The model proposed here gives the required results as it gives us a secure blockchain network to store all of the gathered evidence records. The evidence files are stored in a decentralized manner over the IPFS servers and provide transparency and security to the system. Also, the use of blockchain helps implement immutability to the Evidence Management System thus eliminating any case of evidence tampering.

As seen in graphs above, the Evidence Management System shows optimal performance in Docker Containers [8] running on top of Ubuntu OS running 16 GB RAM and Intel's core i9 processor. It can give more throughput when running on AWS EC2 [2] machines with dedicated 1 GB RAM for each Sawtooth Node. From the above graphs, it can be concluded that for small consortium based Hyperledger Sawtooth Networks, PBFT [5] consensus gives optimal performance. Results of the study also indicate that high reliability could be accomplished with digital data contained in the proposed model.

The blockchain based digital evidence management system tries to employ a lot of the state of the art technologies available readily which are also easy to handle by the users. But it also faces some serious concerns. The EMS database can only be accessed through a web-client on a PC or a desktop at the moment, it does not have mobile compatibility. So, the next step would be to make the web client compatible for mobile phones so the accessing of files becomes easier and handy. This system could also be incorporated in a mobile application to make its functioning on a mobile phone even better. Once it is incorporated as a mobile app, it could make use of the tremendous amount of features in a smartphone which can help the authorities to directly capture the evidence through the phone's camera at the scene of the crime and have it store directly on the database. It could even be paired with various modern day technologies like Augmented Reality (AR) to capture the whole scene of the crime in 3D in order to recreate that scene somewhere or sometime else to help understand and solve the case better. Also, technologies such as Machine Learning and Deep Learning could be implemented, which could allow the system to recognize a face in some CCTV footage or picture from a database directly. ML could also be used to categorize the different evidences automatically once the number of evidence gets very high.

References

1. Ahmad, L., Khanji, S., Iqbal, F., Kamoun, F.: Blockchain- based chain of custody: towards real-time tamper-proof evidence management. In: Proceedings of the 15th International Conference on Availability, Reliability and Security (Virtual Event, Ireland) (ARES 2020). Association for Computing Machinery, New York, NY, USA, Article 48, p. 8 (2020). https://doi.org/10.1145/3407023.3409199
2. Amazon Inc. [n.d.]. Amazon Elastic Compute Cloud (Amazon EC2), A web service that provides secure, resizable compute capacity in the cloud. https://aws.amazon.com/ec2/
3. Anjum, A., Sporny, M., Sill, A.: Blockchain standards for compliance and trust. IEEE Cloud Comput. **4**(4), 84–90 (2017). https://doi.org/10.1109/MCC.2017.3791019
4. Benet, J.: IPFS - Content Addressed, Versioned, P2P File System (2014). arXiv:1407.3561 [cs.NI]
5. Castro, M., Liskov, B., et al.: Practical Byzantine fault tolerance. OSDI **99**, 173–186 (1999)
6. Chen, L., Xu, L., Shah, N., Gao, Z., Lu, Y., Shi, W.: On security analysis of proof-of-elapsed-time (PoET), pp. 282–297 (2017). https://doi.org/10.1007/978-3-319-69084-1_19
7. Chong, C.N., Peng, Z., Hartel, P.H.: Secure audit logging with tamper-resistant hardware. In: Gritzalis, D., De Capitani di Vimercati, S., Samarati, P., Katsikas, S. (eds.) SEC 2003. ITIFIP, vol. 122, pp. 73–84. Springer, Boston, MA (2003). https://doi.org/10.1007/978-0-387-35691-4_7
8. Docker Inc. [n.d.]. Docker, A set of platform as a service products that use OS-level virtualization to deliver software in packages called containers. https://docker.com
9. Giova, G.: Improving chain of custody in forensic investigation of electronic digital systems. Int. J. Comput. Sci. Netw. Secur. **11**(1), 1–9 (2011)
10. Grafana Labs [n.d.]: Grafana, Grafana is the open source analytics and monitoring solution for every database. https://grafana.com
11. Ye, G., Chen, L.: Blockchain application and outlook in the banking industry. Finan. Innov. **2**(1), 24 (2016)
12. Facebook Inc. [n.d.]: Reactjs, A JavaScript library for building user interfaces. https://reactjs.org/
13. InfluxData [n.d.]: Influx DB, An open-source time series database developed by InfluxData. https://www.influxdata.com

14. InfluxData [n.d.]: Telegraf, The open source server agent to help you collect metrics from your stacks, sensors and systems. https://www.influxdata.com/time-seriesplatform/telegraf/

15. Jakobsson, M., Juels, A.: Proofs of Work and Bread Pudding Protocols (Extended Abstract), pp. 258–272. Springer, Boston (1999). https://doi.org/10.1007/978-0-387-35568-9_18

16. Jeong, J., Kim, D., Lee, B., Son, Y.: Design and implementation of a digital evidence management model based on hyperledger fabric. J. Inf. Process. Syst. **16**, 4 (2020)

17. Protocol Labs. IPFS Official Documentation. https://docs.ipfs.ioProtocol Labs (2021). IPFS Official Documentation.https://docs.ipfs.io

18. Lone, A.H., Mir, R.N.: Forensic-chain: blockchain based digital forensics chain of custody with PoC in Hyperledger Composer. Digital Invest. **28**(2019), 44–55 (2019). https://doi.org/10.1016/j.diin.2019.01.002

19. Nakamoto, S., et al.: A peer-to-peer electronic cash system, Bitcoin (2008)

20. Nieto, A., Roman, R., Lopez, J.: Digital witness: safeguarding digital evidence by using secure architectures in personal devices. IEEE Netw. **30**(6), 34–41 (2016). https://doi.org/10.1109/MNET.2016.1600087NM

21. Olson, K., Bowman, M., Mitchell, J., Amundson, S., Middleton, D., Montgomery, C.: Sawtooth: An Introduction. The Linux Foundation (2018)

22. Hyperledger Org. [n.d.]: Hyperledger Sawtooth Explorer, SawtoothExplorer is an application that provides visibility into the Sawtooth Blockchain for Node Operators. https://github.com/hyperledger/sawtooth-explorer

23. Hyperledger Org. 2021. Hyperledger Sawtooth Official Documentation. https://sawtooth.hyperledger.org/docs/core/releases/latest

24. Singh, S., Singh, N.: Blockchain: future of financial and cyber security. In: 2016 2nd International Conference on Contemporary Computing and Informatics (IC3I), pp. 463–467 (2016). https://doi.org/10.1109/IC3I.2016.7918009

25. Tian, Z., Li, M., Qiu, M., Sun, Y., Su, S.: Block- DEF: a secure digital evidence framework using blockchain. Inf. Sci. **491**, 151–165 (2019). https://doi.org/10.1016/j.ins.2019.04.011

26. Vasin, P.: Blackcoin's proof-of-stake protocol v2 (2014). https://blackcoin.co/blackcoin-pos-protocol-v2-whitepaper.pdf71

27. Wang, M., Wu, Q., Qin, B., Wang, Q., Liu, J., Guan, Z.: Lightweight and manageable digital evidence preservation system on bitcoin. J. Comput. Sci. Technol. **33**(3), 568–586 (2018)

28. Wood, G., et al.: Ethereum: a secure decentralised generalised transaction ledger. Ethereum Project Yellow Pap. **151**, 1–32 (2014)

29. Yunianto, E., Prayudi, Y., Sugiantoro, B.: B-DEC: digital evidence cabinet based on blockchain for evidence management. Int. J. Comput. Appl. **975**(2019), 8887 (2019)

30. Zheng, Z., Xie, S., Dai, H., Chen, X., Wang, H.: An overview of blockchain technology: architecture, consensus, and future trends. In: 2017 IEEE International Congress on Big Data (BigData Congress), pp. 557–564 (2017). https://doi.org/10.1109/BigDataCongress.2017.85

Author Index

A
Aeri, Manisha, 244
Aggarwal, Vishali, 236
Agrawal, Aman, 337
Agrawal, Rahul, 12
Ali, Rifaqat, 337
Annappa, B., 34
Antony, Jobin K., 203

B
Bhopale, Amol P., 67
Biswas, Monisha, 22
Borse, Shital S., 160

C
Chandrakar, Preeti, 337
Chaturvedi, Amrita, 175, 304
Chaudhary, Gauri, 268
Chintanpalli, Anantha Krishna, 78
Chopade, Nilkanth B., 253

D
Deshmukh, Smita, 289
Devassy, Binet Rose, 203
Dharavath, Ramesh, 213

E
Edla, Damodar Reddy, 213

G
Gadekar, Aumkar, 314
Gagandeep,, 236
Gandhi, Neel, 95
Geetha, N., 48

G
Giridhar, Ankitha, 85
Goyal, Nitesh, 12
Gupta, Siddharth, 244
Gupta, Sonali, 244

H
Hasija, Yasha, 150
Hoque, Amirul, 128
Hoque, Mohammed Moshiul, 22

J
Jamulkar, Shritesh, 337
Jangpangi, Sachin, 326
Jeevan, Govind, 34
Jha, Namrata, 227

K
Kadroli, Vijayalaxmi, 160
Kamath, Sharanya, 34
Kondaveeti, Hari Kishan, 186
Kshirsagar, Manali, 268
Kumar, Rakesh, 304
Kumar, Saurabh, 12
Kumar, Shailender, 227, 326
Kumar, Shubham, 12
Kumar, Vivek, 12

M
Madarkar, Jitendra, 280
Malladi, Ravisankar, 186
Mana, Suja Cherukullapurath, 1
Manwal, Manika, 244
Mishra, Shakti, 95

© The Editor(s) (if applicable) and The Author(s), under exclusive license
to Springer Nature Switzerland AG 2022
R. Misra et al. (Eds.): ICMLBDA 2021, LNNS 256, pp. 361–362, 2022.
https://doi.org/10.1007/978-3-030-82469-3

N
Naik, Amrita, 213
Narkhede, Manish M., 253
Nimkar, Anant V., 314

O
Oak, Shreya, 314

P
Padmavathi, S., 106
Pandeeswari, S. Thiruchadai, 106
Panwar, Avnish, 244
Patel, Bhavesh, 141
Prakash, Alok, 175
Prasad, Vandana, 78
Prashanth, M. C., 117

R
Raj, Jyoti, 128
Ravikumar, M., 117
Revadekar, Abhishek, 314

S
Sachdeva, Nikhil, 227
Saha, Ashim, 128
Sahani, Manish, 326
Sampathila, Niranjana, 85
Sasipraba, T., 1
Satapathy, Santosh Kumar, 186
Sharif, Omar, 22
Sharma, Arpita, 150
Sharma, Poonam, 280
Shivaprasad, B. J., 117
Singh, Priyadarshan, 326
Singh, Shashank Kumar, 175
Singhal, Palak, 34
Srilakshmi, S. S., 106
Suhirtha, R., 48
Susitha, A., 48
Swetha, A., 48

T
Tiwari, Ashish, 67
Tiwari, Divya, 289
Tiwari, Kartik, 337

Printed in the United States
by Baker & Taylor Publisher Services